Applied RNA Bioscience

Seiji Masuda • Shingo Izawa
Editors

Applied RNA Bioscience

Springer

Editors
Seiji Masuda
Graduate School of Biostudies
Kyoto University
Kyoto, Japan

Shingo Izawa
Applied Biology
Kyoto Institute of Technology
Kyoto, Japan

ISBN 978-981-10-8371-6 ISBN 978-981-10-8372-3 (eBook)
https://doi.org/10.1007/978-981-10-8372-3

Library of Congress Control Number: 2018937410

Printed on acid-free paper

This Springer imprint is published by the registered company Springer Nature Singapore Pte Ltd.
The registered company address is: 152 Beach Road, #21-01/04 Gateway East, Singapore 189721, Singapore

Contents

Chapter 1
Improvement of Yeast Fermentation Efficiency Utilizing mRNAs Preferentially Translated Under Translational Repression

Sae Kato and Shingo Izawa

Abstract As their transcription site (nucleus) and translation site (cytoplasm) are separated by the nuclear membrane, eukaryotic microorganisms such as yeast and *Aspergillus* undergo gene expression regulation differently from prokaryotic microorganisms such as *Escherichia coli*. In eukaryotic microorganisms, mRNAs synthesized in the nucleus are transported outside after intranuclear processing, which includes addition of 5′ caps and 3′ poly(A) tails, splicing, and formation of nuclear export complexes. Additionally, mRNA nuclear export also occurs selectively under certain stress conditions, and this allows efficient gene expression by retaining non-essential mRNAs in the nucleus while preferentially transporting urgent and highly essential mRNAs to the cytoplasm. Recently, it has lately been proven that specific mRNAs are preferentially and selectively translated under stress conditions that repress overall protein synthesis. These facts suggest that, in addition to transcription regulation, gene expression is also precisely regulated at the mRNA nuclear transport stage and cytoplasmic translation stage. This paper focuses on the preferential and selective translation in yeast cells under stress and introduces recent findings regarding applications of this knowledge.

Keywords Translation repression · Preferential translation · Ethanol stress · Lignocellulosic hydrolysates · Vanillin

1.1 Introduction

The budding yeast *Saccharomyces cerevisiae* is a useful microorganism for human life, especially for brewing alcoholic beverages and second-generation bioethanol. However, during the brewing process of wine and Japanese *sake*, yeast cells are

S. Kato · S. Izawa (✉)
Applied Biology, Kyoto Institute of Technology,
Matsugasaki, Sakyo-ku, Kyoto, Japan
e-mail: thioredoxin@kit.ac.jp

S. Masuda, S. Izawa (eds.), *Applied RNA Bioscience*,
https://doi.org/10.1007/978-981-10-8372-3_1

inevitably exposed to various kinds of stress—including ethanol stress (caused by the elevated concentration of ethanol produced by the yeast itself), nitrogen depletion, low temperature, low oxygen, low pH, and high sugar content. Second-generation bioethanol, produced from woody and herbaceous biomass (lignocellulosic biomass) including waste wood and rice straw, may be utilized effectively as a source of energy that does not compete with food resources. It is thus a renewable alternative to petroleum, and its future expansion is anticipated. In the pretreatment process of biomass saccharification, vanillin, furan compounds (furfural and 5-hydroxymethyl furfural), and acetic acid are produced as by-products (Antal et al. 1990; Klinke et al. 2004; Lu et al. 2009; Piotrowski et al. 2014); as they repress and inhibit the growth and fermentability of yeast in the fermenter, the production process of bioethanol is hardly a comfortable environment for yeast. In any case, yeast undergoes alcohol fermentation under a poor environment where various kinds of stressors act in combination.

It has been reported that high concentrations of ethanol, glucose depletion, and lignocellulosic biomass conversion inhibitors such as vanillin strongly repress the translation activity of yeast (Ashe et al. 2000; Kato et al. 2011; Iwaki et al. 2013a, b). Therefore, despite the activation of transcription of most genes, the synthesis of new translation products—namely, proteins—does not occur, and the expression of those genes is repressed. It has been known that, while non-urgent mRNAs such as housekeeping gene mRNAs are sequestered in the cytoplasmic processing bodies (P-bodies) and stress granules (SGs) during the severe stress conditions that cause the pronounced translation repression, some mRNAs required to cope with these stresses are preferentially and selectively translated (Arribere et al. 2011; Zid and O'Shea 2014). By focusing on mRNA flux from transcription to degradation, we analyzed the response mechanism of yeast against stress related to fermentation and brewing. This paper describes mRNAs preferentially translated under the pronounced translational repression and introduces our attempts to improve yeast fermentability by utilizing the promoter regions of the genes encoding these mRNAs.

1.2 Translational Repression and Formation of Cytoplasmic mRNP Granules

Gene expression in eukaryotic cells is influenced by stress and can be regulated not only at the transcriptional stage but also during mRNA nuclear transport, cytoplasmic translation, and degradation. The formation of polysomes, which reflects translation activity, can be analyzed using ribosome profiling (Inada and Aiba 2005): under glucose depletion (a form of nutrient-depletion stress) and severe ethanol stress, the translational activity of yeast is strongly repressed, leading to an observable decrease in polysomes and a corresponding increase in 80S monosomes (Ashe et al. 2000; Lui et al. 2010; Yamauchi and Izawa 2016). This indicates that translation elongation does not occur even though 40S small subunits and 60S large subunits bind to mRNA to form 80S monosomes, also known as the translation initiation

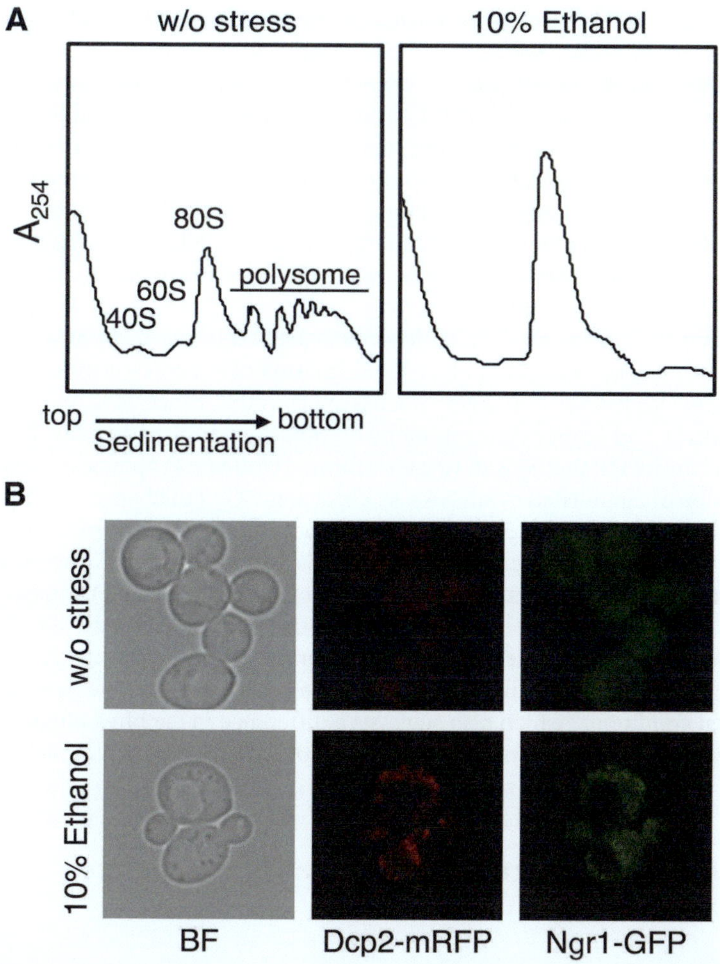

Fig. 1.1 Severe ethanol stress causes the pronounced repression of overall protein synthesis and the formation of mRNP granules. Ribosome profiling indicates that 10% (v/v) ethanol leads to a prominent decrease in polysomes and a corresponding increase in 80S monosomes (**a**). P-bodies (Dcp2-mRFP) and SGs (Ngr1-GFP), in which untranslated mRNAs are segregated, can be observed in yeast cells treated with 10% ethanol (**b**)

complexes (Lui et al. 2010) (Fig. 1.1a). Furthermore, this means that the number of translated mRNAs decreases, while the amount of untranslated mRNAs increases.

Under this kind of pronounced translation repression, untranslated mRNAs dissociate from ribosomes to form mRNP granules called processing bodies (P-bodies) and stress granules (SGs) in the cytoplasm (Fig. 1.1b). P-bodies and SGs are structures without biological membranes and are composed of untranslated mRNAs and various kinds of proteins (Kedersha et al. 2005; Buchan et al. 2008; Balagopal and Parker 2009). Although some of the proteins that make up P-bodies and SGs are common, there is a huge difference between the two. While SGs contain translation initiation factors, poly(A)-binding protein, and 40S ribosome proteins, P-bodies contain the proteins involved in the mRNA decay such as exonuclease and decapping enzymes (Buchan et al. 2008; Grousl et al. 2009). P-bodies and SGs appear to play an important role in translation repression under stress: it is thought that the mRNAs that can be translated under normal conditions are sequestered in P-bodies and SGs in order to repurpose the translation apparatus, allowing preferential and efficient translation of urgently needed mRNAs while under stress (Brengues et al. 2005; Anderson and Kedersha 2006; Balagopal and Parke 2009; Arribere et al. 2011; Zid and O'Shea 2014).

Unlike mammalian cells—in which untranslated mRNAs form SGs first—yeast cells under glucose depletion induce the formation of P-bodies prior to the generation of SGs (Kedersha et al. 2005; Buchan et al. 2008). Since SGs requires severer ethanol stress and higher concentrations of vanillin than P-bodies for their formation, it is considered that SGs in yeast are formed by using P-bodies as a scaffold for the transfer of untranslated mRNAs to emerging SGs (Buchan et al. 2008; Kato et al. 2011; Iwaki et al. 2013a, b). This difference in the sequence of formation may suggest a divergence between mammals and yeast in the "triage," the process that determines the fate of untranslated mRNAs. In yeast, it may be P-bodies—instead of SGs—that determine the future of mRNAs, that is, whether untranslated mRNAs will be degraded, stored temporarily, or returned to ribosomes to be translated. Interestingly, compared to mammalian cells, the formation of SGs in yeast is induced by limited conditions, which also suggests a difference in the physiological roles of SGs between these organisms (Buchan et al. 2008; Balagopal and Parker 2009).

1.3 *HSP26* Is Preferentially Translated Under Glucose Depletion

It has recently been determined that mRNAs transcribed from genes whose promoter regions contain a heat-shock element (HSE)—the recognition sequence of heat-shock-responsive transcriptional activator Hsf1—are preferentially translated in yeast cells under glucose depletion, which otherwise strongly represses translation (Zid and O'Shea 2014) (Fig. 1.2). Although the transcription of various genes whose promoter regions do not contain the HSE was activated by stress-responsive transcription factors such as Msn2/Msn4, general stress-responsive transcription factors, the resulting mRNAs were transported to P-bodies or SGs and were not translated during glucose depletion. In addition, housekeeping gene mRNAs constitutively synthesized before stress treatment were also sequestered in P-bodies and SGs. Meanwhile, mRNAs of *HSP26* and *HSP30*, which carry the HSE in their promoter regions, were not sequestered in SGs and were preferentially translated even under translational repression.

Interestingly, by utilizing the *HSP26* promoter containing the HSE, other genes might also be preferentially expressed under glucose depletion (Zid and O'Shea 2014). These results strongly suggest that preferential and selective translation is

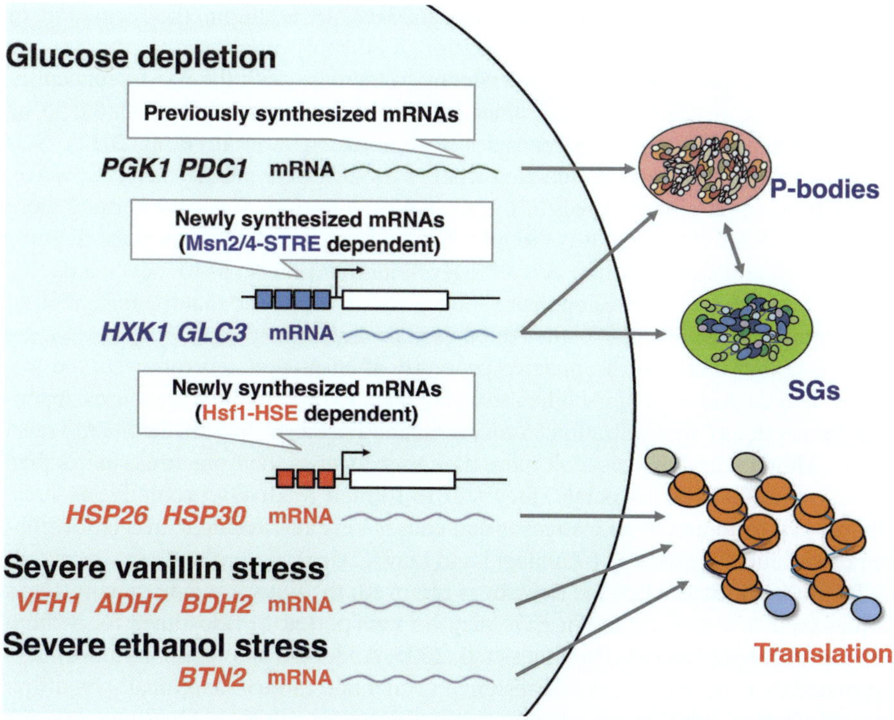

Fig. 1.2 Prioritized translation of mRNAs under the pronounced translation repression. Under glucose depletion stress, mRNAs of *PGK1* and *PDC1*, which are constitutively expressed under non-stressed conditions, and mRNAs of *HXK1* and *GLC3*, which are newly synthesized upon the stress, are not translated and segregated into P-bodies and SGs. On the other hand, *HSP26* and *HSP30* are efficiently translated despite the pronounced translation repression (Zid and O'Shea 2014). Prioritized translation has also been observed in *ADH7* and *BDH2* (severe vanillin stress) and *BTN2* (severe ethanol stress) (Nguyen et al. 2015; Ishida et al. 2016; Yamauchi and Izawa 2016)

not determined by mRNA base sequences, but instead by DNA base sequences in the promoter region. Although it is seemingly incomprehensible that the fate of post-transcriptional mRNAs is dependent upon promoters, this issue may be explained by the intervention of RNA binding proteins (RBPs) between transcription and translation during periods of stress.

1.4 Translational Repression and Induction of mRNP Granule Formation Due to Ethanol Stress

Ethanol produced by yeast through alcoholic fermentation reaches a high concentration at the final stage of the brewing process of wine and Japanese *sake*. Under laboratory conditions, translation activity started to be repressed at an ethanol concentration above 6% (v/v) and was highly repressed in the presence of 10% ethanol

(Iwaki et al. 2013a; Yamauchi and Izawa 2016). In addition, the formation of P-bodies was caused when the concentration of ethanol was above 6%; the size and amount of P-bodies also showed a tendency to increase with the rise in concentration (Izawa et al. 2007). On the other hand, the formation of SGs started to be induced when the ethanol concentration rose above 10% (Kato et al. 2011). SGs were formed very close to the location where P-bodies were synthesized in advance. In addition, under ethanol stress of the same concentration, SGs were formed more slowly than were P-bodies. It is thought that the formation of SGs in yeast requires a higher ethanol concentration and a longer induction time so as to induce a deeper level of translational repression than P-bodies can achieve and that the formation of SGs requires pre-existing P-bodies to be used as scaffolds.

As P-bodies and SGs disappeared promptly after ethanol was removed, the formation and degradation of P-bodies and SGs seem to be reversible responses dependent upon ethanol concentration (Yamamoto and Izawa 2013; Yamauchi and Izawa 2016). Although polysomes took more time to recover to their pre-stress levels than P-bodies and SGs to dissociate, they still re-formed relatively promptly; as such, ethanol is considered to be a stressor that causes very few residual effects for ribosomes and mRNP granules (Yamauchi and Izawa 2016).

It is thought that, when the stressor is removed, the housekeeping gene mRNAs sequestered in P-bodies and/or SGs may be transported to ribosomes for prompt resumption of translation (Brengues et al. 2005; Anderson and Kedersha 2006). It is estimated that translation may be resumed sooner and more economically by utilizing sequestered mRNAs rather than transcribing new mRNAs. Consequently, the resumption of translation following ethanol removal was delayed in yeast strains—such as *pbp1*Δ and *tif4632*Δ—that cannot form SGs (Kato et al. 2011). However, it has also been reported that limited numbers of mRNAs from mRNP granules were re-utilized in translation (Arribere et al. 2011). Thus, further study regarding the extent of mRNA recycling is necessary.

1.5 *BTN2* Is Preferentially Translated Under Translation Repression Due to High Concentrations of Ethanol

Although translation is strongly inhibited in the presence of 10% ethanol concentration, cells do not die immediately, and slight protein synthesis is observed; this observation suggested that some mRNAs are selectively and preferentially translated (Yamauchi and Izawa 2016). Therefore, we tried to identify mRNAs that form polysomes even under high concentrations of ethanol. Ribosome profiling was performed and mRNAs in polysome fractions were collected and analyzed. We identified 27 types of mRNAs—which greatly increased in number in polysomes under high concentrations of ethanol—although very few were found under non-stress conditions. The expression of their corresponding genes was studied, revealing that each responded to 10% ethanol by significantly increasing mRNA levels.

One of those genes, *BTN2*, encodes the binding protein v-SNARE and is involved in intracellular protein transport and sequestration of protein aggregation in the nucleus (Kama et al. 2007; Miller et al. 2015). It has been reported that, as Btn2 plays an important role in the correct localization of various proteins, *btn2Δ* cells show high sensitivity toward various kinds of stress, including ethanol (Chattopadhyay et al. 2000; Chattopadhyay and Pearce 2002; Kim et al. 2005; Espinazo-Romeu et al. 2008; Yang et al. 2011). Although Btn2 proteins are detected at very low levels via Western blot under non-stress conditions, their significant increase was observed after treatment with 8–10% ethanol (Yamauchi and Izawa 2016). These results indicated that, even under high-concentration ethanol stress, which repressed the nuclear export and translation of most mRNAs, *BTN2* mRNAs were actively synthesized and were furthermore preferentially translated. In addition, Btn2 protein levels, which increased due to induction under high-concentration ethanol, decreased smoothly after ethanol was eliminated; they had disappeared within 45 min (Yamauchi and Izawa 2016). Btn2 proteins also disappeared when the ethanol concentration was reduced from 10% to 5%. Since overexpression of *BTN2* under non-stress conditions causes a delay of cell growth (Sopko et al. 2006), it is thought that the expression of *BTN2* is only required under severe stress conditions and is harmful for yeast cells under non-stress conditions.

1.6 Improve Ethanol Tolerance and Fermentability by *BTN2* Promoter

As mentioned earlier, any gene can be expressed under glucose depletion by utilizing the *HSP26* promoter (Zid and O'Shea 2014). Similarly, we studied the possibility of inducing the expression of other non-native genes under high-concentration ethanol stress using the *BTN2* promoter (Fig. 1.3). We constructed an expression plasmid containing the promoter and terminator regions of *BTN2* (pYY2712) flanking the coding regions of other genes, such as *CUR1*, *GIC2*, and *YUR1* (Yamauchi and Izawa 2016). Although these three genes were only slightly expressed under non-stress conditions, their transcription and protein synthesis were remarkably upregulated under high concentrations of ethanol, confirming preferential translation. This result suggests that under the pronounced translation repression caused by severe ethanol as well as glucose depletion, preferential translation is regulated by DNA base sequences in the promoter region.

These findings on *BTN2* led us to the idea that the *BTN2* promoter could be used as a shot in the arm that reactivates yeast cells at the final stage of the brewing process, when fermentability becomes impaired due to the increase in ethanol concentration. We are currently studying the impact of this by inducing the expression of *TRK1* and *PDR18* at the final stage of the brewing process. The formation of a potassium/proton gradient is essential in the maintenance of alcohol tolerance and fermentability, and it has been reported that the overexpression of *TRK1*, which

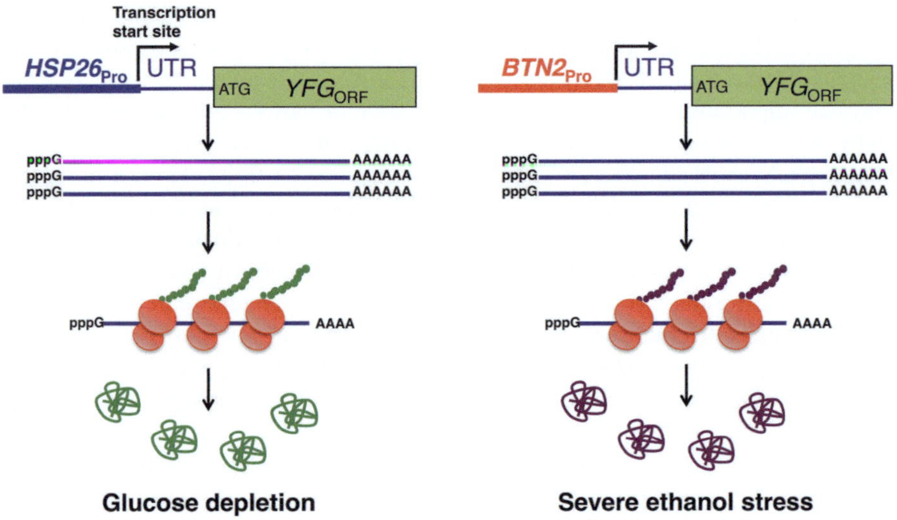

Fig. 1.3 Preferential translation under stress conditions is determined by DNA base sequences in the promoter region. The *HSP26* promoter and the *BTN2* promoter overcome the translation repression and enable protein synthesis under glucose depletion and severe ethanol stress, respectively

encodes a potassium transporter, improves ethanol production (Lam et al. 2014). *PDR18*, which encodes a transporter that promotes extracellular release of ethanol, is also involved in the maintenance of ethanol tolerance and fermentability; the null mutant (*pdr18*Δ) shows accumulation of intracellular ethanol and decreased resistance to ethanol (Teixeira et al. 2012). As the expression level of *TRK1* and *PDR18* decreases with the increase in ethanol concentration during the brewing process of Japanese *sake* (Wu et al. 2006), fermentability at the final stage is anticipated to be improved through the expression of these genes by utilizing the *BTN2* promoter.

1.7 *ADH7* and *BDH2* Are Preferentially Translated Under Translational Repression Due to Vanillin

Lignocellulosic biomass, which is used in the production of second-generation bioethanol, is pretreated with acids and hot-compressed water for saccharification; this process also yields fermentation inhibitors such as vanillin, furfural, and 5-hydroxymethylfurfural (HMF) as by-products (Antal et al. 1990; Klinke et al. 2004; Lu et al. 2009; Piotrowski et al. 2014). They repress the growth and fermentability of yeast in the fermenter, and it is known that vanillin and furfural suppress yeast translation (Iwaki et al. 2013a, b). Severe vanillin stress (>7.5 mM) causes the marked repression of overall protein synthesis (Iwaki et al. 2013a, b; Nguyen et al.

2014, 2015). Similar to *BTN2*, we explored and identified several mRNAs preferentially translated under severe vanillin stress.

ADH7, one of the mRNAs preferentially translated under severe vanillin stress, encodes NADPH-dependent medium-chain alcohol dehydrogenase (MDR) with *ADH6*; the enzyme reduces—and thus detoxifies—vanillin to vanillyl alcohol (Larroy et al. 2002a, b). *ADH7* and *ADH6* are paralogs with highly similar sequences. *ADH6* is constitutively expressed, and its deficient strains (*adh6*Δ) demonstrate sensitivity toward low concentrations of vanillin; on the other hand, *ADH7* expression is not observed under non-stress conditions and low vanillin concentrations, while the sensitivity of *adh7*Δ to low concentrations of vanillin was similar to that of wild-type strains. However, *adh7*Δ demonstrated extreme sensitivity toward high concentrations of vanillin. The expression of both genes was analyzed under such conditions, and while the transcription level of both genes increased, the protein level of Adh6 declined; *ADH6* mRNAs were presumably not translated due to the inhibitory effect of vanillin. Meanwhile, protein synthesis of Adh7 occurred smoothly under vanillin concentrations ranging from 10 to 15 mM, which otherwise repressed overall protein synthesis (Nguyen et al. 2015). These results clearly indicated that *ADH7* mRNAs were preferentially translated even under the marked repression of overall protein synthesis by severe vanillin stress.

Like *ADH7*, *BDH2* was also preferentially translated under translation repression caused by severe vanillin stress (Ishida et al. 2016). *BDH1* and *BDH2* encode NADH-dependent MDR, which also reduces vanillin to vanillyl alcohol (Nording et al. 2002; González et al. 2010). Similar to the relationship between *ADH6* and *ADH7*, *BDH2* expression was not observed under low concentrations of vanillin, whereas *BDH1* was constitutively expressed. However, at high concentrations of vanillin, only *BDH2* was preferentially translated, manifesting in a significant increase in its protein level (Ishida et al. 2016).

Both types of MDR—NADPH-dependent Adh7 and NADH-dependent Bdh2—were preferentially expressed under vanillin-induced translational repression, and this was thought to enable yeast to cope with severe vanillin stress. Since the expression of *ADH7* and *BDH2* was induced neither under non-stress conditions nor under low vanillin concentrations, it is assumed that yeast cells are able to compensate for minor vanillin stress via Adh6 and Bdh1 (Nguyen et al. 2015; Ishida et al. 2016). Conversely, they appear to be unable to combat high concentrations of vanillin without the support of Adh7 and Bdh2.

1.8 Improvement of Tolerance Against Biomass Conversion Inhibitors Using *ADH7* Promoter

We studied whether the expression of other genes could be induced by using *ADH7* and *BDH2* promoters under high vanillin concentrations, as was done with the *HSP26* and *BTN2* promoters. The coding regions of *GPX2*, *ADH6*, *ALD6*, and

ZWF1 were ligated downstream of the *ADH7* and *BDH2* promoters; expression of these genes was not observed under non-stress conditions and low concentrations of vanillin, but their transcriptional activation and increased production of the corresponding proteins were induced by high vanillin concentrations, confirming preferential translation (Nguyen et al. 2015; Ishida et al. 2016). These results indicated that expression of other non-native genes utilizing the *ADH7* and *BDH2* promoters could occur under pronounced translation repression caused by severe vanillin stress. Consequently, we ligated the coding regions of *ADH6* and *ALD6*—which encode aldehyde reductase—downstream of these promoters and introduced them into yeast cells using multicopy vectors; growth was improved significantly under severe stress resulting from a vanillin concentration of 12 mM (Nugyen et al. 2015; Ishida et al. 2016).

The influence of the biomass conversion inhibitors furfural and HMF was also studied. Although the expression of *ADH7* and *BDH2* was induced by HMF, furfural only activated the expression of *BDH2* (Ishida et al. 2016, 2017). These three biomass conversion inhibitors are usually all present in actual biomass hydrolysates (lignocellulosic hydrolysates). Therefore, we investigated the expression of *ADH7* and *BDH2* under the coexistence of vanillin, furfural, and HMF (a condition we call VFH stress) (Ishida et al. 2017). It was confirmed via polysome profiling that the translational activity of yeast was strongly repressed when the three inhibitors coexisted at a concentration level above 7 mM (Ishida et al. 2017). Despite the pronounced translation repression, *ADH7* was not only transcribed but also translated; however, *BDH2* expression was weaker than that of *ADH7*. Furthermore, under translational repression caused by a combination of 4 mM vanillin, 15 mM furfural, and 40 mM HMF—a mixture that approximates actual conditions in lignocellulosic hydrolysates—*ADH7* expression was significantly induced (Ishida et al. 2017). Yeast growth was improved under VFH stress when *ALD6* expression was induced by the *ADH7* promoter (Ishida et al. 2017). These results suggest that the *ADH7* promoter could be utilized for the improvement of not only tolerance of biomass conversion inhibitors but also yeast fermentability in lignocellulosic hydrolysates.

1.9 Characteristics of the Promoters that Enable the Preferential Translation Under the Repression of Overall Protein Synthesis

We very recently found the third gene whose promoter has better capability to induce the expression of other non-native genes than the *ADH7* promoter and the *BDH2* promoter (Nguyen et al. 2018). Beside this gene, *ADH7*, *BDH2*, and *BTN2* contain the HSE in their promoters as well as *HSP26*. contain the HSE in their promoters. The presence of HSE in the promoter regions may contribute to efficient protein synthesis under the pronounced translation repression. Additionally, we found the binding sites of 15 common transcriptional factors in the promoter regions of *ADH7*, *BDH2*, and the newly found gene (Nguyen et al. submitted). Some of these common transcriptional factors including Hsf1 may be crucial for the efficient gene expression under the pronounced translation repression caused by high concentrations of vanillin.

Modifications in the terminator regions of *BTN2*, *ADH7*, and *BDH2* did not result in significant changes in their expression levels. Therefore, it was thought that, under stress-induced translational repression, the base sequences of promoter regions upstream of the transcription start sites would determine whether or not mRNAs would be preferentially translated. Interestingly, although *BTN2*, *ADH7*, and *BDH2* all carry the HSE in their promoter regions, these genes are not expressed during glucose depletion, whereas *HSP26* also contains an HSE in its promoter and is expressed in the absence of glucose. It is suggested that other *cis*-elements in the promoter region may play an important role in preferential translation under various kinds of stress. Although the mechanism that determines the fate of post-transcriptional mRNAs via promoter sequences is still not well understood, the involvement of *trans*-factors—such as mRNA-binding proteins recruited to the promoter region during transcription—has been suspected. During heat-shock response in mammalian cells, translation elongation factor eEF1A is recruited to the HSE of *HSP70*, then binds with *HSP70* mRNA, and promotes nuclear export and translation (Vera et al. 2014). In yeast, it is also possible that one or more factors are recruited to the promoter regions and then bind to mRNAs and cause the selective and preferential translation observed under stress-induced translational repression.

1.10 Conclusion

Under severe stress conditions that cause pronounced translation repression, majority of genes cannot be expressed by simply introducing them into yeast cells and activating their transcription. Therefore, the promoters described in this article might be valuable and expected to become useful tools in the modification of gene expression under severe stress conditions. Induction of the gene expression using the *BTN2* promoter might facilitate the development of yeast strains that maintain high fermentability even at the final stage of the brewing process (Fig. 1.4). The *ADH7* promoter would be useful for improvement of stress tolerance and fermentability of yeast cells in lignocellulosic hydrolysates.

In the real industrial brewing process, specially evolved yeast strains are used for brewing wine and Japanese *sake* or making bioethanol, and these industrial strains have distinct physiological and genetic features to suit the stressful brewing environments (Akao et al. 2011; Noguchi et al. 2012; Marsit and Dequin 2015; Hajar et al. 2017). Indeed, there are various differences in the DNA sequences of the *BTN2* promoter region among laboratory strains and industrial strains (wine yeast EC1118, *sake* yeast Kyokai No. 7 and No. 11, and bioethanol yeast CBS7960). Although laboratory strains have mainly been used in our studies so far, we have also verified that gene expression driven by the *BTN2* promoter was induced in a

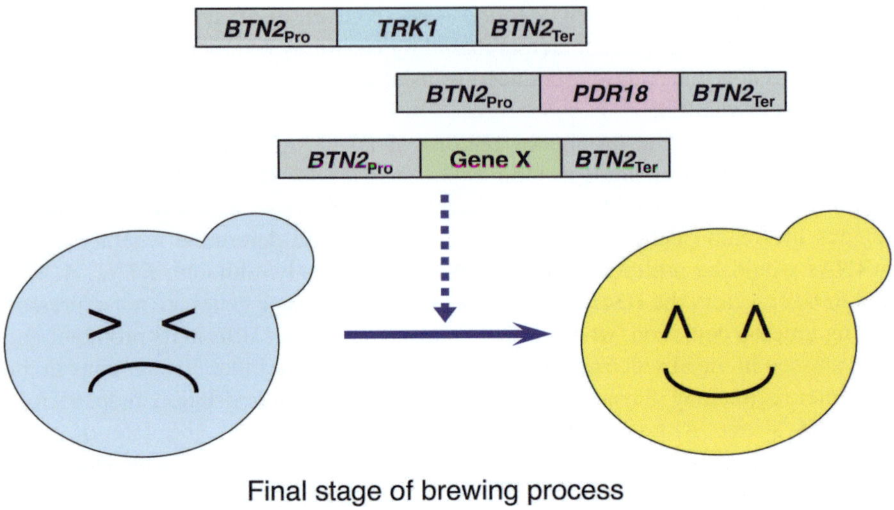

Fig. 1.4 Potential of the *BTN2* promoter as a tool to invigorate yeast cells via the modification of gene expression under the pronounced translation repression. The *BTN2* promoter might enable the induced protein synthesis of key genes, which are involved in stress tolerance and efficient fermentation but usually not expressed at the final stage of brewing process such as *TRK1* and *PDR18* (Teixeira et al. 2012; Lam et al. 2014)

wine yeast strain EC1118 when the ethanol concentration increased during brewing process (Kato et al. 2018). Further studies on the promoters of *BTN2* and *ADH7* of industrial strains and their usefulness under the actual fermentation conditions are underway.

References

Akao T, Yashiro I, Hosoyama A, Kitagaki H, Horikawa H, Watanabe D et al (2011) Whole-genome sequencing of sake yeast *Saccharomyces cerevisiae* Kyokai no. 7. DNA Res 18:423–434. https://doi.org/10.1093/dnares/dsr029

Anderson P, Kedersha N (2006) RNA granules. J Cell Biol 172:803–808. https://doi.org/10.1083/jcb.200512082

Antal MJ, Mok WS, Richards GN (1990) Mechanism of formation of 5-(hydroxymethyl)-2-furaldehyde from D-fructose and sucrose. Carbohydr Res 199:91–109

Arribere JA, Doudna JA, Gilbert WV (2011) Reconsidering movement of eukaryotic mRNAs between polysomes and P bodies. Mol Cell 44:745–758. https://doi.org/10.1016/j.molcel.2011.09.019

Ashe MP, De Long SK, Sachs AB (2000) Glucose depletion rapidly inhibits translation initiation in yeast. Mol Biol Cell 11:833–848. https://doi.org/10.1091/mbc.11.3.833

Balagopal V, Parker R (2009) Polysomes, P bodies and stress granules: states and fates of eukaryotic mRNAs. Curr Opin Cell Biol 21:403–408. https://doi.org/10.1016/j.ceb.2009.03.005

Brengues M, Sheth U, Parker R (2005) Movement of eukaryotic mRNAs between polysomes and cytoplasmic processing bodies. Science 310:486–489. https://doi.org/10.1126/science.1115791

Buchan JR, Muhlrad D, Parker R (2008) P bodies promote stress granule assembly in *Saccharomyces cerevisiae*. J Cell Biol 183:441–455. https://doi.org/10.1083/jcb.200807043
Chattopadhyay S, Pearce DA (2002) Interaction with Btn2p is required for localization of Rsglp: Btn2p-mediated changes in arginine uptake in *Saccharomyces cerevisiae*. Eukaryot Cell 1:606–612. https://doi.org/10.1128/EC.1.4.606-612.2002
Chattopadhyay S, Muzaffar NE, Sherman F, Pearce DA (2000) The yeast model for Batten disease: mutations in btn1, btn2, and hsp30 alter pH homeostasis. J Bacteriol 182:6418–6423. https://doi.org/10.1128/JB.182.22.6418-6423.2000
Espinazo-Romeu M, Cantoral JM, Matallana E, Aranda A (2008) Btn2p is involved in ethanol tolerance and biofilm formation in flor yeast. FEMS Yeast Res 8:1127–1136. https://doi.org/10.1111/j.1567-1364.2008.00397.x
González E, Fernández MR, Marco D, Calam E, Sumoy L, Parés X, Dequin S, Biosca JA (2010) Role of *Saccharomyces cerevisiae* oxidoreductases Bdh1p and Ara1p in the metabolism of acetoin and 2,3-butanediol. Appl Environ Microbiol 76:670–679. https://doi.org/10.1128/AEM.01521-09
Grousl T, Ivanov P, Frýdlova I, Vasicová P, Janda F, Vojtová J, Malinská K, Malcová I, Nováková L, Janosková D, Valásek L, Hasek J (2009) Robust heat shock induces eIF2alpha-phosphorylation-independent assembly of stress granules containing eIF3 and 40S ribosomal subunits in budding yeast, *Saccharomyces cerevisiae*. J Cell Sci 122:2078–2088. https://doi.org/10.1083/jcb.200807043
Hajar S, Azhar M, Abdulla R, Jambo SA, Marbawi H, Gansau JA, Faik AAM, Rodrigues KF (2017) Yeasts in sustainable bioethanol production: a review. Biochem Biophys Rep 10:52–61. https://doi.org/10.1016/j.bbrep.2017.03.003
Inada T, Aiba H (2005) Translation of aberrant mRNAs lacking a termination codon or with a shortened 3'-UTR is repressed after initiation in yeast. EMBO J 24:1584–1595. https://doi.org/10.1038/sj.emboj.7600636
Ishida Y, Nguyen TTM, Kitajima S, Izawa S (2016) Prioritized expression of *BDH2* under bulk translational repression and its contribution to tolerance to severe vanillin stress in *Saccharomyces cerevisiae*. Front Microbiol 7:1059. https://doi.org/10.3389/fmicb.2016.01059
Ishida Y, Nguyen TTM, Izawa S (2017) The yeast *ADH7* promoter enables gene expression under pronounced translation repression caused by the combined stress of vanillin, furfural, and 5-hydroxymethylfurfural. J Biotechnol 252:65. https://doi.org/10.1016/j.jbiotec.2017.04.024
Iwaki A, Kawai T, Yamamoto Y, Izawa S (2013a) Biomass conversion inhibitors, furfural and 5-hydroxymethylfurfural, induce the formation of mRNP granules and attenuate translation activity in yeast. Appl Environ Microbiol 446:225–233. https://doi.org/10.1128/AEM.02797-12
Iwaki A, Ohnuki S, Suga Y, Izawa S, Ohya Y (2013b) Vanillin inhibits translation and induces messenger ribonucleoprotein (mRNP) granule formation in *Saccharomyces cerevisiae*: application and validation of high-content, image-based profiling. PLoS One 8:e61748. https://doi.org/10.1371/journal.pone.0061748
Izawa S, Kita T, Ikeda K, Miki T, Inoue Y (2007) Formation of the cytoplasmic P-bodies in *sake* yeast during Japanese *sake* brewing and wine making. Biosci Biotechnol Biochem 71:2800–2807. https://doi.org/10.1271/bbb.70417
Kama R, Robinson M, Gerst JE (2007) Btn2, a hook1 ortholog and potential batten disease-related protein, mediates late endosome-Golgi protein sorting in yeast. Mol Cell Biol 27:605–621. https://doi.org/10.1128/MCB.00699-06
Kato K, Yamamoto Y, Izawa S (2011) Severe ethanol stress induces assembly of stress granules in *Saccharomyces cerevisiae*. Yeast 28:339–347. https://doi.org/10.1002/yea.1842
Kato S, Yamauchi Y, Izawa S (2018) Pronounced translation repression and the preferential expression of *BTN2* in yeast during the process of alcoholic fermentation and wine making (submitted)
Kedersha N, Stoecklin G, Ayodele M, Yacono P, Lykke-Anderson J, Fritzler MJ, Scheuner D, Kaufman RJ, Golan DE, Anderson P (2005) Stress granules and processing bodies are dynamically linked sites of mRNP remodeling. J Cell Biol 169:871–884. https://doi.org/10.1083/jcb.200502088

Kim Y, Chattopadhyay S, Locke S, Pearce DA (2005) Interaction among Btn1p, Btn2p, and Ist2p reveals potential interplay among the vacuole, amino acid levels, and ion homeostasis in the yeast *Saccharomyces cerevisiae*. Eukaryot Cell 4:281–288. https://doi.org/10.1128/EC.4.2.281-288.2005

Klinke HB, Thomsen AB, Ahring BK (2004) Inhibition of ethanol-producing yeast and bacteria by degradation products produced during pre-treatment of biomass. Appl Microbiol Biotechnol 66:10–26. https://doi.org/10.1007/s00253-004-1642-2

Lam FH, Ghaderi A, Fink GR, Stephanopoulos G (2014) Engineering alcohol tolerance in yeast. Science 346:71–75. https://doi.org/10.1126/scicence.1257859

Larroy C, Fernández MR, González E, Parés X, Biosca JA (2002a) Characterization of the *Saccharomyces cerevisiae YMR318C (ADH6)* gene product as a broad specificity NADPH-dependent alcohol dehydrogenase: relevance in aldehyde reduction. Biochem J 361:163–172. https://doi.org/10.1042/0264-6021:3610163

Larroy C, Parés X, Biosca JA (2002b) Characterization of a *Saccharomyces cerevisiae* NADP(H)-dependent alcohol dehydrogenase (ADHVII), a member of the cinnamyl alcohol dehydrogenase family. Eur J Biochem 269:5738–5745. https://doi.org/10.1046/j.1432-1033.2002.03296.x

Lu X, Yamauchi K, Phaiboonsilpa N, Saka S (2009) Two-step hydrolysis of Japanese beech as treated by semi-flow hot-compressed water. J Wood Sci 55:367–375. https://doi.org/10.1007/s10086-009-1040-6

Lui J, Campbell SG, Ashe MP (2010) Inhibition of translation following glucose depletion in yeast facilitates a rationalization of mRNA content. Biochem Soc Trans 38:1131–1136. https://doi.org/10.1042/BST0381131

Marsit S, Dequin S (2015) Diversity and adaptive evolution of *Saccharomyces* wine yeast: a review. FEMS Yeast Res 15:fov067. https://doi.org/10.1093/femsyr/fov067

Miller SBM, Ho C, Winkler J, Khokhrina M, Neuner A, Mohamed MYH, Guilbride DL, Richiter K, Lisby M, Schiebel E, Mogk A, Bukau B (2015) Compartment-specific aggregases direct distinct nuclear and cytoplasmic aggregate deposition. EMBO J 34:778–797. https://doi.org/10.15252/embj.201489524

Nguyen TTM, Kitajima S, Izawa S (2014) Importance of glucose-6-phosphate dehydrogenase (G6PDH) for vanillin tolerance in *Saccharomyces cerevisiae*. J Biosci Bioeng 118:263–269. https://doi.org/10.1016/j.jbiosc.2014.02.025

Nguyen TTM, Iwaki A, Izawa S (2015) The *ADH7* promoter of *Saccharomyces cerevisiae* is vanillin inducible and enables mRNA translation under severe vanillin stress. Front Microbiol 6:1390. https://doi.org/10.3389/fmicb.2015.01390

Nguyen TTM, Ishida Y, Kato S, Iwaki A, Izawa S (2018) The *VFH1* (*YLL056C*) promoter is vanillin-inducible and enables mRNA translation despite pronounced translation repression caused by severe vanillin stress in *Saccharomyces cerevisiae*. (submitted)

Noguchi C, Watanabe D, Zhou Y, Akao T, Shimoi H (2012) Association of constitutive hyperphosphorylation of Hsf1p with a defective ethanol stress response in *Saccharomyces cerevisiae* sake yeast strains. Appl Environ Microbiol 78:385–392. https://doi.org/10.1128/AEM.06341-11

Nordling E, Jörnvall H, Persson B (2002) Medium-chain dehydrogenases/reductases (MDR) family characterizations including genome comparisons and active site modeling. Eur J Biochem 269:4267–4276. https://doi.org/10.1046/j.1432-1033.2002.03114

Piotrowski JS, Zhang Y, Bates DM, Keating DH, Sato TK, Ong IM, Landick R (2014) Death by a thousand cuts: the challenges and diverse landscape of lignocellulosic hydrolysate inhibitors. Front Microbiol 5:90. https://doi.org/10.3389/fmicb.2014.00090

Sopko R, Huang D, Preston N, Chua G, Papp B, Kafadar K, Snyder M, Oliver SG, Cyert M, Hughes TR, Boone C, Andrews B (2006) Mapping pathways and phenotypes by systematic gene overexpression. Mol Cell 21:319–330. https://doi.org/10.1016/j.molcel.2005.12.011

Teixeira MC, Gosinho CP, Cabrito TR, Mira NP, Sá-Correia I (2012) Increased expression of the yeast multidrug resistance ABC transporter Pdr18 leads to increased ethanol tolerance and ethanol production in high gravity alcoholic fermentation. Microb Cell Fact 11:98. https://doi.org/10.1186/1475-2859-11-98

Vera M, Pani B, Griffiths LA, Muchardt C, Abbott CM, Singer RH, Nudler E (2014) The translation elongation factor eEF1A1 couples transcription to translation during heat shock response. eLife 3:e03164. https://doi.org/10.75554/eLife.03164

Wu H, Zheng X, Araki Y, Sahara H, Takagi H, Shimoi H (2006) Global gene expression analysis of yeast cells during sake brewing. Appl Environ Microbiol 72:7353–7358. https://doi.org/10.1128/AEM.01097-06

Yamamoto Y, Izawa S (2013) Adaptive response in stress granule formation and bulk translational repression upon a combined stress of mild heat shock and mild ethanol stress in yeast. Genes Cells 18:974–984. https://doi.org/10.1111/gtc.12090

Yamauchi Y, Izawa S (2016) Prioritized expression of *BTN2* of *Saccharomyces cerevisiae* under pronounced translation repression induced by severe ethanol stress. Front Microbiol 7:1319. https://doi.org/10.3389/fmicb.2016.01319

Yang J, Bae JY, Lee YM, Kwon H, Moon HY, Kang HA, Yee SB, Kim W, Choi W (2011) Construction of *Saccharomyces cerevisiae* strains with enhanced ethanol tolerance by mutagenesis of the TATA-binding protein gene and identification of novel genes associated with ethanol tolerance. Biotechnol Bioeng 108:1776–1787. https://doi.org/10.1002/bit.23141

Zid BM, O'Shea EK (2014) Promoter sequences direct cytoplasmic localization and translation of mRNAs during starvation in yeast. Nature 514:117–121. https://doi.org/10.1038/nature13578

Chapter 2
Constructing Mutant Ribosomes Containing Mutant Ribosomal RNAs

Kei Kitahara and Kentaro Miyazaki

Abstract The ribosome is the factory for protein biosynthesis, consisting of 3 different ribosomal RNA (rRNA) molecules (16S, 23S, and 5S rRNAs in prokaryotes) and more than 50 different ribosomal proteins. Because almost all organisms have multiple operons for rRNA genes (*rrn* operons), mutational analysis of ribosomes has inevitable technical difficulties, particularly for analyzing the functions of the 16S and 23S rRNAs, which form part of the core structure for the small (30S) and large (50S) subunits, respectively. In this chapter, we introduce six major strategies that allow researchers to perform mutational studies of the prokaryotic ribosome, particularly by focusing on the analysis of the 16S and 23S rRNA molecules. Although conventional mutational studies allow only for a small number of nucleotide changes simultaneously, recent approach developed by our group circumvents this problem in the *Escherichia coli* 16S rRNA gene, allowing for changes of up to 20% of the total nucleotides by interspecies exchange of the gene with that from foreign (non-*E. coli*) bacteria. The outcome of this novel technique has led to the discovery of an unexpected, nontranslational function (ribonuclease inhibitor) in the 16S rRNA molecule. The introduction of such a large sequence perturbation into the central core of the ribosome will open up a new era of ribosomal engineering to create highly functional ribosomes or phenotypic improvements of the host cell, which would be advantageous for biotechnological applications.

Keywords 16S rRNA · Mutational analysis · Functional metagenomics · Orthogonal (O)-translation system

K. Kitahara
Department of Chemistry, Faculty of Science, Hokkaido University, Sapporo, Hokkaido, Japan

K. Miyazaki (✉)
Department of Life Science and Biotechnology, Bioproduction Research Institute, National Institute of Advanced Industrial Science and Technology (AIST), Tsukuba, Ibaraki, Japan

Department of Computational Biology and Medical Sciences, Graduate School of Frontier Sciences, The University of Tokyo, Kashiwa, Chiba, Japan
e-mail: miyazaki-kentaro@aist.go.jp

S. Masuda, S. Izawa (eds.), *Applied RNA Bioscience*,
https://doi.org/10.1007/978-981-10-8372-3_2

2.1 Function and Structure of the Ribosome

The ribosome is the factory responsible for protein biosynthesis, which is omnipresent in every cell or organelle (e.g., mitochondria or chloroplast) of all organisms. The prokaryotic 70S ribosome consists of the small 30S subunit, which contains the 16S rRNA (1542 bases in *Escherichia coli*) as well as about 20 different ribosomal proteins, and the large 50S subunit, which contains the 23S and 5S rRNAs (2904 and 121 bases, respectively, in *E. coli*) as well as about 35 different ribosomal proteins, where S is defined as the Svedberg unit or sedimentation coefficient. The 70S ribosome (or simply the ribosome), which is formed by the association of both the 30S and 50S subunits, has a total molecular mass of 2.3 MDa, with a diameter of roughly 26 nm (Schuwirth et al. 2005; Schmeing and Ramakrishnan 2009). We can obtain an intuitive grasp for the size of the ribosome by imagining that approximately 20 ribosomes aligned in a row will sum up to the length of the short axis of an *E. coli* (0.5 μm). The ribosome is dominated by RNA with a molecular mass ratio of RNAs to proteins of about 2:1 (Milo et al. 2010). The ribosome is known to be densely packed inside the cell (Klumpp et al. 2013); in dividing *E. coli* cells, the number of the ribosomes per cell is about 18,800, which corresponds to about 25% of the dry mass of a cell (Neidhardt 1987; Kitahara and Suzuki 2009). The number of mRNAs in a cell at any given time is about 2400–7800, and a single mRNA is translated by multiple ribosomes on average, forming structures called polysomes (Milo et al. 2010).

At the beginning of translation, the small and large subunits associate to make the 70S ribosome (Voorhees and Ramakrishnan 2013; Ramakrishnan 2014). The role of the small subunit is to rigorously monitor the recognition of codons on the mRNA by the anticodons on the tRNAs (Voorhees and Ramakrishnan 2013), whereas that of the large subunit is to catalyze the polymerization of L-amino acids attached to the 3′ ends (CCA) of the tRNAs via a peptidyl transfer reaction (Rodnina 2013). It should be noted that aminoacyl-tRNAs that are erroneously mischarged by D-amino acids do not serve as good substrates for the peptidyl transfer reaction (Englander et al. 2015). Newly synthesized polypeptides thus consist solely of L-amino acids that are correctly aligned as directed by the mRNAs.

The biological significance of the ribosome is obvious because proteins and the translation system for their synthesis are essential to sustain life. In particular, rRNAs (16S, 23S, and 5S), which are the main components of the ribosome, are regarded as one of the most ancient molecules; all prokaryotic organisms (i.e., bacteria and archaea) possess these molecules, suggesting that rRNAs appeared well before the emergence of the last universal common ancestor of life (Forterre 2015). The genes for these rRNAs are nevertheless among the most highly conserved genes, thus suggesting their biological importance. T. Steitz and P. B. Moore have postulated a bold hypothesis—that the ancient ribosome was composed entirely of RNA(s) (Moore and Steitz 2002), which advocates that the first translated protein was synthesized by ribosomes composed entirely of RNA, and thus the origin of rRNAs is older than that of proteins.

The tertiary structures of the large and small subunits were revealed one after another before and after the year 2000, respectively. The first atomic structures were determined for those derived from extremophiles, such as the thermophilic bacterium *Thermus thermophilus* (Clemons et al. 1999; Tocilj et al. 1999; Schluenzen et al. 2000; Wimberly et al. 2000), the radiation-resistant bacterium *Deinococcus radiodurans* (Harms et al. 2001; Schlunzen et al. 2001), and the halophilic archaeon *Haloarcula marismortui* (Ban et al. 2000), while recent studies also revealed high-resolution structures for the *E. coli* 70S ribosome (Noeske et al. 2015). Importantly, it was shown that the functional core of the ribosome is occupied mainly by rRNAs, and ribosomal proteins on the other hand are located along the periphery of both subunits (Noeske et al. 2015). The abovementioned hypothesis by Steitz and Moore (Moore and Steitz 2002) was inevitably derived from structural insights. These structural insights, together with a large amount of biochemical data from the past, have clarified the detailed functions and movements of the ribosome during the steps of translation. Recent reviews show how the three processes of translation (i.e., initiation, elongation, and termination) proceed according to these structural analyses (Voorhees and Ramakrishnan 2013; Ramakrishnan 2014).

Recently, crystal structures of eukaryotic ribosomes and high-resolution structures of mitochondrial ribosomes were clarified using cryo-electron microscopy (Ben-Shem et al. 2011; Klinge et al. 2011; Rabl et al. 2011; Brown et al. 2014; Greber et al. 2014, 2015; Kaushal et al. 2014, 2015; Ramakrishnan 2014; Yusupova and Yusupov 2014; Amunts et al. 2015; Ott et al. 2016; Frank 2017). It was revealed that, although rRNAs have significant varieties in length between prokaryotes, eukaryotes, and organelles, every organismal ribosome contains essentially an identical core structure.

2.2 Constructing Mutant Ribosomes Containing Modified rRNAs

To clarify the function of rRNAs, we have long conducted mutational analyses by constructing a series of mutant ribosomes using the *E. coli* ribosome as a model. However, there is an intrinsic difficulty in performing these experiments, more specifically in obtaining a pure population of mutant ribosome in the cells. This is primarily because most bacterial genomes, including that of *E. coli*, contain multiple copies of the *rrn* operon (rRNA operon that usually encodes the 16S, 23S, and 5S rRNAs as well as several tRNAs as shown in Fig. 2.1) whose sequences are slightly different from each other. Because of this intrinsic heterogeneity, even the "wild-type" ribosome constitutes a mixture of various "wild-type" ribosomes that are composed of slightly different rRNAs. It is thus difficult to observe a clear phenotypic effect of mutant rRNAs due to the existence of a large population of endogenous, heterogeneous "wild-type" ribosomes if a null mutant is not used (Nomura 1999). To overcome this obstacle, researchers should choose the most appropriate methodology for their purposes from (1) to (5) as described below or the in vitro reconstitution method described in (6) (Table 2.1).

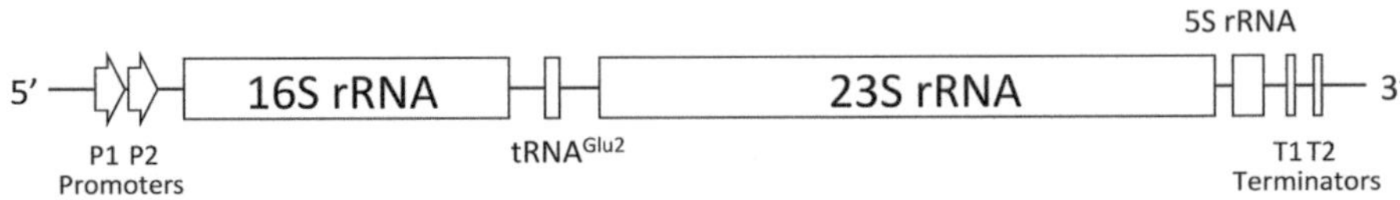

Fig. 2.1 Detailed structure of the *rrnB* operon in *Escherichia coli*. Transcription of rRNAs starts from either P1 or P2 promoter to generate a long precursor containing the 16S rRNA, $tRNA^{Glu2}$, 23S rRNA, and 5S rRNA in this order. Depending on the operons (*E. coli* contains seven *rrn* operons), there are some variations in sequences or arrangements in regulatory sequences, rRNA genes, and the types of tRNAs

Table 2.1 Six methodologies introduced in this chapter

Methods	Notes
1. Use of non-model microorganisms	Use an organism carrying a single copy of an *rrn* operon in the genome to prepare uniform ribosome species
2. Overexpression	Use a plasmid overexpression system for rRNAs to be investigated. Minimize the background derived from the genomic *rrn* operons by introducing antibiotic-resistant (e.g., spectinomycin resistance for 16S rRNA, thiostrepton resistance for 23S rRNA) point mutations into the rRNA genes to be investigated
3. Affinity-tag purification	Introduce affinity-tag sequence in the rRNA sequence to be investigated and separate the tagged RNA for the use of in vitro reconstitution study
4. O-Ribosome	Use orthogonal translational system by reverting the SD-anti-SD sequence pair. The orthogonal translation system would work independent from the normal translational system. Researchers can use standard *E. coli* (but not Δ7) strain as a host
5. Use of null mutant of rrn operon (Δ7 prrn)	Specialized *E. coli* strains have been developed that lack all the seven chromosomally encoded *rrn* operons. Researchers use a plasmid system to investigate the functionality of the rRNAs based on the survivability (or phenotypes)
6. In vitro reconstitution	e.g. iSAT (integrated synthesis, assembly, and translation) technology

Genetic approaches as such are powerful because the function of a mutant ribosome can be simply observed through the phenotypes of the mutant cells; researchers can also sometimes reveal unexpected functions of rRNAs that cannot be deduced from structural data alone. We introduce here several approaches to construct mutant prokaryotic ribosomes.

2.2.1 The Use of Non-model Microorganisms

One of the simplest ways to obtain a pure population of a ribosome containing mutated rRNA is to use a microorganism that naturally contains only one copy of *rrn* operon in the genome (Leviev et al. 1994; Nomura 1999). For example, by using microorganisms such as the archaeon *Halobacterium halobium*, various spontaneous point mutations have been discovered in the rRNA that render the cell, and therefore the ribosomes, resistant to specific antibiotics (Kloss et al. 1999). There are, however, some technical demerits in using such non-model organisms; a long time is often required to cultivate the cells due to slow growth. Difficulties in genetic manipulations of those organisms are also disadvantageous.

If a researcher attempts to introduce mutations in the rRNAs of the *E. coli* 16S rRNA, the following methods (2)–(5) are recommended.

2.2.2 Classical Overexpression Method

This method uses a multi-copy vector (e.g., pKK3535, pLK35), which harbors a single *rrnB* operon from *E. coli* (Brosius et al. 1981a, b). Mutations to be investigated can be introduced site-specifically into the plasmid-born 16S rRNA gene using a conventional oligonucleotide primer-based method such as QuikChange mutagenesis. It is good practice to also simultaneously introduce a known spectinomycin resistance mutation (C1192T) in the same gene as a phenotypic marker for exogenous 16S rRNA to minimize the background of wild-type endogenous 16S rRNAs as described below; otherwise, the mutations to be investigated would not display a clear phenotype (Green and Noller 1997; Moine et al. 2000). After introducing the vector carrying the mutated 16S rRNA genes into an *E. coli* host, the mutant 16S rRNA is overexpressed. In this state, both the wild-type (seven copies from the genome) and mutant 16S rRNAs are expressed in the same cell. The trick here is to inactivate the wild-type 16S rRNA by the addition of spectinomycin into the cultivation medium so that only 16S rRNA carrying the C1192U mutation can function in the cell. The phenotype of the cell is thus determined solely by the function of the plasmid-born, mutant 16S rRNA gene. Likewise, when we attempt to introduce mutations into the 23S rRNA gene, it is advisable to co-introduce a thiostrepton resistance mutation (A1068T) or macrolide resistance mutation (A2058G) into the gene carrying the mutations to be investigated (Moine et al. 2000). Thus, the background derived from the wild-type 23S rRNAs can be minimized by addition of thiostrepton or erythromycin in the medium. Moine et al. introduced simultaneous random mutations into three nucleotides in the binding site of the ribosomal protein S8 of the 16S rRNA, in which the C1192U mutation was co-introduced (Moine et al. 2000). By selecting on spectinomycin-containing plates, they successfully obtained functional sequences for the S8-binding site, including the wild-type *E. coli* sequence as well as naturally occurring non-*E. coli*-type sequences and unnatural sequences, suggesting that there are many alternative functional sequences for protein-binding nucleotides. Interestingly, they also reported that some mutant sequences were as functional as that of the wild-type counterpart as judged by the doubling times of the wild-type and mutant strains.

2.2.3 Affinity-Tag Purification

This method is used for purification of ribosomes with a mutant rRNA for in vitro biochemical experiments rather than to directly observe the phenotype of the cell containing the mutant ribosome. Affinity purification is a frequently used method for purification of recombinant proteins fused to an affinity tag (e.g., polyhistidine-tag). To date, some affinity tags have been developed for RNA molecules through which ribosome-containing mutant rRNA with an affinity tag can be purified in a manner similar to protein purification schemes. More specifically, by inserting a streptavidin-binding aptamer (Golovina et al. 2010) or MS2 stem loop (Gupta and Culver 2014; Terasaka et al. 2014) sequences into a plasmid-born rRNA gene to be mutated, only the ribosomes containing mutant rRNAs would be affinity-purified. We can use *E. coli* with wild-type *rrn* operons in the genome as a host; however, when researchers attempt to extract the ribosome for biochemical assays, it is strongly recommended to use a knockout mutant strain for ribonuclease (RNase) I because the periplasmic RNase has strong non-specific RNase activity that can degrade rRNAs upon cell disruption (Kitahara and Miyazaki 2011).

2.2.4 O-Ribosome

The O-ribosome system, where the letter O refers to "orthogonal" (Rackham and Chin 2005), is an in vivo translation system, which is becoming a popular replacement for tedious biochemical characterizations of the ribosome. Generally, there is a purine-rich region (e.g., AGGAGG) several bases upstream from the start codon of an mRNA (Shine and Dalgarno 1975), which is known as the ribosomal binding site (RBS) or Shine-Dalgarno (SD) sequence. On the other hand, at the 3′ end of the 16S rRNA, there is a pyrimidine-rich sequence (e.g., CCUCCU), which is complementary to the SD sequence, called the anti-SD sequence. When base pairs are formed between the SD and anti-SD, the ribosome 30S subunit is placed around the start codon of the mRNA to be translated. Native mRNAs contain various SD sequences, whose translational efficiency is highly dependent on the sequence. Such sequence variation is important for tuning the expression levels of each individual mRNA in a cell. Based on the variable nature of the SD sequences, it is also possible to rationally design the SD sequence so that translational levels can be regulated as desired. Unlike the case for SD sequences, however, anti-SD is functionally important since its modification can perturb the cell's proteomic profile. Nevertheless, if the anti-SD in the 16S rRNA and SD in mRNA sequences are simultaneously altered, such pair maintains the ability to form a complex, while the 16S rRNA becomes less reactive (or near completely inactive) with native mRNAs. This system, the so-called the "orthogonal translation system", was established after rigorous screening of nontoxic combinations of SD/anti-SD sequence to minimize the cross-reactivity with native mRNAs and render specific recognitions between the modified 16S rRNA and its cognate mRNA (Rackham and Chin 2005; Wang et al. 2007). The mutated SD thus selected is called the O-SD, and the mRNA with O-SD is called the O-mRNA, whereas the mutated anti-SD sequence is called the O-anti-SD, and the 16S rRNA with O-anti-SD is called the O-16S rRNA (Rackham and Chin 2005; Wang et al. 2007). By using the O-ribosome system, we can virtually separate two translational systems in the same cell; while the normal translation by the wild-type ribosome is used to sustain cell growth, the second artificial O-ribosome system can be exclusively used for another task such as the expression of a specific gene of interest. P. R. Cunningham's group has succeeded in selecting functional 16S rRNA sequences by randomizing particular areas (containing seven to nine bases) of the gene. They used a chloramphenicol acetyl transferase (CAT) gene as a reporter for O-mRNA, and functional 16S rRNAs that are capable of translating the CAT gene were selected on chloramphenicol-containing plates (Lee et al. 1996; Morosyuk et al. 2001; Saraiya et al. 2008; Baker et al. 2016). The O-ribosome system was also applied to establish proteins with an unnatural amino acid, which would be highly toxic if incorporated into the natural translational system. More specifically, Chin's group developed a mutant ribosome that efficiently incorporated an unnatural amino acid allocated to the amber codon. This was conducted by introducing mutations into the O-16S rRNA, which reduced the affinity of the ribosome to RF1 (Wang et al. 2007).

However, there are some limitations to the O-ribosome system. We can introduce mutations only into the 16S rRNA gene but not 23S rRNA gene, which is unlinked to the O-SD sequence. However, it was reported independently by two groups that the 16S and 23S rRNAs could be fused into a single chain (Fried et al. 2015; Orelle et al. 2015). Specifically, a circular permuted construct of the 23S rRNA was made and inserted into a specific region of the 16S rRNA via relatively short linkers. As the result, both subunits of the ribosome containing the tethered rRNAs were connected by these linkers just like a pair of castanets. If an O-anti-SD sequence was introduced into the tethered rRNA, it is possible to create a special O-ribosome that includes both the 16S and 23S rRNAs that would specifically translate O-mRNA. By using this experimental system, we can introduce, for example, mutations into the peptidyl-transferase center domain of the 23S rRNA without any resulting growth defects of the cell, which was difficult to achieve in conventional systems (Fried et al. 2015; Orelle et al. 2015). Thus, this system will help develop ribosomes that can catalyze the incorporation of unnatural amino acids that are usually not incorporated efficiently into polypeptides.

2.2.5 *Use of an* rrn *Operon Null Mutant (Δ7 prrn)*

A series of mutant *E. coli* strains known as Δ7 prrn is a series in which all seven chromosomal *rrn* operons are knocked out. Among them are, for example, TA647 developed by C. Squires (Asai et al. 1999) and SQ171 developed by S. Quan (Quan et al. 2015) (similar strains are also available for *Bacillus subtilis* (Yano et al. 2015)

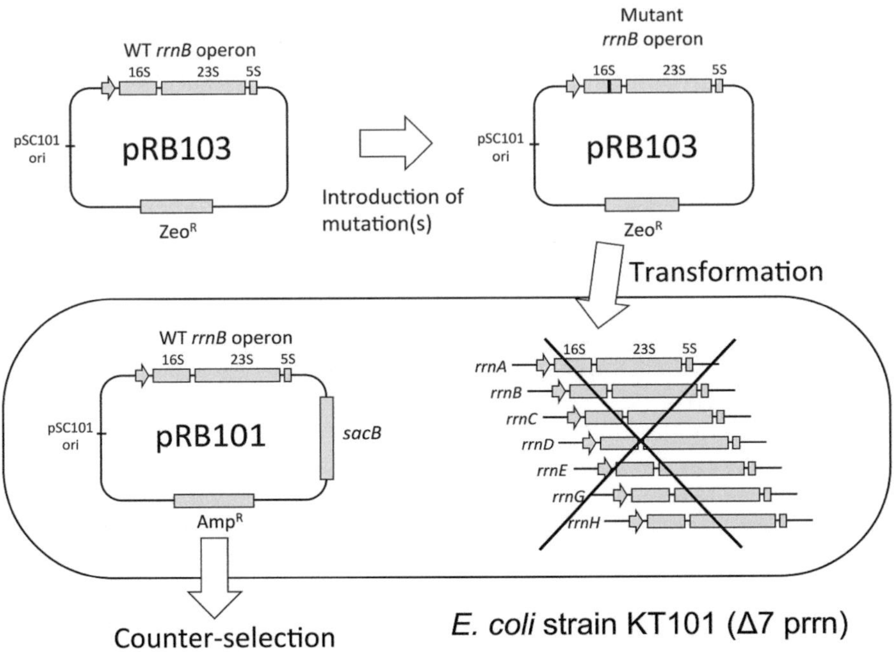

Fig. 2.2 Preparing a pure population of mutant ribosome using the Δ7 system. The *E. coli* strain KT101 has no intact *rrn* operons in its genome. Alternatively, it contains pRB101, a growth rescue plasmid, which contains the wild-type *rrnB* operon. We use the plasmid pRB103 to introduce mutations into rRNAs by means of, for example, QuikChange mutagenesis. The mutated plasmid is then introduced into KT101 cells. After transforming the KT101 cells, we transfer the Zeocin-resistant colony (containing both pRB101 and pRB103) onto sucrose-containing LB plate. Through this process, we can obtain cells that have completely lost pRB101

and yeast (Oakes et al. 1998; Rakauskaite and Dinman 2008). Despite the lack of all *rrn* operons, the Δ7 prrn strains can still grow because they possess an exo-chromosomal copy of the *rrn* operon by supplying a complementation (or growth rescue) plasmid harboring the *E. coli rrnB* operon as well as a counterselection marker (Fig. 2.2). To introduce mutations into the rRNA genes, we prepared another *rrn* operon-containing vector having different antibiotic resistance genes (Kitahara and Miyazaki 2011). The mutant plasmid carrying a mutated rRNA gene, which can be conducted using the QuikChange mutagenesis protocol, for example, is used to transform a Δ7 prrn strain. Subsequent counterselection to eliminate the preexisting complementation plasmid allows for the isolation of mutants whose growth is supported solely by the exogenous mutant rRNAs. The merit of this system is that we can completely exclude wild-type rRNAs from the cell; the replacement of the complementation plasmid with the mutant plasmid can be done by taking advantage of a counterselectable marker, such as *sacB*, which is toxic to the cell in the presence of sucrose, in the complementation plasmid (Kitahara et al. 2007). If mutations render the ribosome nonfunctional, the host *E. coli* would not grow on the sucrose plate.

Likewise, if mutations render the ribosome less functional, the host *E. coli* would grow slower compared to that complemented by the self (wild-type *E. coli*) rRNA. We can also assume that, if mutations render the ribosome more functional, the host *E. coli* would grow faster than the wild-type counterpart. Such fast growers should be readily selected by the continuous cultivation of the mixed population of the library. We can thus judge whether or not mutant rRNA is functional or how active the mutant rRNA is using the host's growth phenotype as a readout.

In the newest Δ7 system, KT101 (Kitahara and Suzuki 2009) is a derivative of SQ171, in which the complementation plasmid pRB101 (pSC101 ori) encoding the *rrnB* operon, *sacB*, and ampicillin resistance (Ap^R) gene was introduced (SQ171 and KT101 also harbor pTRNA67 [p15A ori] to supply tRNAs that were co-eliminated during the complete deletion of the *rrn* operons from the genome). For the introduction of mutations, there are *rrnB*-encoding plasmids such as pRB102 (pSC101 ori, Km^R) and pRB103 (pSC101 ori, $Zeocin^R$) (Fig. 2.2). Another system uses MY201 rna^- (a derivative of KT101) as a host that contains pMY201 (encoding *rrnB* operon, *sacB*, Ap^R gene, and p15A ori) (Miyazaki et al. 2017). In this system, pMY201 can be replaced with a combination of two plasmids (pML103Δ and pMS205aTp1) in which artificial mutations can be introduced. The pML103Δ (pSC101 ori) encodes the 23S and 5S rRNAs as well as some tRNA genes, and the pMS205aTp1 (p15A ori) encodes the 16S rRNA gene. The two-plasmid system lowers the probability of introducing undesired mutations, which can be generated during the course of engineering a rather large plasmid carrying the entire *rrn* operon.

2.2.6 *In Vitro Reconstitution*

The ribosome has long been known for its ability to be reconstituted from its components (three rRNAs and total ribosomal proteins); these purified components can be mixed to form functional subunits or 70S ribosome (Traub and Nomura 1968; Nierhaus and Dohme 1974). Using a system called iSAT, Jewett et al. have shown that functional ribosomes can be assembled from rRNAs that are transcribed from their genes. By changing the genetic information, it was possible to construct mutant ribosomes with mutated rRNAs (Jewett et al. 2013).

2.3 Constructing Ribosomes with Hypermutations in 16S rRNA

Although the introduction of multiple mutations into an rRNA may appear to be a good strategy for the genetic analysis of rRNA function, the fact is that we can introduce only about ten random mutations (far less than 1% of total length, e.g., 1542 bases for *E. coli* 16S rRNA) at once into the 16S rRNA (Laios et al. 2004). This result implies that the ribosome is such a sophisticated molecule such that point mutations are generally deleterious. It is thus difficult to construct a functional ribosome with multiple mutations in rRNAs. Despite this, we have succeeded in introducing hypermutations into the 16S rRNA.

Because both ends of the 16S rRNA genes are highly conserved among many species of bacteria (Cannone et al. 2002), various 16S rRNA gene fragments can be amplified by PCR using the same set of universal primers. In our previous study (Kitahara et al. 2012; Kitahara and Miyazaki 2014) (Fig. 2.3), we used metagenomic DNA extracted from various environmental specimens (sea water, soils, and fermented foods) and amplified various 16S rRNA gene fragments using the primer pair. The fragments were cloned into pRB103 (Zeocin^R, pSC101 ori) to generate a plasmid library containing 16S rRNAs from various species whose sequences were highly divergent, yet the total lengths were nearly the same. Next, we took advantage of the Δ7 prrn strain KT101 (Kitahara and Suzuki 2009) and tested whether or not foreign 16S rRNA can function in the *E. coli* ribosome. As the result, we found that the gammaproteobacterial *E. coli* 16S rRNA can be replaced, for example, by the 16S rRNA from a different class of bacteria (betaproteobacteria). Interestingly, by comparing approximately 30 different sequences that were functional in *E. coli*, we found that as many as 628 (40.7%) bases out of 1542 bases can be mutated (Fig. 2.4), the number of which would be further increased as the number of functional sequences increases. This method can be used for hyper-mutagenesis of the 16S rRNA. By using this method, we have successfully found that the 16S rRNA is a specific inhibitor of RNase I in *E. coli*, and further mutational analysis showed that it is the helix-41 in 16S rRNA that is involved in the specific interaction between the 16S rRNA and RNase I (Kitahara and Miyazaki 2011).

2.4 Improving the Method to Construct Hypermutation Libraries of 16S rRNA

In general, when we express recombinant proteins from an expression vector, we can clearly designate both the N and C termini by allocating the start and stop codons for both ends of an ORF, respectively. On the other hand, for RNA molecules and particularly for functional RNA molecules in prokaryotes, transcription starts from the transcription start point under the promoter and ends at the terminator. This transcript is a precursor that is then processed by RNases to be converted into a mature molecule. In the case of the ribosome, ribosomal proteins bind to rRNAs while they are being transcribed, thus making RNP particles (ribosome assembly) (Shajani et al. 2011; Davis and Williamson 2017). Many assembly factors (RNA helicase, RNA chaperones, ribosome-dependent GTPase, and RNA modification enzyme) promote the assembly during the process as well. In the case of the 16S rRNA, although both 5′ and 3′ termini of pre-16S rRNA form a

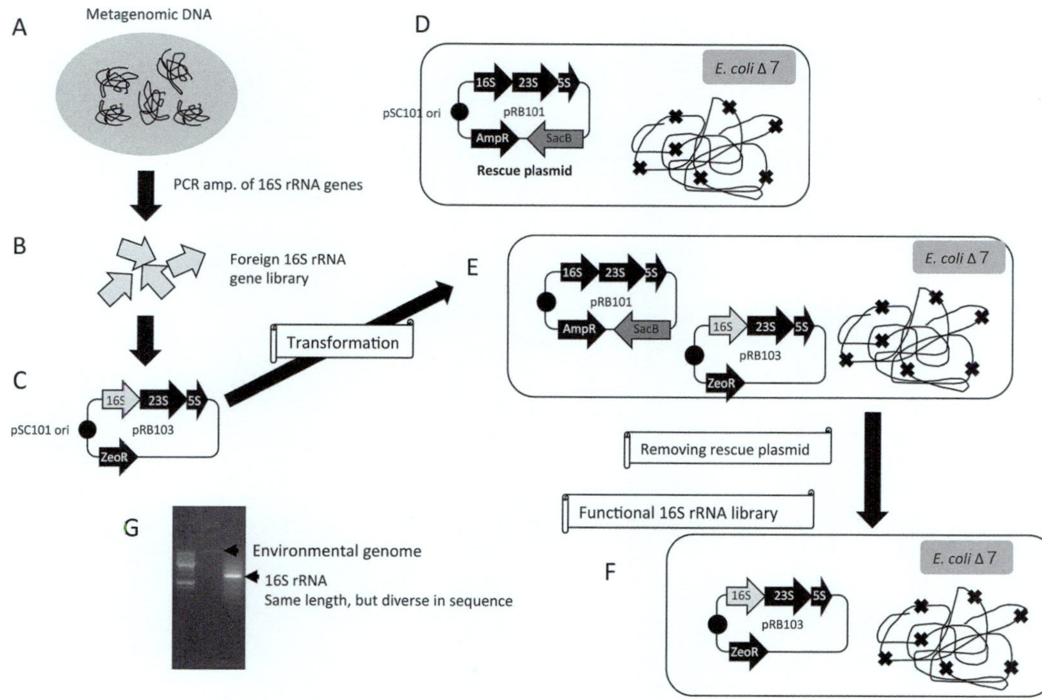

Fig. 2.3 Systematic selection of functional 16S rRNA sequences from the metagenome in *E. coli.* Metagenomic DNA samples that were extracted from various environmental specimens (**a**) were used as a source for the 16S rRNA gene. By using a set of universal primers, full-length (except small fragments on both ends) 16S rRNA genes were amplified (**b**). These fragments were used to replace the corresponding sequences of pRB103, thus generating a plasmid library containing 16S rRNA genes from various species (**c**). These plasmids were then introduced into a Δ7 prrn strain KT101 (**d**, **e**). Transformants that were selected on Zeocin plate containing both pRB101 (rescue plasmid) and pRB103 were isolated, resuspended with liquid medium, and transferred onto a sucrose-containing plate to eliminate pRB101 (**f**). Cell growth on the sucrose plate suggests that the cells contain a functional sequence. (**g**) DNA fragments of various 16S rRNA genes amplified from metagenomic DNA

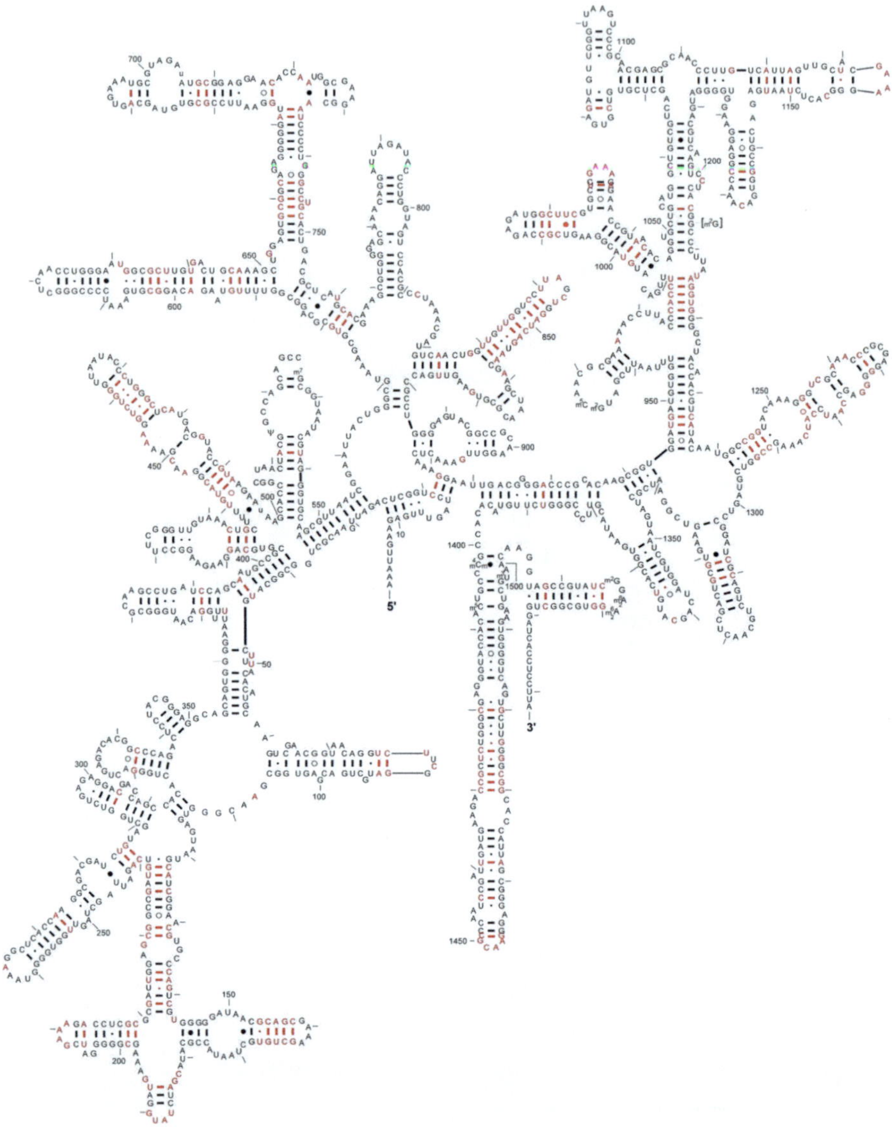

Fig. 2.4 A betaproteobacterial 16S rRNA sequence that is functional in *E. coli.* A functional sequence (B07: only 80.9% identical to *E. coli* sequence) found from a metagenomic library. Bases that are different from the *E. coli* sequence are highlighted in red

processing stem, this stem and other neighboring regions are excised by multiple RNases, resulting in the mature 16S rRNA (1542 bases in *E. coli*). After processing, the 5′ and 3′ termini of the new molecules are no longer adjacent (Kitahara and Suzuki 2009; Shajani et al. 2011). Our research has shown that, when a foreign (non-*E. coli*) 16S rRNA is expressed in *E. coli*, *E. coli* sequences for both 5′ and 3′

termini should be strictly conserved; otherwise, the ribosomal 30S subunit does not function correctly (Kitahara et al. 2012; Miyazaki et al. 2017). When researchers introduce hypermutations into 16S rRNA by substitution with metagenome-derived 16S rRNA, we recommend that they use primer pairs Bac1f and UN1542r because they can minimize the risk of interfering with the ribosomal assembly process as well as inter-RNA helix misinteraction of helix 2 or central pseudoknots (Miyazaki et al. 2017; Tsukuda et al. 2017).

2.5 Conclusion

Although numerous structural studies have been performed for the ribosome, functional analyses are still under way, which should be performed by mutational approaches combined with biochemical characterizations. The main reason for the scarcity of mutational studies on the ribosome is primarily because the construction of a mutant ribosome requires a lot of effort as well as some technical tips as described in this chapter. Although an in vitro mutagenesis and reconstitution method exists (6), it has not been widely used; genetic approaches described in the present chapter using live *E. coli* or other prokaryotes should continue to be used as standard methods for constructing mutant ribosomes. The demand to utilize such technologies will continue to increase in order to better elucidate the functions of rRNAs as well as to obtain ribosomes with improved function for biotechnological or synthetic biological purposes.

References

Amunts A, Brown A, Toots J, Scheres SH, Ramakrishnan V (2015) Ribosome. The structure of the human mitochondrial ribosome. Science 348:95–98

Asai T, Zaporojets D, Squires C, Squires CL (1999) An *Escherichia coli* strain with all chromosomal rRNA operons inactivated: complete exchange of rRNA genes between bacteria. Proc Natl Acad Sci U S A 96:1971–1976

Baker KA, Lamichhane R, Lamichhane T, Rueda D, Cunningham PR (2016) Protein-RNA dynamics in the central junction control 30S ribosome assembly. J Mol Biol 428:3615–3631

Ban N, Nissen P, Hansen J, Moore PB, Steitz TA (2000) The complete atomic structure of the large ribosomal subunit at 2.4 A resolution. Science 289:905–920

Ben-Shem A, Garreau de Loubresse N, Melnikov S, Jenner L, Yusupova G, Yusupov M (2011) The structure of the eukaryotic ribosome at 3.0 A resolution. Science 334:1524–1529

Brosius J, Dull TJ, Sleeter DD, Noller HF (1981a) Gene organization and primary structure of a ribosomal RNA operon from *Escherichia coli*. J Mol Biol 148:107–127

Brosius J, Ullrich A, Raker MA, Gray A, Dull TJ, Gutell RR, Noller HF (1981b) Construction and fine mapping of recombinant plasmids containing the rrnB ribosomal RNA operon of *E. coli*. Plasmid 6:112–118

Brown A, Amunts A, Bai XC, Sugimoto Y, Edwards PC, Murshudov G, Scheres SH, Ramakrishnan V (2014) Structure of the large ribosomal subunit from human mitochondria. Science 346:718–722

Cannone JJ, Subramanian S, Schnare MN, Collett JR, D'Souza LM, Du Y, Feng B, Lin N, Madabusi LV, Müller KM, Pande N, Shang Z, Yu N, Gutell RR (2002) The comparative RNA web (CRW) site: an online database of comparative sequence and structure information for ribosomal, intron, and other RNAs. BMC Bioinformatics 3:2
Clemons WM Jr, May JL, Wimberly BT, Mccutcheon JP, Capel MS, Ramakrishnan V (1999) Structure of a bacterial 30S ribosomal subunit at 5.5 A resolution. Nature 400:833–840
Davis JH, Williamson JR (2017) Structure and dynamics of bacterial ribosome biogenesis. Philos Trans R Soc Lond Ser B Biol Sci 372:pii: 20160181
Englander MT, Avins JL, Fleisher RC, Liu B, Effraim PR, Wang J, Schulten K, Leyh TS, Gonzalez RL Jr, Cornish VW (2015) The ribosome can discriminate the chirality of amino acids within its peptidyl-transferase center. Proc Natl Acad Sci U S A 112:6038–6043
Forterre P (2015) The universal tree of life: an update. Front Microbiol 6:717
Frank J (2017) The mechanism of translation. F1000Res 6:198
Fried SD, Schmied WH, Uttamapinant C, Chin JW (2015) Ribosome subunit stapling for orthogonal translation in *E. coli*. Angew Chem Int Ed 54:12791–12794
Golovina AY, Bogdanov AA, Dontsova OA, Sergiev PV (2010) Purification of 30S ribosomal subunit by streptavidin affinity chromatography. Biochimie 92:914–917
Greber BJ, Boehringer D, Leibundgut M, Bieri P, Leitner A, Schmitz N, Aebersold R, Ban N (2014) The complete structure of the large subunit of the mammalian mitochondrial ribosome. Nature 515:283–286
Greber BJ, Bieri P, Leibundgut M, Leitner A, Aebersold R, Boehringer D, Ban N (2015) Ribosome. The complete structure of the 55S mammalian mitochondrial ribosome. Science 348:303–308
Green R, Noller HF (1997) Ribosomes and translation. Annu Rev Biochem 66:679–716
Gupta N, Culver GM (2014) Multiple in vivo pathways for *Escherichia coli* small ribosomal subunit assembly occur on one pre-rRNA. Nat Struct Mol Biol 21:937–943
Harms J, Schluenzen F, Zarivach R, Bashan A, Gat S, Agmon I, Bartels H, Franceschi F, Yonath A (2001) High resolution structure of the large ribosomal subunit from a mesophilic eubacterium. Cell 107:679–688
Jewett MC, Fritz BR, Timmerman LE, Church GM (2013) In vitro integration of ribosomal RNA synthesis, ribosome assembly, and translation. Mol Syst Biol 9:678
Kaushal PS, Sharma MR, Booth TM, Haque EM, Tung CS, Sanbonmatsu KY, Spremulli LL, Agrawal RK (2014) Cryo-EM structure of the small subunit of the mammalian mitochondrial ribosome. Proc Natl Acad Sci U S A 111:7284–7289
Kaushal PS, Sharma MR, Agrawal RK (2015) The 55S mammalian mitochondrial ribosome and its tRNA-exit region. Biochimie 114:119–126
Kitahara K, Miyazaki K (2011) Specific inhibition of bacterial RNase T2 by helix 41 of 16S ribosomal RNA. Nat Commun 2:549
Kitahara K, Miyazaki K (2014) Revisiting bacterial phylogeny. Mob Genet Elem 3(1):e24210
Kitahara K, Suzuki T (2009) The ordered transcription of RNA domains is not essential for ribosome biogenesis in *Escherichia coli*. Mol Cell 34:760–766
Kitahara K, Kajiura A, Sato NS, Suzuki T (2007) Functional genetic selection of Helix 66 in *Escherichia coli* 23S rRNA identified the eukaryotic-binding sequence for ribosomal protein L2. Nucleic Acids Res 35:4018–4029
Kitahara K, Yasutake Y, Miyazaki K (2012) Mutational robustness of 16S ribosomal RNA, shown by experimental horizontal gene transfer in *Escherichia coli*. Proc Natl Acad Sci U S A 109:19220–19225
Klinge S, Voigts-Hoffmann F, Leibundgut M, Arpagaus S, Ban N (2011) Crystal structure of the eukaryotic 60S ribosomal subunit in complex with initiation factor 6. Science 334:941–948
Kloss P, Xiong L, Shinabarger DL, Mankin AS (1999) Resistance mutations in 23 S rRNA identify the site of action of the protein synthesis inhibitor linezolid in the ribosomal peptidyl transferase center. J Mol Biol 294:93–101
Klumpp S, Scott M, Pedersen S, Hwa T (2013) Molecular crowding limits translation and cell growth. Proc Natl Acad Sci U S A 110:16754–16759

Laios E, Waddington M, Saraiya AA, Baker KA, O'Connor E, Pamarathy D, Cunningham PR (2004) Combinatorial genetic technology for the development of new anti-infectives. Arch Pathol Lab Med 128:1351–1359
Lee K, Holland-Staley CA, Cunningham PR (1996) Genetic analysis of the Shine-Dalgarno interaction: selection of alternative functional mRNA-rRNA combinations. RNA 2:1270–1285
Leviev IG, Rodriguez-Fonseca C, Phan H, Garrett RA, Heilek G, Noller HF, Mankin AS (1994) A conserved secondary structural motif in 23S rRNA defines the site of interaction of amicetin, a universal inhibitor of peptide bond formation. EMBO J 13:1682–1686
Milo R, Jorgensen P, Moran U, Weber G, Springer M (2010) BioNumbers—the database of key numbers in molecular and cell biology. Nucleic Acids Res 38:D750–D753
Miyazaki K, Sato M, Tsukuda M (2017) PCR primer design for 16S rRNAs for experimental horizontal gene transfer test in *Escherichia coli*. Front Bioeng Biotechnol 5:14
Moine H, Squires CL, Ehresmann B, Ehresmann C (2000) In vivo selection of functional ribosomes with variations in the rRNA-binding site of Escherichia coli ribosomal protein S8: evolutionary implications. Proc Natl Acad Sci U S A 97:605–610
Moore PB, Steitz TA (2002) The involvement of RNA in ribosome function. Nature 418:229–235
Morosyuk SV, Santalucia J Jr, Cunningham PR (2001) Structure and function of the conserved 690 hairpin in *Escherichia coli* 16 S ribosomal RNA. III. Functional analysis of the 690 loop. J Mol Biol 307:213–228
Neidhardt FC (1987) *Escherichia coli* and *Salmonella typhimurium*, cellular and molecular biology. ASM Press, Washington DC
Nierhaus KH, Dohme F (1974) Total reconstitution of functionally active 50S ribosomal subunits from Escherichia coli. Proc Natl Acad Sci U S A 71:4713–4717
Noeske J, Wasserman MR, Terry DS, Altman RB, Blanchard SC, Cate JH (2015) High-resolution structure of the Escherichia coli ribosome. Nat Struct Mol Biol 22:336–341
Nomura M (1999) Engineering of bacterial ribosomes: replacement of all seven *Escherichia coli* rRNA operons by a single plasmid-encoded operon. Proc Natl Acad Sci U S A 96:1820–1822
Oakes M, Aris JP, Brockenbrough JS, Wai H, Vu L, Nomura M (1998) Mutational analysis of the structure and localization of the nucleolus in the yeast *Saccharomyces cerevisiae*. J Cell Biol 143:23–34
Orelle C, Carlson ED, Szal T, Florin T, Jewett MC, Mankin AS (2015) Protein synthesis by ribosomes with tethered subunits. Nature 524:119–124
Ott M, Amunts A, Brown A (2016) Organization and regulation of mitochondrial protein synthesis. Annu Rev Biochem 85:77–101
Quan S, Skovgaard O, Mclaughlin RE, Buurman ET, Squires CL (2015) Markerless *Escherichia coli rrn* deletion strains for genetic determination of ribosomal binding sites. G3 5:2555–2557
Rabl J, Leibundgut M, Ataide SF, Haag A, Ban N (2011) Crystal structure of the eukaryotic 40S ribosomal subunit in complex with initiation factor 1. Science 331:730–736
Rackham O, Chin JW (2005) A network of orthogonal ribosome x mRNA pairs. Nat Chem Biol 1:159–166
Rakauskaite R, Dinman JD (2008) rRNA mutants in the yeast peptidyltransferase center reveal allosteric information networks and mechanisms of drug resistance. Nucleic Acids Res 36:1497–1507
Ramakrishnan V (2014) The ribosome emerges from a black box. Cell 159:979–984
Rodnina MV (2013) The ribosome as a versatile catalyst: reactions at the peptidyl transferase center. Curr Opin Struct Biol 23:595–602
Saraiya AA, Lamichhane TN, Chow CS, Santalucia J Jr, Cunningham PR (2008) Identification and role of functionally important motifs in the 970 loop of *Escherichia coli* 16S ribosomal RNA. J Mol Biol 376:645–657
Schluenzen F, Tocilj A, Zarivach R, Harms J, Gluehmann M, Janell D, Bashan A, Bartels H, Agmon I, Franceschi F, Yonath A (2000) Structure of functionally activated small ribosomal subunit at 3.3 angstroms resolution. Cell 102:615–623

Schlünzen F, Zarivach R, Harms J, Bashan A, Tocilj A, Albrecht R, Yonath A, Franceschi F (2001) Structural basis for the interaction of antibiotics with the peptidyl transferase centre in eubacteria. Nature 413:814–821

Schmeing TM, Ramakrishnan V (2009) What recent ribosome structures have revealed about the mechanism of translation. Nature 461:1234–1242

Schuwirth BS, Borovinskaya MA, Hau CW, Zhang W, Vila-Sanjurjo A, Holton JM, Cate JH (2005) Structures of the bacterial ribosome at 3.5 A resolution. Science 310:827–834

Shajani Z, Sykes MT, Williamson JR (2011) Assembly of bacterial ribosomes. Annu Rev Biochem 80:501–526

Shine J, Dalgarno L (1975) Determinant of cistron specificity in bacterial ribosomes. Nature 254:34–38

Terasaka N, Hayashi G, Katoh T, Suga H (2014) An orthogonal ribosome-tRNA pair via engineering of the peptidyl transferase center. Nat Chem Biol 10:555–557

Tocilj A, Schlünzen F, Janell D, Glühmann M, Hansen HA, Harms J, Bashan A, Bartels H, Agmon I, Franceschi F, Yonath A (1999) The small ribosomal subunit from *Thermus thermophilus* at 4.5 A resolution: pattern fittings and the identification of a functional site. Proc Natl Acad Sci U S A 96:14252–14257

Traub P, Nomura M (1968) Structure and function of *E. coli* ribosomes. V. Reconstitution of functionally active 30S ribosomal particles from RNA and proteins. Proc Natl Acad Sci U S A 59:777–784

Tsukuda M, Nakashima N, Miyazaki K (2015) Counterselection method based on conditional silencing of antitoxin genes in *Escherichia coli*. J Biosci Bioeng 120:591–595

Tsukuda M, Kitahara K, Miyazaki K (2017) Comparative RNA function analysis reveals high functional similarity between distantly related bacterial 16 S rRNAs. Sci Rep 7(1):9993

Voorhees RM, Ramakrishnan V (2013) Structural basis of the translational elongation cycle. Annu Rev Biochem 82:203–236

Wang K, Neumann H, Peak-Chew SY, Chin JW (2007) Evolved orthogonal ribosomes enhance the efficiency of synthetic genetic code expansion. Nat Biotechnol 25:770–777

Wimberly BT, Brodersen DE, Clemons WM Jr, Morgan-Warren RJ, Carter AP, Vonrhein C, Hartsch T, Ramakrishnan V (2000) Structure of the 30S ribosomal subunit. Nature 407:327–339

Yano K, Masuda K, Akanuma G, Wada T, Matsumoto T, Shiwa Y, Ishige T, Yoshikawa H, Niki H, Inaoka T, Kawamura F (2015) Growth and sporulation defects in *Bacillus subtilis* mutants with a single *rrn* operon can be suppressed by amplification of the *rrn* operon. Microbiology 162:35–45

Yusupova G, Yusupov M (2014) High-resolution structure of the eukaryotic 80S ribosome. Annu Rev Biochem 83:467–486

Chapter 3
Development and Application of a Highly Efficient Protein Synthesis Technique Using Riboswitches in Microorganisms

Takahiro Yamauchi and Naoki Sugimoto

Abstract Riboswitches are conserved RNA elements that precisely and specifically target small molecules to control gene expression. For example, a thiamine pyrophosphate (TPP)-dependent riboswitch downregulates gene expression upon binding TPP in *Aspergillus oryzae*. Here we found that physiological concentrations of Mg^{2+} play a critical role in the regulation of TPP binding by the TPP-dependent riboswitch in *A. oryzae*. We further showed that the wild-type TPP-dependent riboswitch located in a 5′-untranslated region (UTR) intron of the *thiA* gene directly interacted with TPP and led to improper splicing, resulting in the retention of a remnant of the intron. Based on this mechanism, we constructed an "on riboswitch," in which the remnant sequence inhibiting the gene expression was truncated from the intron, leading to proper splicing on TPP binding. The riboswitch developed here should be a useful tool for gene regulation.

Keywords Riboswitch · Thiamine pyrophosphate · *Aspergillus oryzae* · Biotechnology · Splicing

3.1 What Is a Riboswitch?

In the twenty-first century, a gene expression regulation mechanism in which a part of mRNA directly recognizes the metabolite and regulates downstream gene expression was discovered in bacteria (Winkeler et al. 2002). Because this unique

T. Yamauchi (✉)
Research Department, Hakutsuru Sake Brewing Co. Ltd., Kobe, Japan
e-mail: yamauchi@hakutsuru.co.jp

N. Sugimoto
Frontier Institute for Biomolecular Engineering Research (FIBER), Graduate School of Frontiers of Innovative Research in Science and Technology (FIRST), Konan University, Kobe, Japan
e-mail: sugimoto@konan-u.ac.jp

S. Masuda, S. Izawa (eds.), *Applied RNA Bioscience*,
https://doi.org/10.1007/978-981-10-8372-3_3

mRNA acts as the gene expression switch, it has been called "riboswitch." At present, riboswitches that recognize vitamins, amino acids, and purine bases have been discovered, and they have been classified into more than ten classes of riboswitches based on the types of signal molecules to be recognized (Breaker 2012). Riboswitches have been discovered in bacteria as well as in fungi and plants, and they have become widely recognized as functional RNA present in the living world. Interestingly, only thiamine pyrophosphate (TPP)-dependent riboswitches widely exist from bacteria to fungi as well as in plants.

The TPP-dependent riboswitch is present in the untranslated region (UTR) of mRNA encoding TPP synthase. If the intracellular TPP is deficient, TPP synthase is translated and stimulates the production of TPP. However, when the amount of TPP becomes excessive in the cell, it directly binds to the riboswitch as a signal molecule without any protein factors. Because the structure of mRNA is changed by TPP binding, it inhibits the translation of TPP synthase and consequently suppresses the production of TPP. Thus, the TPP-dependent riboswitch keeps the amount of TPP in the cell constant by controlling the production of its own TPP synthase.

The secondary structure and higher-order structure at the TPP recognition site of the TPP riboswitch are common to the biological species; however, the mechanism for controlling gene expression is entirely different among organisms. For example, protein translation in bacteria begins with ribosomal binding to the Shine-Dalgarno (SD) sequence, which is the ribosome-binding site on mRNA. However, when TPP binds to the riboswitch, the structure of mRNA changes and the ribosome is unable bind to the SD sequence (Fig. 3.1). On the other hand, riboswitches of filamentous fungi, including *Aspergillus oryzae*, are present in the introns. Riboswitches of filamentous fungi have a gene expression regulation mechanism specific to eukaryotes in that introns do not exist in bacteria.

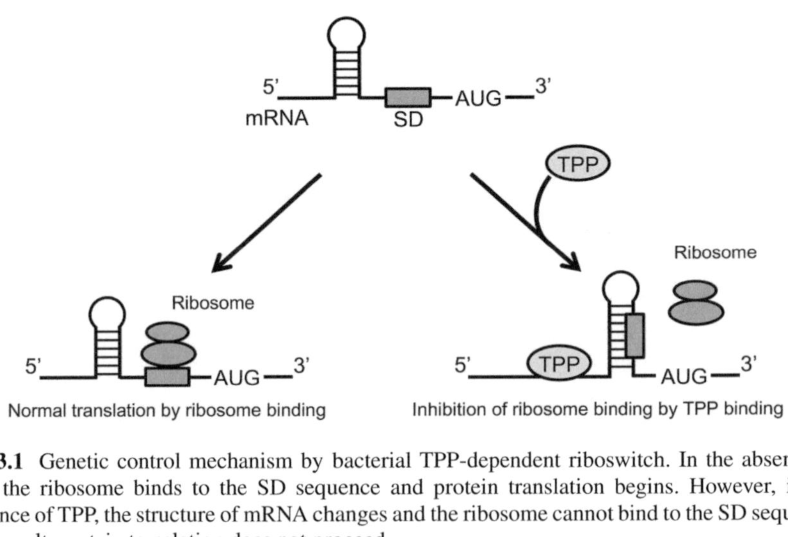

Fig. 3.1 Genetic control mechanism by bacterial TPP-dependent riboswitch. In the absence of TPP, the ribosome binds to the SD sequence and protein translation begins. However, in the presence of TPP, the structure of mRNA changes and the ribosome cannot bind to the SD sequence. As a result, protein translation does not proceed

3.2 Riboswitch of *A. oryzae*

Figure 3.2 shows a photomicrograph of "koji," which is *A. oryzae* cultured with steamed rice. Sake breweries use rice starch as the raw material; rice starch is saccharified by several amylases from *A. oryzae* and further converted to alcohol by yeast. Therefore, one of the most important roles of *A. oryzae* is the production of amylase. The method used to produce *A. oryzae* amylases for sake is very unique. The brewers start culturing *A. oryzae* by adhering its spores to steamed rice. This solid culture method is different from liquid cultures in that the raw material has less moisture and the microorganisms being cultured are in direct contact with the air. *A. oryzae* is cultured for about 40–50 h, during which they produce a large amount of enzymes, such as proteases, lipases, and amylases. Since mass production of enzymes by *A. oryzae* cannot be seen in liquid culture, such production and external release of enzymes is specific to solid cultures. Figure 3.3 shows the differences in the methods used for brewing sake, beer, and wine. In wine brewing, there is no need to saccharify the raw material because yeast alcohol fermentation is carried out using glucose contained in grapes. In beer brewing, barley starch is converted to maltose by malt, and maltose is converted to alcohol by yeast. In this process, saccharification and alcohol fermentation are carried out in separate tanks. On the other hand, sake undergoes both saccharification and alcoholic fermentation simultaneously in the same tank. Hence, *A. oryzae* are indispensable to the production of sake.

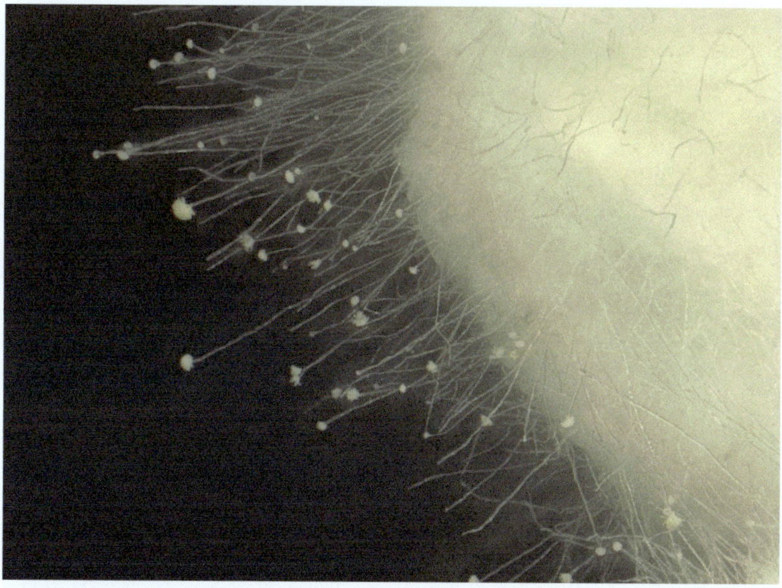

Fig. 3.2 Microphotograph of *A. oryzae* growing on steamed rice. In Japanese sake brewing, spores of *A. oryzae* sprinkled on steamed rice grow in 2 days, during which various enzymes are released outside the cells. Among these enzymes, some amylases play an important role in converting rice starch into sugar

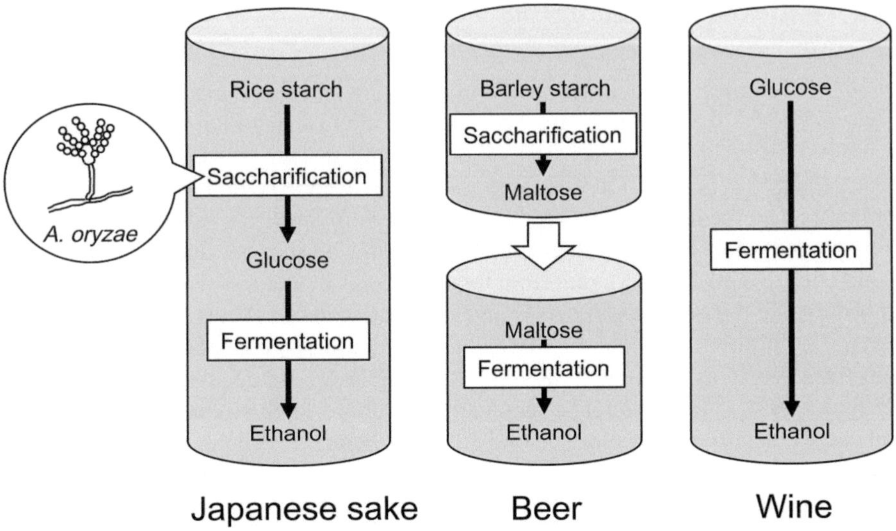

Fig. 3.3 Differences in fermentation type with alcoholic beverages. In Japanese sake brewing, saccharification of rice starch and alcohol fermentation are performed in parallel. The features of this fermentation are that the saccharification progresses gradually and the resulting glucose is also consumed sequentially by fermentation. In beer brewing, barley starch is decomposed with malt saccharification enzyme to prepare wort and ferment using wort. Grape juice glucose is used in winemaking; therefore, only fermentation is performed without saccharification

We found a consensus sequence with the bacterial TPP-dependent riboswitch from the intron in the 5′-UTR of *thiA* encoding the *A. oryzae* putative thiamine synthesis gene (Kubodera et al. 2003). RNA was synthesized from the putative TPP-binding domain of *thiA* by in vitro transcription, and the binding property between the synthesized RNA and TPP was confirmed using circular dichroism (CD) spectroscopy (Yamauchi et al. 2005). Figure 3.4a shows the CD spectra of the synthesized RNA in buffers containing 10 mM NaCl, 50 mM Tris-HCl (pH 7.0), 1 mM $MgCl_2$, and various concentrations of TPP at 20 °C. The CD spectra have a large positive peak near 265 nm and a small negative peak around 240 nm, indicating that the synthesized RNA contains an RNA/RNA A-form helix (Saenger 1984). The peak intensities decreased without a peak shift by the titration of TPP; therefore, CD intensity at 265 nm is suitable to estimate equilibrium parameters of TPP binding to the synthesized RNA. Figure 3.4b shows the CD intensity changes at 265 nm of the synthesized RNA with TPP concentration with or without 1 mM Mg^{2+}. The width of the CD intensity change by TPP titration in the presence of 1 mM Mg^{2+} is four times larger than that in the absence of Mg^{2+}, indicating that Mg^{2+} is important for the structural changes in the synthesized RNA induced by TPP binding. The estimated value of the x-axis of a Scatchard plot (normalized CD intensity/[TPP] versus normalized CD intensity) of the CD intensity change during TPP titration was 1. This indicated the presence of one TPP-binding site in the riboswitch (data not

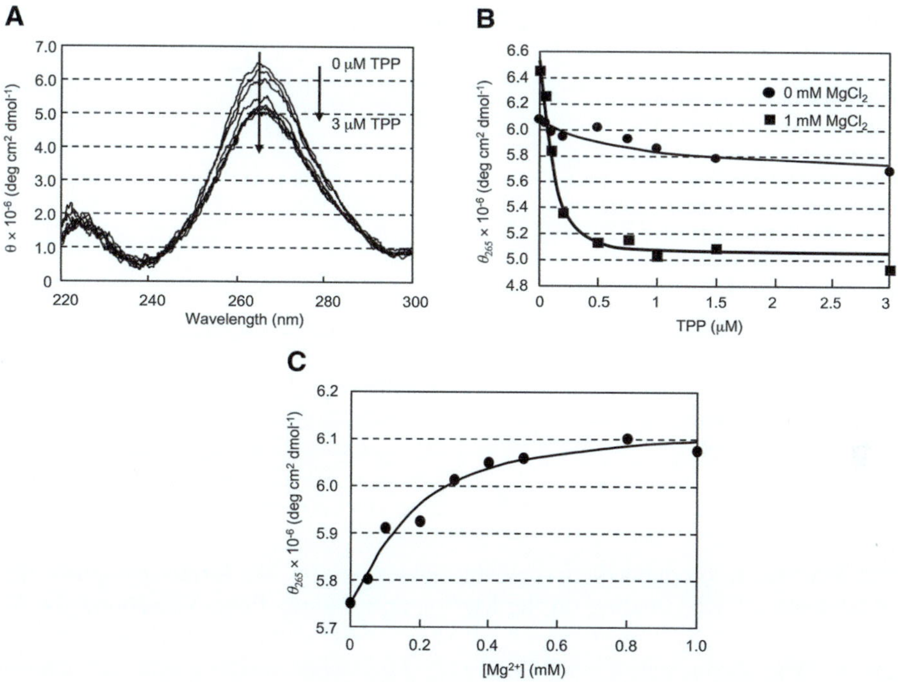

Fig. 3.4 (**a**) CD spectra of the *thiA* riboswitch in 50 mM Tris-HCl (pH 7.0) containing 10 mM NaCl, 1 mM Mg^{2+}, and 0–3 μM TPP (from upper to lower at 265 nm) at 20 °C. (**b**) CD intensities of the *thiA* riboswitch at 265 nm in 50 mM Tris-HCl (pH 7.0) containing 10 mM NaCl and various concentrations of TPP with (*squares*) or without (*circles*) 1 mM Mg^{2+} at 20 °C. (**c**) CD intensities of the *thiA* riboswitch at 265 nm in 50 mM Tris-HCl (pH 7.0) containing 10 mM NaCl and various concentrations of $MgCl_2$ 20 °C

shown). Therefore, the data were fitted to Eq. (3.1) on the basis of a model that postulates one binding site to estimate the equilibrium parameters (Rippe 1997):

$$\theta = a\left(K_a[\mathrm{RNA}] + K_a[\mathrm{TPP}] + 1 - \left(\left(K_a[\mathrm{RNA}] + K_a[\mathrm{TPP}] + 1\right)^2 - 4K_a^2[\mathrm{RNA}][\mathrm{TPP}]\right)^{1/2}\right) / 2K_a[\mathrm{RNA}] + b, \quad (3.1)$$

where a is the scale factor, b is the initial CD value, [RNA] is the molar concentration of RNA, [TPP] is the molar concentration of TPP, and K_a is the apparent association constant for TPP binding. The estimated K_a value at 1 mM Mg^{2+} was 50×10^6 M^{-1} at 20 °C, indicating that TPP was directly bound to the synthesized RNA. In other words, it was suggested that the consensus sequence with the bacterial TPP riboswitch found in *A. oryzae* acts as a TPP riboswitch. More interestingly, it was found that the binding constant of the *thiA* riboswitch and TPP decreased to 1.2×10^6 M^{-1} in the absence of Mg^{2+} (Figure 3.4b). The K_a with 1 mM Mg^{2+} was about 40-fold higher than that without Mg^{2+} at 20 °C, indicating that TPP binding to

Table 3.1 (**A**) Association constants (K_a) of thiamine pyrophosphate binding to the *thiA* riboswitch with various concentrations of Mg^{2+}. (**B**) Association constants (K_a) of thiamine pyrophosphate binding to the *thiA* riboswitch with various concentrations of Na^+ in the presence/absence of 1 mM Mg^{2+}

(A)

Mg^{2+} (mM)	$K_a \times 10^{-6}$ (M^{-1})
0	1.2 ± 0.6
0.2	5.0 ± 1.7
0.5	28 ± 6.2
1.0	50 ± 34
10	50 ± 16

(B)

	Concentration of Mg^{2+} (mM)	
Condition	0	1.0
10 mM Na^+	1.2 ± 0.6	50 ± 34
100 mM Na^+	3.7 ± 1.4	51 ± 41
1 M Na^+	2.2 ± 2.0	29 ± 13

All experiments in (**A**) and (**B**) were carried out in buffer containing 50 mM Tris-HCl (pH 7.0) at 20 °C. (B) Values presented as the $K_a \times 10^{-6}$ M^{-1}

the *thiA* riboswitch depends on the Mg^{2+} concentration. We further examined the dependence of TPP binding on the Mg^{2+} concentration. Table 3.1a shows the K_a values with various concentrations of Mg^{2+} in the presence of 10 mM Na^+ at 20 °C. The K_a increased with increasing Mg^{2+} concentration and was almost saturated at 1 mM Mg^{2+}, demonstrating that a physiological concentration of Mg^{2+} (≤1 mM) is sufficient for TPP binding. Generally, Mg^{2+} binding to RNA is required for the structural stability of RNAs, facilitating the formation of active structures (Zheng et al. 2015; Cunha and Bussi 2017). Furthermore, previous studies have shown that the binding constant of a metabolite for a riboswitch increases with increasing Mg^{2+} concentration (Batey et al. 2004; McPhie et al. 2016). These results combined with the quantitative parameters determined in the present study suggest that Mg^{2+} controls metabolite binding to riboswitches.

Since Na^+ alone can sometimes stabilize specific subdomains of RNA and induce RNA activity (Murray et al. 1998; Vieregg et al. 2007), we examined the effect of high concentrations of Na^+ on TPP binding to the *thiA* riboswitch. Table 3.1b shows the estimated K_a values for TPP binding to the *thiA* riboswitch in 50 mM Tris-HCl (pH 7.0) containing 10, 100, or 1000 mM NaCl in the absence of Mg^{2+}. The results showed that this Na^+ concentration range did not affect TPP binding, suggesting that even a high Na^+ concentration cannot induce formation of the active *thiA* riboswitch structure that is required for tight TPP binding. Moreover, the K_a values of TPP binding in the presence of 10, 100, or 1000 mM Na^+ and 1 mM Mg^{2+} are almost identical within the error range (Table 3.1b), indicating that the tight TPP binding induced by Mg^{2+} binding to the riboswitch was not significantly inhibited by high concentrations of Na^+. These results demonstrated that specific Mg^{2+} binding to the riboswitch plays a critical role in the ability of the riboswitch to form an active structure that is pivotal for binding TPP.

Since Mg^{2+} is required for TPP binding to the *thiA* riboswitch, we quantified the amount of Mg^{2+} bound to the riboswitch. Figure 3.4c shows the CD intensity of the *thiA* riboswitch at 265 nm in 50 mM Tris-HCl (pH7.0) containing 10 mM NaCl and 0.0–1.0 mM Mg^{2+} at 20 °C. The CD spectra of the *thiA* riboswitch had a large positive peak near 265 nm, and a peak shift was not evident at any Mg^{2+} concentration tested. Therefore, the CD intensity at 265 nm could be used to estimate the equilibrium parameters of Mg^{2+} binding to the riboswitch. The data were analyzed using a model that postulates cooperative Mg^{2+} binding to estimate the Mg^{2+} binding parameters (Cantor and Schimmel 1980); the curve fit is given by Eq. (3.2):

$$\theta = aK_{\mathrm{a}}^{\alpha}\left[\mathrm{Mg}^{2+}\right]^{\alpha} / \left(1 + K_{\mathrm{a}}^{\alpha}\left[\mathrm{Mg}^{2+}\right]^{\alpha}\right) + b \tag{3.2}$$

where a is the scale factor, b is the initial CD value, α is the Hill coefficient, $[Mg^{2+}]$ is the molar concentration of Mg^{2+}, and K_a is the apparent association constant of Mg^{2+} binding. The estimated parameters at 10 mM Na^+ were $K_a = (6.1 \pm 1.7) \times 10^3\ M^{-1}$ and $\alpha = 1.3 \pm 0.4$ at 20 °C. The Hill coefficient indicated that the *thiA* riboswitch has at least one Mg^{2+} binding site and that Mg^{2+} binding is independent even if several Mg^{2+} binding sites are available. Since a long RNA molecule harbors many Mg^{2+} binding sites, the estimated K_a might represent the average of several levels of affinity for a number of RNA-binding sites. The crystal structure of bacterial and plant TPP riboswitches were respectively analyzed. It has been reported that the phosphate moiety of TPP is bound to the TPP riboswitch via Mg^{2+} (Serganov et al. 2006; Thore et al. 2006). These results suggest that Mg^{2+} plays an important role in the binding of TPP to the *thiA* riboswitch. Therefore, riboswitch function may be controlled by metabolites, as well as the Mg^{2+} concentration, under physiological conditions. By understanding the binding mechanism of the TPP riboswitch and TPP, it will be possible to tune the TPP riboswitch artificially. Flavin mononucleotide (FMN) riboswitches have been extensively studied for the binding between the riboswitch and FMN, and it is now possible to tune the gene regulation ability of the FMN riboswitches (Rode et al. 2015a, b).

Unlike bacterial TPP-dependent riboswitches, the *thiA* riboswitch was predicted to control splicing of an intron because it is present in the intron of UTR of *thiA*. Therefore, we analyzed how the *thiA* riboswitch is involved in intron control by reverse transcriptase-polymerase chain reaction (RT-PCR) and DNA sequencing (Yamauchi et al. 2008). We constructed a promoter-β-glucuronidase (GUS) reporter gene plasmid containing 5′-UTR of *thiA*. We introduced this plasmid into *A. oryzae* and then treated the transformed cells with or without 10 μM thiamine. These cells take up thiamine and spontaneously convert it into TPP (Kubodera et al. 2003), allowing the activation of the TPP-dependent riboswitch by the application of thiamine as an external stimulus.

Total RNA was extracted from the cells, and pre-mRNA and spliced mRNA were amplified by RT-PCR. Figure 3.5a shows the native polyacrylamide gel electrophoresis (PAGE) results of the RT-PCR products. DNA sequencing revealed that mature mRNA was produced only in the absence of thiamine, whereas the level of immature mRNA increased in the presence of thiamine. Moreover, sequencing of the immature mRNA clearly showed that the 5′-splicing site in the absence of thiamine

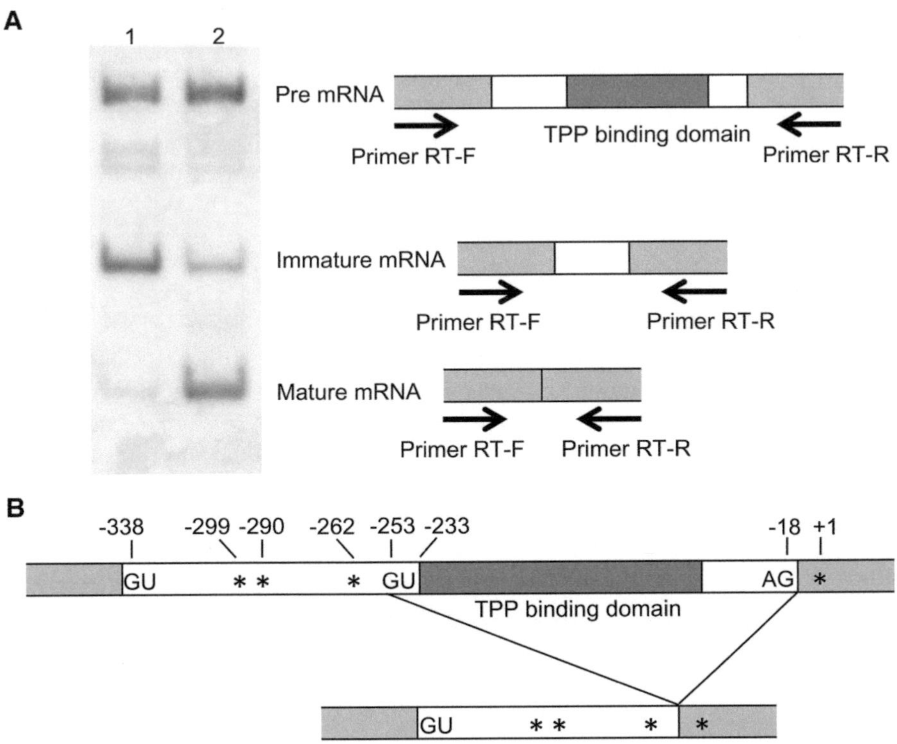

Fig. 3.5 (**a**) Pre-mRNA and spliced mRNA were amplified by RT-PCR with primer RT-F and RT-R. Light gray, white, and dark gray rectangles indicate exons, introns, and TPP-binding domain, respectively. RT-PCR products were separated by native PAGE on a 5% acrylamide gel. Lanes 1 and 2 indicate the PCR products in the absence and presence of thiamine, respectively. (**b**) The 5′-UTR of pre-mRNA and immature mRNA produced from the *thiA* gene. Asterisks indicate putative start codons, and numbers indicate the sequence position relative to the translation initiation site (+1)

was at position −338 relative to the translation initiation site (+1) (Figure 3.5b). In contrast, in the presence of thiamine, the 5′-splicing site was at position −253, demonstrating that immature mRNA contains an 85-base remnant of the intron. In general, in an intron splicing reaction, the guanosine residue at the 5′-splicing site of the intron and the adenosine residue at the branch site inside the intron are combined to form a lariat structure as an intermediate. The data suggested that the intron containing the *thiA* riboswitch forms a lariat structure with the guanosine residue at position −338 in the absence of TPP and at position −253 in the presence of TPP. It has been reported that the TPP-dependent riboswitch from *Neurospora crassa* uses a long-distance base-pairing interaction to regulate alternative splicing; therefore, the *thiA* riboswitch may use the same mechanism (Li and Breaker 2013). The remnant sequence of the intron has three potential translation initiation codons at positions −262, −290, and −299 (Fig. 3.5b), which can induce frameshifts in the codons, leading to protein inactivation. These results are consistent with the findings that the alternative translation initiation codons observed in the TPP-dependent riboswitch system in *N. crassa* can reduce protein expression (Cheah 2007). These results confirm that the *thiA* riboswitch alters the 5′-splicing site on TPP binding.

3.3 Functional Modification of *thiA* Riboswitch

A. oryzae has been used for sake brewing as well as for traditional Japanese brewing of other foods such as miso and soy sauce for a long time. The main role of *A. oryzae* in these brewing processes is the production of enzymes such as amylase and protease, and it is known that *A. oryzae* has an excellent protein production capacity (Ichishima 2016). Therefore, it is very advantageous to use *A. oryzae* as a host for producing useful proteins in large quantities. However, during the production of these proteins, one must be aware of the toxicity of the proteins to the host. In particular, when the desired useful protein inhibits the growth of the host cell, it is difficult to produce the protein. To overcome this problem, it is conceivable to grow the host microorganisms while suppressing the production of the useful protein and to produce the protein after the host has grown sufficiently. In other words, it is preferable to be able to start protein production at any time by artificially using a "switch" for protein production.

If the *thiA* riboswitch is used for such a switch, the production of any useful protein can be controlled by *A. oryzae*. However, the *thiA* riboswitch is an "off riboswitch" that negatively regulates the gene expression by TPP binding. If it is possible to construct an "on riboswitch," which positively regulates gene expression by binding of TPP, the application range of the riboswitch as a gene expression control tool will be widened. Therefore, we focused on the gene regulatory mechanism of the *thiA* riboswitch and attempted to engineer an "on riboswitch" that can upregulate gene expression in response to TPP (Yamauchi et al. 2008).

The 85-base remnant of the intron is produced by splicing in the presence of TPP. We suspected that removing the remnant sequence would prevent improper splicing from proceeding in the presence of TPP (Fig. 3.6). Initially, we constructed an engineered *thiA* riboswitch E1 by removing the 85 bases from the 5′-splicing site of the intron. We then constructed a promoter-GUS reporter gene plasmid containing the E1 riboswitch and analyzed it in the absence and presence of thiamine. The GUS activities of the original *thiA* riboswitch in the absence and presence of thiamine were 3708 ± 15 and 106 ± 1.3 unit/mg protein, respectively, demonstrating the downregulation of the wild-type riboswitch. On the other hand, the GUS activities of the engineered E1 riboswitch in the absence and presence of thiamine were 357 ± 6.2 and 450 ± 12 unit/mg protein, respectively. Although the response to exogenous thiamine was altered, the difference in GUS activities of the E1 riboswitch was insufficient for use in gene regulation.

Further, we attempted to improve the extent of upregulation mediated by the riboswitch. In fungi, there is a very strong correlation between the total length of an intron and the length from the 5′-splicing site to the branch point (Kupfer et al.

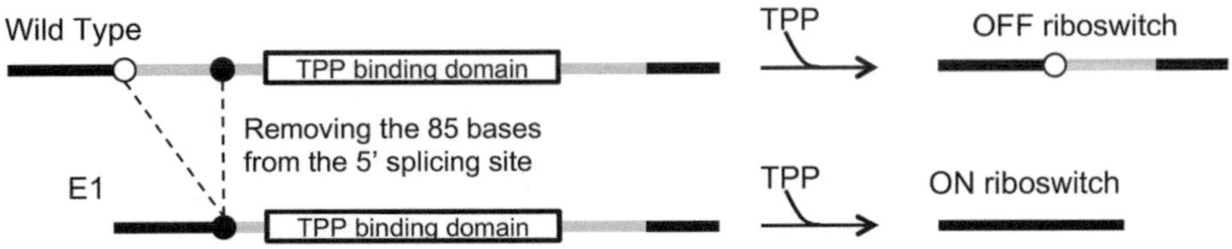

Fig. 3.6 Strategy for controlling gene expression by the engineering of the *thiA* riboswitch. Black and gray lines indicate exons and introns, respectively. White and black circles indicate the 5′-splicing sites in the absence and presence of TPP, respectively

2004). Because the TPP-binding domain of the *thiA* riboswitch is located between the 5′-splicing site and the branch point, the length between the 5′-splicing site and the TPP-binding domain could be important for the splicing reaction (Fig. 3.7a). In the absence of TPP, the length between the 5′-splicing site and the TPP-binding domain of the *thiA* riboswitch was 105 bases (Fig. 3.7b). On the other hand, in the presence of TPP, this length was 20 bases. From this observation, we can conclude that the main 5′-splicing site in the presence of TPP was close to the TPP-binding domain. We therefore truncated the sequence between the 5′-splicing site and the TPP-binding domain of the E1 riboswitch (Fig. 3.7a), generating the E2, E3, and E4 riboswitches. We then inserted these new riboswitches into the promoter-GUS reporter gene plasmids. Surprisingly, the GUS activities of the E2 and E3 riboswitches in the presence of thiamine were 4.7 and 4.3 times higher, respectively, than those in the absence of thiamine, whereas the E4 riboswitch construct had almost no GUS activity (Fig. 3.7b). In addition, the splicing products were confirmed by RT-PCR and DNA sequencing. As a result, the splicing products in the E2 and E3 riboswitches in the presence of thiamine were higher than those in the absence of thiamine, indicating that these RT-PCR results are consistent with the results of GUS activity. Although the length between the 5′-splicing site and the TPP-binding domain of the *thiA* riboswitch should be further optimized, these alterations in gene expression are comparable with the differences in gene expression mediated by the naturally occurring TPP-dependent riboswitch in *N. crassa*, where the gene expression in the presence of thiamine is approximately four times higher than that in the absence of thiamine (Cheah 2007).

To further confirm gene regulation mediated by the engineered TPP-dependent riboswitches, it is important to compare their binding affinities for TPP. Figure 3.7c shows the normalized GUS activities of *thiA* and E2 riboswitches with various concentrations of thiamine. Because one TPP molecule, generated spontaneously from thiamine in vivo, binds to each riboswitch, the data from the *thiA* and E2 riboswitches were fitted to Eq. (3.3) to estimate the equilibrium parameters based on a model that assumes a single binding site (Rippe 1997):

$$\theta = aK_a\left[thiamine\right] / \left(1 + K_\mathrm{a}\left[thiamine\right]\right) + b \qquad (3.3)$$

where θ is the normalized GUS activity, K_a is the apparent association constant of thiamine binding, [thiamine] is the molar concentration of thiamine, a is a scale

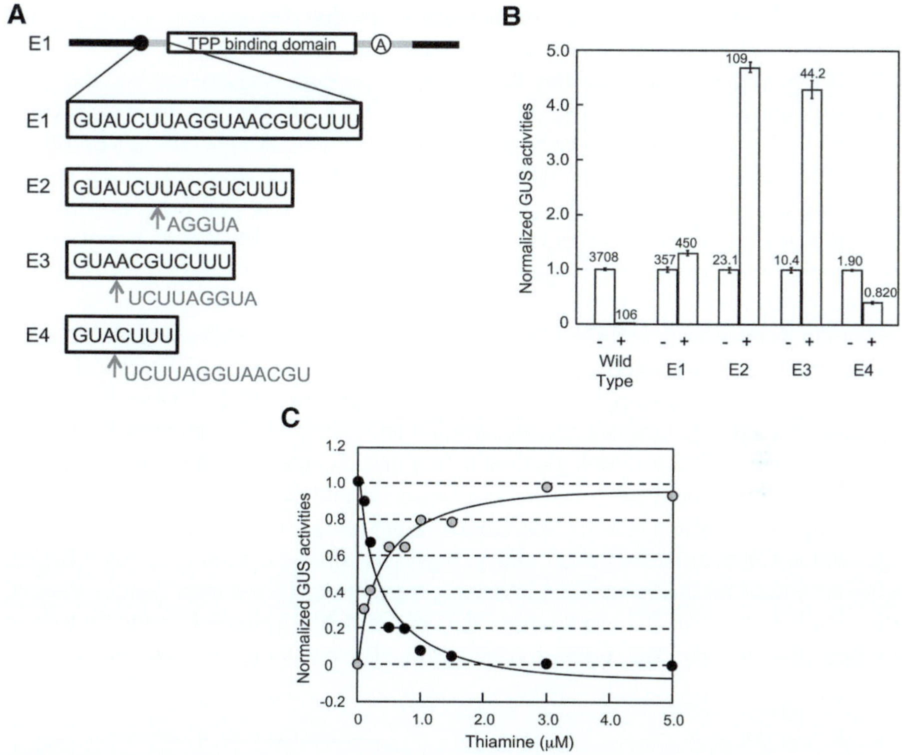

Fig. 3.7 (**a**) E2–E4 riboswitches were designed by truncating the E1 riboswitch. The gray sequences and arrows indicate the truncated sequence and position, respectively. The circled "A" indicates the branch point of splicing. (**b**) GUS activity of the engineered riboswitches in the absence (−) and presence (+) of thiamine. The GUS activity of the riboswitches in the absence of thiamine was normalized to 1.0. The actual GUS activities (unit/mg protein) are indicated at the top of each bar. (**c**) GUS activity of the *thiA* (black) and E2 (gray) riboswitches at different thiamine concentrations

factor, and b is the initial θ value. The apparent K_a values for these riboswitches at 30 °C were $(3.2 \pm 1.1) \times 10^6$ and $(3.5 \pm 0.6) \times 10^6$ M^{-1}, respectively. Thus, the affinities of these two riboswitches for thiamine were essentially identical, suggesting that the engineered TPP-dependent riboswitch retained normal TPP binding.

3.4 Future Prospects

We successfully engineered the *thiA* riboswitch from its natural state as an "off riboswitch" into an artificial "on riboswitch." On the other hand, TPP riboswitches that induce gene expression in bacteria have been developed, and the research application of riboswitches is rapidly progressing (Nomura and Yokobayashi 2007). Currently, wild-type riboswitches functioning with *A. oryzae* are only those that bind to TPP, whereas bacterial riboswitches are diverse. Henceforth, it can be expected to combine the riboswitch expression platform of *A. oryzae* with bacterial riboswitch aptamers and to be able to regulate gene expression by the new riboswitches using various molecules with *A. oryzae*. In addition, because introns present in *A. oryzae* are also present in mammals, it is thought that gene control using riboswitches in various higher organisms is possible. In other words, in the near future, it is likely that the production of any protein can be controlled by riboswitches using any host organism and any signal molecule. For that purpose, it is important to know the higher-order structure of the riboswitch and the thermodynamic stability energy related to the physical properties of the riboswitch in detail (Nakano et al. 2014).

A. oryzae is recognized not only in Japan but also in the US Food and Drug Administration (FDA) as a safe food, and it is sometimes called a "national fungus" because it is a microorganism representing Japan. Therefore, it can safely be used as a host microorganism producing not only food but also useful protein. *A. oryzae* can be cultured in a liquid medium, such as bacteria and yeast, as well as in a solid medium, such as wheat bran or rice. Solid culture has an advantage that cells can be cultured at a higher density than that in liquid culture, and *A. oryzae* has a higher efficiency than bacteria and yeast in producing protein. To produce useful proteins, the concept of a "cell factory" where the host cell is considered as the factory is present. To increase the protein production efficiency of this cell factory, the

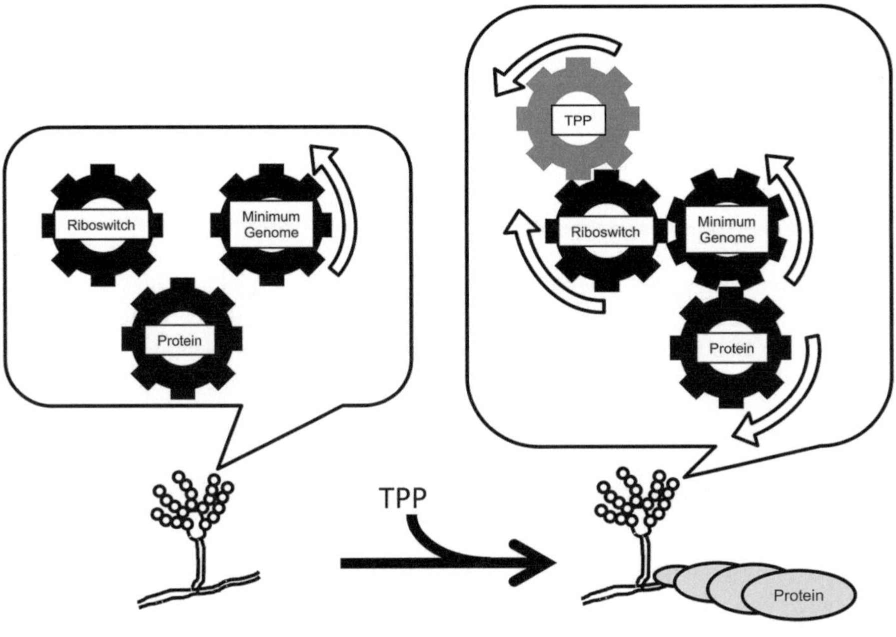

Fig. 3.8 Scheme of a "cell factory" that combines the TPP riboswitch with minimal genome and TPP and produces the target protein with high efficiency using *A. oryzae*

development of microorganisms having the minimum genome, i.e., lacking the genes that are unrelated to proliferation or inhibit proliferation, has been advanced in bacteria and yeast. If *A. oryzae* having the minimum genome is engineered, it can be developed as the ultimate cell factory (Fig. 3.8). Thus, as research on each of the riboswitches and *A. oryzae* develops, it is expected that the riboswitch and *A. oryzae* will be a very effective combination for useful protein production.

References

Batey RT, Gilbert SD, Montange RK (2004) Structure of a natural guanine-responsive riboswitch complexed with the metabolite hypoxanthine. Nature 432:411–415

Breaker RR (2012) Riboswitches and the RNA world. Cold Spring Harb Perspect Biol 4(2):a003566

Cantor CR, Schimmel RR (1980) Biophysical chemistry part III: the behavior of biological macromolecules. W. H. Freeman, San Francisco, pp 863–866

Cheah MT (2007) Control of alternative RNA splicing and gene expression by eukaryotic riboswitches. Nature 447:497–500

Cunha RA, Bussi G (2017) Unraveling Mg^{2+}-RNA binding with atomistic molecular dynamics. RNA 23:628–638

Ichishima E (2016) Development of enzyme technology for Aspergillus oryzae, A. Sojae, and A. Luchuensis, the national microorganisms of Japan. Biosci Biotechnol Biochem 80:1681–1692

Kubodera T, Watanabe M, Yoshiuchi K, Yamashita N, Nishimura A, Nakai S, Gomi K, Hanamoto H (2003) Thiamine-regulated gene expression of Aspergillus oryzae thiA requires splicing of the intron containing a riboswitch-like domain in the 5′-UTR. FEBS Lett 555:516–520

Kupfer DM, Drabenstot SD, Buchanan KL, Lai H, Zhu H, Dyer DW, Roe BA, Murphy JW (2004) Introns and splicing elements of five diverse fungi. Eukaryot Cell 3:1088–1100

Li S, Breaker RR (2013) Eukaryotic TPP riboswitch regulation of alternative splicing involving long-distance base pairing. Nucleic Acids Res 41:3022–3031

McPhie P, Brown P, Chen B, Dayie TK, Minton AP (2016) Modulation of conformational equilibria in the S-adenosylmethionine (SAM) II riboswitch by SAM, Mg^{2+} and trimethylamine N-oxide. Biochemistry 55:5010–5020

Murray JB, Seyhan AA, Walter NG, Burke JM, Scott WG (1998) The hammerhead, hairpin and VS ribozymes are catalytically proficient in monovalent cations alone. Chem Biol 5:587–595

Nakano S, Miyoshi D, Sugimoto N (2014) Effects of molecular crowding on the structures, interactions, and functions of nucleic acids. Chem Rev 114:2733–2758

Nomura Y, Yokobayashi Y (2007) Reengineering a natural riboswitch by dual genetic selection. J Am Chem Soc 129:13814–13815

Rippe K (1997) Analysis of protein-DNA binding at equilibrium. BIF Futura 12:20–26

Rode AB, Endoh T, Sugimoto N (2015a) Key tertiary interactions in FMN riboswitch aptamers required for ligand binding. Bull Chem Soc Jpn 88:946–948

Rode AB, Endoh T, Sugimoto N (2015b) Tuning riboswitch-mediated gene regulation by rational control of aptamer ligand binding properties. Angew Chem Int Ed 54:905–909

Saenger W (1984) Water and nucleic acids. In: Principles of nucleic acid structure. Springer, New York, pp 368–384

Serganov A, Polonskaia A, Phan AT, Breaker RR, Patel DJ (2006) Structural basis for gene regulation by a thiamine pyrophosphate-sensing riboswitch. Nature 441:1167–1171

Thore S, Leibundgut M, Ban N (2006) Structure of the eukaryotic thiamine pyrophosphate riboswitch with its regulatory ligand. Science 312:1208–1211

Vieregg J, Cheng W, Bustamante C, Tinoco I Jr (2007) Measurement of the effect of monovalent cations on RNA hairpin stability. J Am Chem Soc 129:14966–14973
Winkeler W, Nahvi A, Breaker RR (2002) Thiamine derivatives bind messenger RNAs directly to regulate bacterial gene expression. Nature 419:952–956
Yamauchi T, Miyoshi D, Kubodera T, Nishimura A, Nakai S, Sugimoto N (2005) Roles of Mg^{2+} in TPP-dependent riboswitch. FEBS Lett 579:2583–2588
Yamauchi T, Miyoshi D, Kubodera T, Ban M, Nishimura A, Sugimoto N (2008) Riboswitches for enhancing target gene expression in eukaryotes. Chembiochem 9:1040–1043
Zheng H, Shabalin IG, Handing KB, Bujnicki JM, Minor W (2015) Magnesium-binding architectures in RNA crystal structures: validation, binding preferences, classification and motif detection. Nucleic Acids Res 43:3789–3801

Chapter 4
Reprogramming of Cells by Lactic Acid Bacteria

Naofumi Ito and Kunimasa Ohta

Abstract Living organisms have been classified into three domains—archaea, eukaryota, and prokaryota—based on their cell structure and genetic evolution (Woese CR, Kandler O, Wheelis ML. Proc Natl Acad Sci USA 87:4576–4579, 1990). The eukaryotic cells have organelles that originated from prokaryotes living within these cells as endosymbionts (Martin W, Hoffmeister M, Rotte C, Henze K. Biol Chem 382:1521–1539. https://doi.org/10.1515/BC.2001.187, 2001). Endosymbionts affected the evolution and diversity of living organisms by horizontal gene transfer (Woese CR. Proc Natl Acad Sci USA 99:8742–8747. https://doi.org/10.1073/pnas.132266999, 2002; Timmis JN, Ayliffe MA, Huang CY, Martin W. Nat Rev Genet 5:123–135. https://doi.org/10.1038/nrg1271, 2004). The origin of eukaryotic cells was explained by the endosymbiotic theory, which has been advanced and substantiated with microbiological evidence (Margulis L. Origin of eukaryotic cells: evidence and research implications for a theory of the origin and evolution of microbial, plant and animal cells on the precambrian earth. Yale University Press, New Heaven, 1970). The partnership between a primitive anaerobic eukaryotic predator cell and an aerobic bacterial cell was potentially established about 1.5 billion years ago. At present, it is widely believed that eubacteria infected archaebacteria, leading to the translocation of genomic DNA and the evolution of

N. Ito
Department of Developmental Neurobiology, Graduate School of Life Sciences, Kumamoto University, Kumamoto, Japan

Program for Leading Graduate Schools "HIGO Program", Kumamoto University, Kumamoto, Japan

K. Ohta (✉)
International Research Core for Stem Cell-based Developmental Medicine, Kumamoto University, Kumamoto, Japan

Japan Agency for Medical Research and Development (AMED), Tokyo, Japan

Department of Developmental Neurobiology, Graduate School of Life Sciences, Kumamoto University, Kumamoto, Japan

Program for Leading Graduate Schools "HIGO Program", Kumamoto University, Kumamoto, Japan
e-mail: ohta9203@gpo.kumamoto-u.ac.jp

S. Masuda, S. Izawa (eds.), *Applied RNA Bioscience*,
https://doi.org/10.1007/978-981-10-8372-3_4

eukaryotic cells (Hartman H, Fedorov A. Proc Natl Acad Sci USA 99:1420–1425. https://doi.org/10.1073/pnas.032658599, 2002). Over time, endosymbiotic interactions and genomic scrambling in various organisms contributed to the generation of new organisms.

To examine whether a bacterial infection can alter cell fate, human dermal fibroblast (HDF) cells were co-cultured with lactic acid bacteria (LAB), which are known to have beneficial effects on the physiology of the host. We previously showed that when HDF cells were incorporated with LAB, the LAB-incorporated HDF cells formed clusters and expressed a subset of common pluripotent markers. Moreover, the LAB-incorporated cell clusters could differentiate into cells of any of the three germ layers, indicating successful reprogramming of the host HDF cells by LAB. In this review, we discuss the nuclear reprogramming mechanisms in the existing examples of cellular reprogramming by bacteria.

Keywords Reprogramming · Multipotency · Lactic acid bacteria · Eukaryotic cells

4.1 Introduction

In general, cell fate of the terminally differentiated somatic cells was believed to be irreversible. However, this belief was challenged with the introduction of the somatic cell nuclear transfer (SCNT) technique in 1952, to study early embryonic development in frogs. Although various mammalian species have been successfully cloned using SCNT, the cloning efficiency of this technique was very low. Although cell fate was thought to be restricted during development, the generation of animals using the SCNT technique by genetically manipulating the ova demonstrated that the epigenome of differentiated cells can be reset to a pluripotent state (Briggs and King 1952; Gurdon 1962).

Pluripotent stem cells are the undifferentiated cells that have the ability to proliferate for an indefinite period, divide to generate daughter cells by self-renewal, and differentiate into various types of specialized cells to fulfill the developmental requirements. Stem cells are categorized into three groups: somatic stem cells, embryonic stem (ES) cells, and the induced pluripotent stem (iPS) cells. ES cells were the first pluripotent cells to be isolated from an embryo. They were highly valuable because of their potential to differentiate into more than 260 types of cells (Srivastava and DeWitt 2016). In 2006, Takahashi and Yamanaka first reported the successful generation of human iPS cells in mice by retroviral transduction of the Yamanaka factors (OCT4, SOX2, Klf4, and c-Myc) (Takahashi and Yamanaka 2006). The generation of iPS cells was a breakthrough in stem cell research, the use of which has solved problems such as induction efficiency of pluripotent stem cells and cell fate controls. Since then, these cells have been utilized in several clinical applications, including disease investigation and drug screening.

Microbial colonization of the human gut is an evolution-driven process that has beneficial effects on the host physiology, immunity, and nutrient intake. In some cases, these effects persist throughout life (Ley et al. 2008). The human fetus exists in a sterile state inside the mother's body. It is exposed to bacteria for the first time at birth (Aagaard et al. 2014). After birth and until death, bacteria continuously interact with the human body via the gut and/or at body surfaces. In total, over 500 kinds of bacteria reside in the human body without being excluded by the immune system (Kawamoto et al. 2014). While much remains unclear about the influence of these microorganisms on human cells, recent studies have revealed their effects on immunology, ecology, and cancer therapy (Kawashima et al. 2013; Collins 2014; Louis et al. 2014). The results of these studies have shown that the human–microbe interactions are not only involved in disease pathogenesis but also in the maintenance of homeostasis.

Recently, the phenomenon of somatic cell reprograming by bacteria has been reported (Ohta et al. 2012; Masaki et al. 2013). In the current review, we discuss this newly discovered reprogramming technique for the generation of stem cell-like cells by the incorporation of bacteria into host cells.

4.2 Microbial Communities

In nature, host–microbe interactions exist in all kinds of organisms (Kleiner et al. 2012; Matsuura et al. 2012). When multiple species live in sympatry, their interaction with each other is called symbiosis. In case of symbiosis, the host–bacteria interactions are strictly limited by the signal molecule cascade, specifically in the inducted organs (Cao and Goodrich-Blair 2017). Alternatively, there are symbiotic relationships that do not require the involvement of specific organs (Fesel and Zuccaro 2016). In that case, although limited, a broader range of microorganisms is established in vivo. The extent of the physiological activity of individual microorganisms is also weaker than symbiosis, but it has a big influence as a whole. In these examples, the endosymbiotic bacteria act as an additional organ and play a critical role in the functioning of host metabolic system.

In humans, the presence of microbial communities has been identified as being either beneficial or harmful. In humans, studies have been conducted on the effect of microbiota on homeostasis and metabolism (Consortium 2012). It has been found that the host–microbe balance may be maintained by factors such as the body mass index and/or age of the host (Costello et al. 2009). Intestinal microbial dysbiosis in individuals may be linked with disease pathogenesis (Holmes et al. 2012).

Surprisingly, microbial communities also affect their host's epigenetic status by producing specific compounds. *Shigella flexneri* is a pathogenic bacterium that causes fever and diarrhea in humans via toxic protein circulation in the blood (Ashida et al. 2011). *S. flexneri* infects the host and disseminates after repressing the host's innate immune system by producing the protein OspF, which in turn interacts with the NF-κB-responsive genes (Arbibe et al. 2007). The pathogenic bacterium

Chlamydia trachomatis causes host epigenetic dysregulation via its proteins (Lad et al. 2007; Pennini et al. 2010). *Listeria monocytogenes*, a pathogenic bacterium found in unpasteurized and/or raw milk, vegetables, and meats, also interacts with the human epigenome (Lebreton et al. 2011). Virulence factor LntA, produced by *L. monocytogenes*, binds to the chromatin repressor BAHD1 in the host nuclei and activates the interferon-stimulated genes and chromatin remodeling. Species of *Clostridium*, an anaerobic commensal bacterial genus, also stimulate host cell epigenetic regulation by secreting butyrate (Furusawa et al. 2013).

4.3 Modulation of Cellular Homeostasis, Cell Lineage Fate, and Stem Cell Development by Endosymbiotic Microbial Community

All multicellular organisms contain microbial communities on their body surfaces and in their internal organs. A very good example of this is the presence of intestinal microflora in human beings (Browne et al. 2017). Multicellular organisms have developed complex ecosystems in their bodies by acquiring ectodermal nerves and mesodermal muscles from the last common ancestor who possessed endodermal intestine-like structures (Sebe-Pedros et al. 2017). Endosymbiotic bacteria are involved in the host's metabolic processes by helping in essential amino acid synthesis, thereby maintaining a symbiotic relationship due to their metabolic dependence. This results in the evolution of both, the symbiotic bacteria and the host (Douglas 2016).

Bacterial metabolites alter host development. Certain organisms harbor commensal bacteria, which produce harmful substances like butyrate (Barcenilla et al. 2000). However, such organisms have structures called crypts, which protect the cells from being damaged by butyrate (Kaiko et al. 2016). Interestingly, crypt-less organisms, such as zebra fish, do not harbor butyrate-producing bacteria. Further, the bacterial peptidoglycan motif confers protection against oxidation-mediated cell death via activation of the innate immune sensor Nod2 in the stem cells of the intestinal crypts (Nigro et al. 2014). The enteric bacterial protein AvrA from *Salmonella typhimurium* increases tumor formation by activating the host's β-catenin signaling pathway (Lu et al. 2014).

Commensal bacteria also interfere with and modulate the mammalian homeostasis regulatory pathways. Intestinal microbiota change distal autoimmune regulator expression in mice (Nakajima et al. 2014). Toxin B, produced by *Clostridium difficile*, acts as a Wnt signaling antagonist in the host by binding to the Frizzled 1, 2, and 7 receptors (Tao et al. 2016). Infectious *Mycobacterium* spp. express pre-microRNAs, which are small RNAs with a stem–loop structure that binds to the RNA-induced silencing complex. This microRNA of bacterial origin regulates cellular homeostasis in the host (Furuse et al. 2014). Non-coding RNA transcribed from *Salmonella* spp. was processed by Argonaut-2 in the host, which caused a reduction in the host's Sal1 expression. This facilitated the infection of host cells by *Salmonella* spp. (Gu et al. 2017). These results suggest that the commensal bacteria may regulate the gene and protein expression profiles of the host, in order to optimize the habitat to suit their survival.

Commensal bacteria affect the development of multicellular organisms; this was observed irrespective of these bacteria forming a symbiotic or parasitic relationship. Symbiotic bacteria in *Caenorhabditis elegans* produce folates that play a critical role in germline stem cell development (Chaudhari et al. 2016). Unknown proteins in the endosymbiotic microbiota promote beta-cell proliferation in the pancreas of zebra fish (Hill et al. 2016). Infection by the commensal bacterium *Enterococcus faecalis* caused cellular transformation and alterations in the transcriptome profile, resulting in induction of the stem/progenitor cell markers Ly6A and Ly6E and the tumorigenic marker Dclk1 (Wang et al. 2015). Infection by the obligate intracellular gram-negative bacterium *Chlamydia pneumoniae* causes defects in host vascularization via the bacterial heat shock protein GroEL1 (Lin et al. 2013). The symbiotic *Wolbachia* spp. of bacteria that are present in arthropods code for genes involved in reprogramming the efficiency of iPS cell generation, indicating that products derived from commensal bacteria stimulate stem cell development (Ikeda et al. 2017).

Symbiosis between microorganisms and multicellular organisms enables us to understand concepts such as ecological adaptation, niche construction, epigenetic inheritance, and developmental plasticity. Ecological, evolutionary, and developmental biologists advocated the concept of "Eco–Evo–Devo" for a deeper understanding of the epigenetic effects and environmental pressures associated with the evolution of organisms (Okada 2004; Gilbert et al. 2015). It is well-known that metabolism by intestinal symbionts affects host development (Moran and Yun 2015; Ziegler et al. 2017). We propose that understanding how the relationship between the microbiome and multicellular organisms affects genetics, metabolism, immunity, and evolution would help us to understand the ecosystems, especially their evolutionary and developmental aspects.

4.4 Cell Reprogramming by Lactic Acid Bacteria (LAB)

LAB form a group of related low-GC-content gram-positive bacteria that are traditionally used to produce fermented foods (Taguchi et al. 2015). LAB have been empirically awarded the "generally recognized as safe" (GRAS) status. GRAS bacteria are used in a variety of therapeutic strategies (Shin et al. 2016). In humans, despite microbial species being stable over a long period of time after their establishment in the host gut, differences in gut microbiota are often observed among individuals (Schloissnig et al. 2013). LAB occupy important niches in the gastrointestinal tract of humans and are responsible for several health-promoting probiotic effects, including a positive influence on the normal microflora, competitive exclusion of pathogens, and stimulation/modulation of mucosal immunity (Gordon 2008). To exert beneficial effects on human physiology, probiotic LAB adhere to the

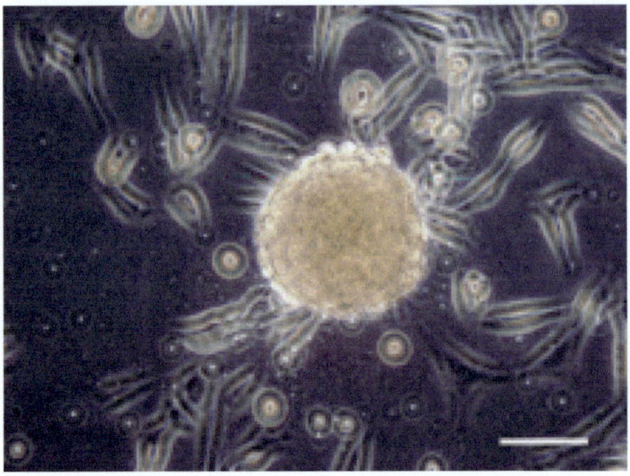

Fig. 4.1 Cell clusters induced by lactic acid bacteria. After reaching confluency, human dermal fibroblast cells were trypsinized. Lactic acid bacteria in logarithmic phase were collected by centrifugation and mixed with human dermal fibroblasts. Notably, cell clusters were not formed by the addition of lactic acid bacteria into the culture dish directly. Bar = 100 μm

surface of the intestine through mucin, extracellular matrix, and lectin (Yan et al. 2013; Nishiyama et al. 2016). Although the role of LAB in gastrointestinal microecology has been a subject of extensive research, the effect of LAB infection in cells is unclear. This is because the mucosal layer provides a protective barrier between the underlying epithelium and lumen, which contains noxious agents and microorganisms (Gouyer et al. 2015; Lechuga and Ivanov 2017).

Our group conducted an experiment in which human cells were artificially infected with nonpathogenic and noninfectious LAB; changes in stemness characteristics were observed (Ohta et al. 2012; Ito and Ohta 2015). LAB-mediated reprograming takes place in a manner which is completely different from that by conventional methods, such as those for iPS cell production. However, both methods employ the addition of exogenous substances to the host cells.

LAB has been successfully incorporated into human dermal fibroblast (HDF) cells by trypsinization and cocultivation (Figs. 4.1 and 4.2). This effect was achieved without the use of genetic modification or pluripotency-inducing chemicals. The LAB-treated HDF cells were observed to form clusters, like embryoid spheres, and had lost their ability for self-renewal. In these clustered cells, LAB were localized in the internal membranes of the host cell organelles. LAB-incorporated cell clusters also expressed a subset of pluripotent stem cell marker genes, such as *NANOG*, *OCT3/4*, and *SOX2*. The expression of *HOX*, which controls embryonic development, was notably decreased. The LAB-incorporated cell clusters could transform into derivatives of all three germ layers in vivo and in vitro. When these cells were implanted into an immunodeficient mouse testis, the cells survived after 3 months without forming a teratoma. Among these cells, there were some existing anti-α1-fetoprotein (endoderm), neurofilament (ectoderm), or smooth muscle actin (mesoderm) antibody immunoreactive cells, thereby indicating that the implanted LAB-incorporated cells had a capacity to differentiate into derivatives of all three germ layers. In an in vitro experiment, the LAB-incorporated cell clusters could differentiate into cells derived from the three germ layers, when cultured with

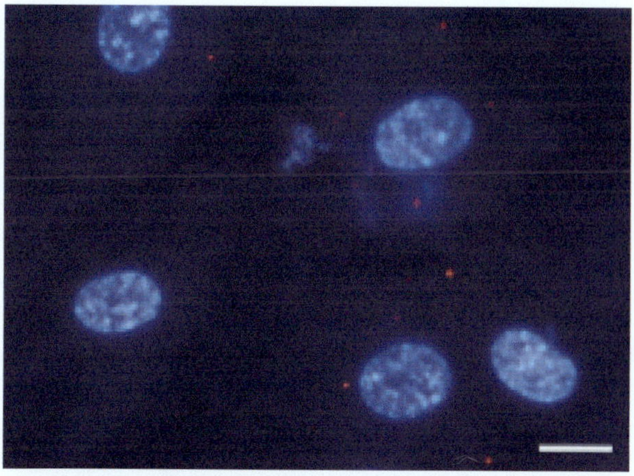

Fig. 4.2 Immunocytochemistry for lactic acid bacteria -induced cell cluster. Lactic acid bacteria incorporation into human cells (host cells) was detected using the anti-gram-positive bacteria marker antibody (sc-58135, Santa Cruz Biotechnology, Inc.10410 Finnell Street, Dallas, Texas 75220, USA) (Magenta). Nucleic acids were stained using Hoechst 33342 (Cyan). Bar = 10 μm

lineage-specific culture media. The LAB-induced cell clusters did not divide, although proliferation of differentiated cells was observed (Fig. 4.3).

LAB do not possess homologs of *cagA* found in *H. pylori*, which injects the pathogenic CagA protein into human stomach via the type IV secretion system (Tegtmeyer et al. 2017). Thus, the acquisition of stemness is induced by LAB via a mechanism different from that of *H. pylori*-induced intestinal metaplasia. Activation of innate immunity accelerates nuclear reprogramming via epithelial–mesenchymal transition (Bessede et al. 2014; Lee et al. 2014). Innate immune activation stimulated by bacteria might be pertinent in bacterial reprogramming. Direct in vivo reprogramming by using iPS cell technology for ubiquitously expressing *Oct4*, *Sox2*, *Klf4*, and *c-Myc*, results in totipotency with teratoma formation (Abad et al. 2013). In vitro cellular lineage transdifferentiation increases the therapeutic potential by directly treating the target tissues in patients (Guo et al. 2014; Song et al. 2016).

Self-renewal is an important property for tissue regeneration and pluripotency; however, it is also a double-edged sword because it may result in teratoma formation and tumorigenesis (Ben-David and Benvenisty 2011). The problem of tumor formation, following teratoma formation, can be resolved in some cells. Efforts are being undertaken to reduce tumor formation (Itakura et al. 2015; Bedel et al. 2017; Itakura et al. 2017). The LAB-reprogrammed cell mass does not show the property of self-renewal. LAB can also be used in vivo as transfer vehicles of the therapeutic agent for therapeutic treatment owing to their safety toward humans (Taniguchi et al. 2010). Therefore, we suggest that the LAB-induced spheres can be potentially used in in vivo cell-based therapies, thereby avoiding the risk of teratoma formation. Further studies about the reprogramming process activated by LAB and the comparison of dedifferentiated cells would yield novel insights into their therapeutic effects and mechanisms of action. Determining the LAB-derived reprogramming factor(s) would open a new avenue for the acquisition of pluripotency at the molecular level.

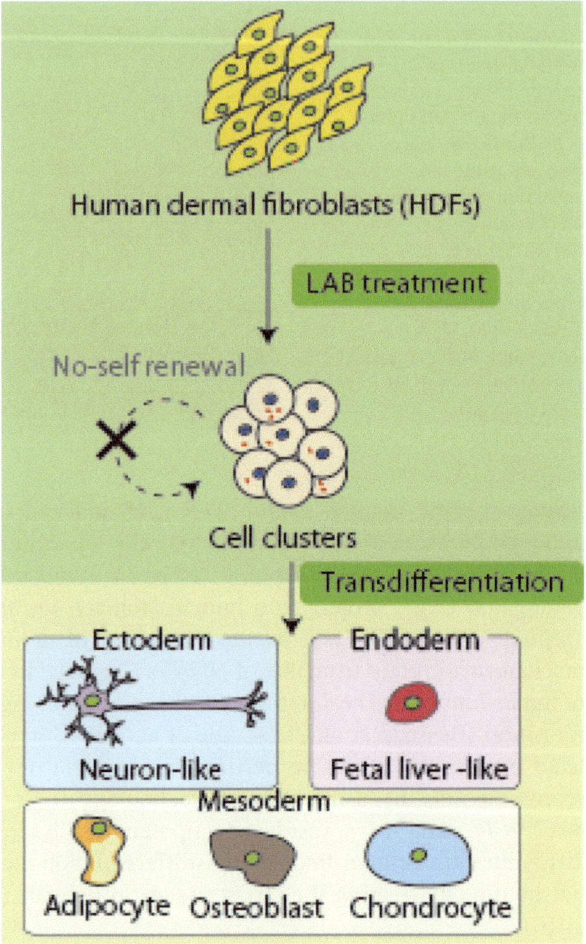

Fig. 4.3 A schematic representation of an in vitro cell cluster formed by lactic acid bacteria. Lactic acid bacteria -incorporated cell clusters are able to transdifferentiate into other cell types when cultured in appropriate differentiation media

4.5 Cell Reprogramming by Members of *Mycobacteriaceae*

Mycobacterium leprae is an acid-fast gram-positive bacterium belonging to the genus *Mycobacterium*, which includes *M. tuberculosis* (Lienhardt et al. 2012). It mainly infects macrophages and Schwann cells and results in peripheral neuropathy designated as Hansen's disease (Moura et al. 2013). *M. leprae* infects humans, proliferates slowly over 20 years, and finally causes painless burns, ulcers, blindness, corneal damage, nerve paralysis, loss of corneal sensation, and iritis (Smith et al. 2017). *Mycobacterium* infections can elicit acute immunological reactions that can cause inflammatory and edematous skin lesions and also result in further impairment of nerve functions.

M. leprae infects macrophages and/or Schwann cells lining the nerve fibers, in the peripheral nervous system. For their propagation, *Mycobacterium* actively infect the permissive macrophages and dodge the macrophages with bactericidal activity (Cambier et al. 2014). *M. leprae* is an obligate parasite occurring in nature. It has been isolated only in humans and armadillos and can grow only in infected cells. This implies that it cannot grow in any laboratory media (Lagier et al. 2015).

Members of *Mycobacteriaceae* (gram-positive, aerobic, and nonmotile bacteria) cause critical diseases in mammals (Fernando and Britton 2006). For its propagation in humans, *M. leprae* hijacks the cellular lineage control system. It infects the Schwann cells and transforms them into stem-cell-like progenitors (Masaki et al. 2013, 2014). The *M. leprae*-infected stem-cell-like progenitors migrate and spread to the smooth muscle-like cells, thereby increasing bacterial infection and cellular transformation. *Mycobacterium tuberculosis*, known to be a serious pathogen and the causative agent of tuberculosis in humans, infects macrophages via E-cadherin-dependent mesenchymal–epithelial transition (Cronan et al. 2016). Formation of granulomas, due to *M. tuberculosis* infection, leads to the destruction of cytoskeletal structure and a shape change in the host cells. Disruption of E-cadherin function in granuloma macrophages increases immune cell access, host survival, and protection against bacterial infection and decreases the bacterial load. These results suggest that *Mycobacterium* spp. can control cell lineage-fate in their hosts, in order to survive and expand their habitat.

4.6 Conclusions

In this chapter, we described how LAB and *Mycobacterium* spp. reprogram mammalian cells. However, these bacteria, which facilitate cellular reprogramming, share less phenotypic and phylogenetic similarities, in terms of the presence or absence of pathogenicity, gram status, and the phyla to which they belong (namely, *Firmicutes* or *Actinobacteria*). The uptake of bacteria by human cells seems to be a critical event for the acquisition of transdifferentiation. It is suggested that cellular reprogramming by bacteria is a general outcome of human–microbe interactions.

In humans, dedifferentiated cells can occur in different states of differentiation, depending on their individual reprogramming (Gafni et al. 2013; Takashima et al. 2014; Zhao et al. 2015). Cells reprogrammed by bacteria cannot be classified into any of these states because of the complexity of their expressed pluripotency markers (Tanabe 2013).

Induction of pluripotency or an undifferentiated state in host cells by microorganisms involves different genes that are common with the naive and primed pluripotent stem cells. Other common genes include those involved in regulating cell structure, adhesion, and body segmentation. This means that if one gene responsible for maintaining the dedifferentiated state undergoes a change due to mutation or environmental stress, the permanent gene expression in the cell can complement this loss or gain of gene expression. This has also been considered to be a mechanism for preventing irregular reprogramming in cells and maintaining homeostasis.

The phenomenon of human cell reprogramming by bacteria is a recently discovered aspect of human–microbe interactions. By introducing a reprogramming gene-containing plasmid into cells, direct reprogramming can be accomplished in vivo; however, at present, this technique is not well established for the therapeutic treatment. Understanding the mechanism of direct reprogramming of human cells in vivo by microorganisms occurring in nature will clarify the phenomenon caused by artificial direct reprogramming of human living body. The molecular mechanisms and signaling pathways involved in cellular reprogramming by bacteria remain unknown; however, recently developed techniques (such as genomics and microbiome analysis) can provide valuable insights into the underlying mechanisms. Thus, detailed knowledge about bacteria-mediated reprogramming will help in understanding human–microbe interactions and somatic lineage reprogramming in depth.

Acknowledgments We thank all the members of our laboratory for their support and insightful discussions. This work was supported by KAKENHI (16674994, 16690582, 25650082), AMED-CREST (15652254), Kumamoto University's Advanced Research Project "Stem Cell-Based Tissue Regeneration Research and Education Unit", Mitsubishi Foundation, and Institute for Fermentation, Osaka.

References

Aagaard K, Ma J, Antony KM, Ganu R, Petrosino J, Versalovic J (2014) The placenta harbors a unique microbiome. Sci Transl Med 6:237ra265. doi:6/237/237ra6510.1126/scitranslmed.3008599

Abad M, Mosteiro L, Pantoja C, Canamero M, Rayon T, Ors I, Grana O, Megias D, Dominguez O, Martinez D, Manzanares M, Ortega S, Serrano M (2013) Reprogramming in vivo produces teratomas and iPS cells with totipotency features. Nature 502:340–345. https://doi.org/10.1038/nature12586

Arbibe L, Kim DW, Batsche E, Pedron T, Mateescu B, Muchardt C, Parsot C, Sansonetti PJ (2007) An injected bacterial effector targets chromatin access for transcription factor NF-kappaB to alter transcription of host genes involved in immune responses. Nat Immunol 8:47–56. https://doi.org/10.1038/ni1423

Ashida H, Ogawa M, Kim M, Suzuki S, Sanada T, Punginelli C, Mimuro H, Sasakawa C (2011) Shigella deploy multiple countermeasures against host innate immune responses. Curr Opin Microbiol 14:16–23. https://doi.org/10.1016/j.mib.2010.08.014

Barcenilla A, Pryde SE, Martin JC, Duncan SH, Stewart CS, Henderson C, Flint HJ (2000) Phylogenetic relationships of butyrate-producing bacteria from the human gut. Appl Environ Microbiol 66:1654–1661

Bedel A, Beliveau F, Lamrissi-Garcia I, Rousseau B, Moranvillier I, Rucheton B, Guyonnet-Dupérat V, Cardinaud B, de Verneuil H, Moreau-Gaudry F, Dabernat S (2017) Preventing pluripotent cell teratoma in regenerative medicine applied to hematology disorders. Stem Cells Transl Med 6:382–393. https://doi.org/10.5966/sctm.2016-0201

Ben-David U, Benvenisty N (2011) The tumorigenicity of human embryonic and induced pluripotent stem cells. Nat Rev Cancer 11:268–277

Bessede E, Staedel C, Acuna Amador LA, Nguyen PH, Chambonnier L, Hatakeyama M, Belleannee G, Megraud F, Varon C (2014) Helicobacter pylori generates cells with cancer stem cell properties via epithelial-mesenchymal transition-like changes. Oncogene 33:4123–4131. https://doi.org/10.1038/onc.2013.380

Briggs R, King TJ (1952) Transplantation of living nuclei from blastula cells into enucleated frogs' eggs. Proc Natl Acad Sci U S A 38:455–463

Browne HP, Neville BA, Forster SC, Lawley TD (2017) Transmission of the gut microbiota: spreading of health. Nat Rev Microbiol. https://doi.org/10.1038/nrmicro.2017.50

Cambier CJ, Takaki KK, Larson RP, Hernandez RE, Tobin DM, Urdahl KB, Cosma CL, Ramakrishnan L (2014) Mycobacteria manipulate macrophage recruitment through coordinated use of membrane lipids. Nature 505:218–222. https://doi.org/10.1038/nature12799

Cao M, Goodrich-Blair H (2017) Ready or not: microbial adaptive responses in dynamic symbiosis environments. J Bacteriol 199:e00883–e00816. https://doi.org/10.1128/JB.00883-16

Chaudhari SN, Mukherjee M, Vagasi AS, Bi G, Rahman MM, Nguyen CQ, Paul L, Selhub J, Kipreos ET (2016) Bacterial folates provide an exogenous signal for *C. elegans* germline stem cell proliferation. Dev Cell 38:33–46. https://doi.org/10.1016/j.devcel.2016.06.013

Collins SM (2014) A role for the gut microbiota in IBS. Nat Rev Gastroenterol Hepatol 11:497–505. https://doi.org/10.1038/nrgastro.2014.40

Costello EK, Lauber CL, Hamady M, Fierer N, Gordon JI, Knight R (2009) Bacterial community variation in human body habitats across space and time. Science 326:1694–1697. https://doi.org/10.1126/science.1177486

Cronan MR, Beerman RW, Rosenberg AF, Saelens JW, Johnson MG, Oehlers SH, Sisk DM, Jurcic Smith KL, Medvitz NA, Miller SE, Trinh LA, Fraser SE, Madden JF, Turner J, Stout JE, Lee S, Tobin DM (2016) Macrophage epithelial reprogramming underlies mycobacterial granuloma formation and promotes infection. Immunity 45:861–876. https://doi.org/10.1016/j.immuni.2016.09.014

Douglas AE (2016) How multi-partner endosymbioses function. Nat Rev Microbiol 14:731–743. https://doi.org/10.1038/nrmicro.2016.151

Fernando SL, Britton WJ (2006) Genetic susceptibility to mycobacterial disease in humans. Immunol Cell Biol 84:125–137. https://doi.org/10.1111/j.1440-1711.2006.01420.x

Fesel PH, Zuccaro A (2016) Dissecting endophytic lifestyle along the parasitism/mutualism continuum in *Arabidopsis*. Curr Opin Microbiol 32:103–112. https://doi.org/10.1016/j.mib.2016.05.008

Furusawa Y, Obata Y, Fukuda S, Endo TA, Nakato G, Takahashi D, Nakanishi Y, Uetake C, Kato K, Kato T, Takahashi M, Fukuda NN, Murakami S, Miyauchi E, Hino S, Atarashi K, Onawa S, Fujimura Y, Lockett T, Clarke JM, Topping DL, Tomita M, Hori S, Ohara O, Morita T, Koseki H, Kikuchi J, Honda K, Hase K, Ohno H (2013) Commensal microbe-derived butyrate induces the differentiation of colonic regulatory T cells. Nature 504:446–450. https://doi.org/10.1038/nature12721

Furuse Y, Finethy R, Saka HA, Xet-Mull AM, Sisk DM, Smith KL, Lee S, Coers J, Valdivia RH, Tobin DM, Cullen BR (2014) Search for microRNAs expressed by intracellular bacterial pathogens in infected mammalian cells. PLoS One 9:e106434. https://doi.org/10.1371/journal.pone.0106434

Gafni O, Weinberger L, Mansour AA, Manor YS, Chomsky E, Ben-Yosef D, Kalma Y, Viukov S, Maza I, Zviran A, Rais Y, Shipony Z, Mukamel Z, Krupalnik V, Zerbib M, Geula S, Caspi I, Schneir D, Shwartz T, Gilad S, Amann-Zalcenstein D, Benjamin S, Amit I, Tanay A, Massarwa R, Novershtern N, Hanna JH (2013) Derivation of novel human ground state naive pluripotent stem cells. Nature 504:282–286. https://doi.org/10.1038/nature12745

Gilbert SF, Bosch TC, Ledon-Rettig C (2015) Eco-Evo-Devo: developmental symbiosis and developmental plasticity as evolutionary agents. Nat Rev Genet 16:611–622. https://doi.org/10.1038/nrg3982

Gordon S (2008) Elie metchnikoff: father of natural immunity. Eur J Immunol 38:3257–3264. https://doi.org/10.1002/eji.200838855

Gouyer V, Dubuquoy L, Robbe-Masselot C, Neut C, Singer E, Plet S, Geboes K, Desreumaux P, Gottrand F, Desseyn JL (2015) Delivery of a mucin domain enriched in cysteine residues strengthens the intestinal mucous barrier. Sci Rep 5:9577. https://doi.org/10.1038/srep09577

Gu H, Zhao C, Zhang T, Liang H, Wang XM, Pan Y, Chen X, Zhao Q, Li D, Liu F, Zhang CY, Zen K (2017) Salmonella produce microRNA-like RNA fragment Sal-1 in the infected cells to facilitate intracellular survival. Sci Rep 7:2392. https://doi.org/10.1038/s41598-017-02669-1

Guo Z, Zhang L, Wu Z, Chen Y, Wang F, Chen G (2014) In vivo direct reprogramming of reactive glial cells into functional neurons after brain injury and in an Alzheimer's disease model. Cell Stem Cell 14:188–202. https://doi.org/10.1016/j.stem.2013.12.001

Gurdon JB (1962) The developmental capacity of nuclei taken from intestinal epithelium cells of feeding tadpoles. J Embryol Exp Morphol 10:622–640

Hartman H, Fedorov A (2002) The origin of the eukaryotic cell: a genomic investigation. Proc Natl Acad Sci U S A 99:1420–1425. https://doi.org/10.1073/pnas.032658599

Hill JH, Franzosa EA, Huttenhower C, Guillemin K (2016) A conserved bacterial protein induces pancreatic beta cell expansion during zebrafish development. eLife 5:e20145. https://doi.org/10.7554/eLife.20145

Holmes E, Li JV, Marchesi JR, Nicholson JK (2012) Gut microbiota composition and activity in relation to host metabolic phenotype and disease risk. Cell Metab 16:559–564. https://doi.org/10.1016/j.cmet.2012.10.007

Ikeda T, Uchiyama I, Iwasaki M, Sasaki T, Nakagawa M, Okita K, Masui S (2017) Artificial acceleration of mammalian cell reprogramming by bacterial proteins. Genes Cells. https://doi.org/10.1111/gtc.12519

Itakura G, Kobayashi Y, Nishimura S, Iwai H, Takano M, Iwanami A, Toyama Y, Okano H, Nakamura M (2015) Controlling immune rejection is a fail-safe system against potential tumorigenicity after human iPSC-derived neural stem cell transplantation. PLoS One 10:e0116413. https://doi.org/10.1371/journal.pone.0116413

Itakura G, Kawabata S, Ando M, Nishiyama Y, Sugai K, Ozaki M, Iida T, Ookubo T, Kojima K, Kashiwagi R, Yasutake K, Nakauchi H, Miyoshi H, Nagoshi N, Kohyama J, Iwanami A, Matsumoto M, Nakamura M, Okano H (2017) Fail-safe system against potential tumorigenicity after transplantation of iPSC Derivatives. Stem Cell Rep 8:673–684. https://doi.org/10.1016/j.stemcr.2017.02.003

Ito N, Ohta K (2015) Reprogramming of human somatic cells by bacteria. Develop Growth Differ 57:305–312. https://doi.org/10.1111/dgd.12209

Kaiko GE, Ryu SH, Koues OI, Collins PL, Solnica-Krezel L, Pearce EJ, Pearce EL, Oltz EM, Stappenbeck TS (2016) The colonic crypt protects stem cells from microbiota-derived metabolites. Cell 165:1708–1720. https://doi.org/10.1016/j.cell.2016.05.018

Kawamoto S, Maruya M, Kato LM, Suda W, Atarashi K, Doi Y, Tsutsui Y, Qin H, Honda K, Okada T, Hattori M, Fagarasan S (2014) Foxp3(+) T cells regulate immunoglobulin a selection and facilitate diversification of bacterial species responsible for immune homeostasis. Immunity 41:152–165. https://doi.org/10.1016/j.immuni.2014.05.016

Kawashima T, Kosaka A, Yan H, Guo Z, Uchiyama R, Fukui R, Kaneko D, Kumagai Y, You DJ, Carreras J, Uematsu S, Jang MH, Takeuchi O, Kaisho T, Akira S, Miyake K, Tsutsui H, Saito T, Nishimura I, Tsuji NM (2013) Double-stranded RNA of intestinal commensal but not pathogenic bacteria triggers production of protective interferon-beta. Immunity 38:1187–1197. https://doi.org/10.1016/j.immuni.2013.02.024

Kleiner M, Wentrup C, Lott C, Teeling H, Wetzel S, Young J, Chang YJ, Shah M, NC VB, Zarzycki J, Fuchs G, Markert S, Hempel K, Voigt B, Becher D, Liebeke M, Lalk M, Albrecht D, Hecker M, Schweder T, Dubilier N (2012) Metaproteomics of a gutless marine worm and its symbiotic microbial community reveal unusual pathways for carbon and energy use. Proc Natl Acad Sci U S A 109:E1173–E1182. https://doi.org/10.1073/pnas.1121198109

Lad SP, Li J, da Silva Correia J, Pan Q, Gadwal S, Ulevitch RJ, Li E (2007) Cleavage of p65/RelA of the NF-kappaB pathway by *Chlamydia*. Proc Natl Acad Sci U S A 104:2933–2938. https://doi.org/10.1073/pnas.0608393104

Lagier J-C, Edouard S, Pagnier I, Mediannikov O, Drancourt M, Raoult D (2015) Current and past strategies for bacterial culture in clinical microbiology. Clin Microbiol Rev 28:208–236. https://doi.org/10.1128/CMR.00110-14

Lebreton A, Lakisic G, Job V, Fritsch L, Tham TN, Camejo A, Mattei PJ, Regnault B, Nahori MA, Cabanes D, Gautreau A, Ait-Si-Ali S, Dessen A, Cossart P, Bierne H (2011) A bacterial protein targets the BAHD1 chromatin complex to stimulate type III interferon response. Science 331:1319–1321. https://doi.org/10.1126/science.1200120

Lechuga S, Ivanov AI (2017) Disruption of the epithelial barrier during intestinal inflammation: quest for new molecules and mechanisms. Biochim Biophys Acta 1864:1183–1194. https://doi.org/10.1016/j.bbamcr.2017.03.007

Lee DG, Kim HS, Lee YS, Kim S, Cha SY, Ota I, Kim NH, Cha YH, Yang DH, Lee Y, Park GJ, Yook JI, Lee YC (2014) Helicobacter pylori CagA promotes Snail-mediated epithelial-mesenchymal transition by reducing GSK-3 activity. Nat Commun 5:4423. https://doi.org/10.1038/ncomms5423

Ley RE, Hamady M, Lozupone C, Turnbaugh PJ, Ramey RR, Bircher JS, Schlegel ML, Tucker TA, Schrenzel MD, Knight R, Gordon JI (2008) Evolution of mammals and their gut microbes. Science 320:1647–1651. https://doi.org/10.1126/science.1155725

Lienhardt C, Glaziou P, Uplekar M, Lonnroth K, Getahun H, Raviglione M (2012) Global tuberculosis control: lessons learnt and future prospects. Nat Rev Microbiol 10:407–416. https://doi.org/10.1038/nrmicro2797

Lin YW, Huang CY, Chen YH, Shih CM, Tsao NW, Lin CY, Chang NC, Tsai CS, Tsai HY, Tsai JC, Huang PH, Li CY, Lin FY (2013) GroEL1, a heat shock protein 60 of Chlamydia pneumoniae, impairs neovascularization by decreasing endothelial progenitor cell function. PLoS One 8:e84731. https://doi.org/10.1371/journal.pone.0084731

Louis P, Hold GL, Flint HJ (2014) The gut microbiota, bacterial metabolites and colorectal cancer. Nat Rev Microbiol 12:661–672. https://doi.org/10.1038/nrmicro3344

Lu R, Wu S, Zhang YG, Xia Y, Liu X, Zheng Y, Chen H, Schaefer KL, Zhou Z, Bissonnette M, Li L, Sun J (2014) Enteric bacterial protein AvrA promotes colonic tumorigenesis and activates colonic beta-catenin signaling pathway. Oncogene 3:e105. https://doi.org/10.1038/oncsis.2014.20

Margulis L (1970) Origin of eukaryotic cells: evidence and research implications for a theory of the origin and evolution of microbial, plant and animal cells on the precambrian earth. Yale University Press, New Heaven

Martin W, Hoffmeister M, Rotte C, Henze K (2001) An overview of endosymbiotic models for the origins of eukaryotes, their ATP-producing organelles (mitochondria and hydrogenosomes), and their heterotrophic lifestyle. Biol Chem 382:1521–1539. https://doi.org/10.1515/BC.2001.187

Masaki T, Qu J, Cholewa-Waclaw J, Burr K, Raaum R, Rambukkana A (2013) Reprogramming adult Schwann cells to stem cell-like cells by leprosy bacilli promotes dissemination of infection. Cell 152:51–67. https://doi.org/10.1016/j.cell.2012.12.014

Masaki T, McGlinchey A, Cholewa-Waclaw J, Qu J, Tomlinson SR, Rambukkana A (2014) Innate immune response precedes Mycobacterium leprae-induced reprogramming of adult Schwann cells. Cell Reprogram 16:9–17. https://doi.org/10.1089/cell.2013.0064

Matsuura Y, Kikuchi Y, Hosokawa T, Koga R, Meng XY, Kamagata Y, Nikoh N, Fukatsu T (2012) Evolution of symbiotic organs and endosymbionts in lygaeid stinkbugs. ISME J 6:397–409. https://doi.org/10.1038/ismej.2011.103

Moran NA, Yun Y (2015) Experimental replacement of an obligate insect symbiont. Proc Natl Acad Sci U S A 112:2093–2096. https://doi.org/10.1073/pnas.1420037112

Moura ML, Dupnik KM, Sampaio GA, Nobrega PF, Jeronimo AK, do Nascimento-Filho JM, Miranda Dantas RL, Queiroz JW, Barbosa JD, Dias G, Jeronimo SM, Souza MC, Nobre ML (2013) Active surveillance of Hansen's disease (leprosy): importance for case finding among extra-domiciliary contacts. PLoS Negl Trop Dis 7:e2093. https://doi.org/10.1371/journal.pntd.0002093

Nakajima A, Negishi N, Tsurui H, Kadowaki-Ohtsuji N, Maeda K, Nanno M, Yamaguchi Y, Shimizu N, Yagita H, Okumura K, Habu S (2014) Commensal bacteria regulate thymic Aire expression. PLoS One 9:e105904. https://doi.org/10.1371/journal.pone.0105904

Nigro G, Rossi R, Commere PH, Jay P, Sansonetti PJ (2014) The cytosolic bacterial peptidoglycan sensor Nod2 affords stem cell protection and links microbes to gut epithelial regeneration. Cell Host Microbe 15:792–798. https://doi.org/10.1016/j.chom.2014.05.003

Nishiyama K, Sugiyama M, Mukai T (2016) Adhesion properties of lactic acid bacteria on intestinal mucin. Microorganisms 4:E34. https://doi.org/10.3390/microorganisms4030034

Ohta K, Kawano R, Ito N (2012) Lactic acid bacteria convert human fibroblasts to multipotent cells. PLoS One 7:e51866. https://doi.org/10.1371/journal.pone.0051866

Okada TS (2004) From embryonic induction to cell lineages: revisiting old problems for modern study. Int J Dev Biol 48:739–742. https://doi.org/10.1387/ijdb.041918to

Pennini ME, Perrinet S, Dautry-Varsat A, Subtil A (2010) Histone methylation by NUE, a novel nuclear effector of the intracellular pathogen Chlamydia trachomatis. PLoS Pathog 6:e1000995. https://doi.org/10.1371/journal.ppat.1000995

Schloissnig S, Arumugam M, Sunagawa S, Mitreva M, Tap J, Zhu A, Waller A, Mende DR, Kultima JR, Martin J, Kota K, Sunyaev SR, Weinstock GM, Bork P (2013) Genomic variation landscape of the human gut microbiome. Nature 493:45–50. https://doi.org/10.1038/nature11711

Sebe-Pedros A, Degnan BM, Ruiz-Trillo I (2017) The origin of Metazoa: a unicellular perspective. Nat Rev Genet 18:498–512. https://doi.org/10.1038/nrg.2017.21

Shin JM, Gwak JW, Kamarajan P, Fenno JC, Rickard AH, Kapila YL (2016) Biomedical applications of nisin. J Appl Microbiol 120:1449–1465. https://doi.org/10.1111/jam.13033

Smith CS, Aerts A, Saunderson P, Kawuma J, Kita E, Virmond M (2017) Multidrug therapy for leprosy: a game changer on the path to elimination. Lancet Infect Dis. https://doi.org/10.1016/s1473-3099(17)30418-8

Song G, Pacher M, Balakrishnan A, Yuan Q, Tsay HC, Yang D, Reetz J, Brandes S, Dai Z, Putzer BM, Arauzo-Bravo MJ, Steinemann D, Luedde T, Schwabe RF, Manns MP, Scholer HR, Schambach A, Cantz T, Ott M, Sharma AD (2016) Direct reprogramming of hepatic myofibroblasts into hepatocytes in vivo attenuates liver fibrosis. Cell Stem Cell 18:797–808. https://doi.org/10.1016/j.stem.2016.01.010

Srivastava D, DeWitt N (2016) In vivo cellular reprogramming: the next generation. Cell 166:1386–1396. https://doi.org/10.1016/j.cell.2016.08.055

Taguchi S, Ooi T, Mizuno K, Matsusaki H (2015) Advances and needs for endotoxin-free production strains. Appl Microbiol Biotechnol 99:9349–9360. https://doi.org/10.1007/s00253-015-6947-9

Takahashi K, Yamanaka S (2006) Induction of pluripotent stem cells from mouse embryonic and adult fibroblast cultures by defined factors. Cell 126:663–676. https://doi.org/10.1016/j.cell.2006.07.024

Takashima Y, Guo G, Loos R, Nichols J, Ficz G, Krueger F, Oxley D, Santos F, Clarke J, Mansfield W, Reik W, Bertone P, Smith A (2014) Resetting transcription factor control circuitry toward ground-state pluripotency in human. Cell 158:1254–1269. https://doi.org/10.1016/j.cell.2014.08.029

Tanabe S (2013) Perspectives of gene combinations in phenotype presentation. World J Stem Cells 5:61–67. https://doi.org/10.4252/wjsc.v5.i3.61

Taniguchi S, Fujimori M, Sasaki T, Tsutsui H, Shimatani Y, Seki K, Amano J (2010) Targeting solid tumors with non-pathogenic obligate anaerobic bacteria. Cancer Sci 101:1925–1932. https://doi.org/10.1111/j.1349-7006.2010.01628.x

Tao L, Zhang J, Meraner P, Tovaglieri A, Wu X, Gerhard R, Zhang X, Stallcup WB, Miao J, He X, Hurdle JG, Breault DT, Brass AL, Dong M (2016) Frizzled proteins are colonic epithelial receptors for *C. difficile* toxin B. Nature 538:350–355. https://doi.org/10.1038/nature19799

Tegtmeyer N, Neddermann M, Asche CI, Backert S (2017) Subversion of host kinases: a key network in cellular signaling hijacked by *Helicobacter pylori* CagA. Mol Microbiol 105:358–372. https://doi.org/10.1111/mmi.13707

The Human Microbiome Project Consortium (2012) A framework for human microbiome research. Nature 486:215–221. https://doi.org/10.1038/nature11209

Timmis JN, Ayliffe MA, Huang CY, Martin W (2004) Endosymbiotic gene transfer: organelle genomes forge eukaryotic chromosomes. Nat Rev Genet 5:123–135. https://doi.org/10.1038/nrg1271

Wang X, Yang Y, Huycke MM (2015) Commensal bacteria drive endogenous transformation and tumour stem cell marker expression through a bystander effect. Gut 64:459–468. https://doi.org/10.1136/gutjnl-2014-307213

Woese CR (2002) On the evolution of cells. Proc Natl Acad Sci U S A 99:8742–8747. https://doi.org/10.1073/pnas.132266999

Woese CR, Kandler O, Wheelis ML (1990) Towards a natural system of organisms: proposal for the domains archaea, bacteria, and eucarya. Proc Natl Acad Sci U S A 87:4576–4579

Yan H, Ohno N, Tsuji NM (2013) The role of C-type lectin receptors in immune homeostasis. Int Immunopharmacol 16:353–357. https://doi.org/10.1016/j.intimp.2013.04.013

Zhao Y, Zhao T, Guan J, Zhang X, Fu Y, Ye J, Zhu J, Meng G, Ge J, Yang S, Cheng L, Du Y, Zhao C, Wang T, Su L, Yang W, Deng H (2015) A XEN-like state bridges somatic cells to pluripotency during chemical reprogramming. Cell 163:1678–1691. https://doi.org/10.1016/j.cell.2015.11.017

Ziegler M, Seneca FO, Yum LK, Palumbi SR, Voolstra CR (2017) Bacterial community dynamics are linked to patterns of coral heat tolerance. Nat Commun 8:14213. https://doi.org/10.1038/ncomms14213

Chapter 5
Generation of Aptamers Against Natural Toxins and Their Application as Biosensors

Yuji Morita and Daisuke Fujiwara

Abstract RNA or ssDNA aptamers, which are capable of binding to target molecules with high affinity and specificity, are selected in vitro from large combinatorial nucleic acid libraries by a process known as systematic evolution of ligands by exponential enrichment (SELEX). Using this SELEX technology, many aptamers have been generated against a wide range of target molecules, including proteins, nucleic acids, small molecules, and whole cells. In this chapter, we described various methods for generating aptamers, including methods that do not require target immobilization. Among these aptamers, ones that are specific for natural toxins, such as mycotoxins, are of great interest to the food industry, as they can used in developing tools (biosensors) for ensuring food safety. We also summarized several aptamer-based detection strategies. Lastly, we described biosensor applications of aptamers for natural toxins.

Keywords RNA · Aptamer · SELEX · Aptasensor · Natural toxin

5.1 Introduction

Aptamers are single-stranded RNA or DNA (ssDNA) molecules that can bind to target molecules with high affinity and specificity (Bunka and Stockley 2006). Importantly, they are attractive alternatives to antibodies for the detection of various molecules, including nucleic acids, proteins, small molecules (Jenison et al. 1994), and whole cells (Mayer et al. 2010). Compared to antibodies, aptamers are relatively cost-effective to produce and can have longer shelf life. Additionally,

Y. Morita (✉)
Research Laboratories for Health Science & Food Technologies, Research and Development Division, Kirin Company Limited, Yokohama, Japan
e-mail: Yuji_Morita@kirin.co.jp

D. Fujiwara
Central Laboratories for Key Technologies, Research and Development Division, Kirin Company Limited, Yokohama, Japan

S. Masuda, S. Izawa (eds.), *Applied RNA Bioscience*,
https://doi.org/10.1007/978-981-10-8372-3_5

aptamers can be easily labeled with a wide range of molecules, such as fluorescent dyes and enzymes, leading to the development of a variety of detection tools. Due to these advantages, there is an increased interest in developing aptamer-based sensors for various targets, including natural toxins. In 1990, three groups independently developed methods to generate catalytic or ligand binding RNA molecules, including aptamers, from RNA libraries (Ellington and Szostak 1990; Robertson and Joyce 1990; Tuerk and Gold 1990). The process, known as systematic evolution of ligands by exponential enrichment (SELEX), is a powerful tool for in vitro selection of nucleic acids with catalytic activity or ligand binding affinity from very large random RNA or ssDNA libraries containing approximately 10^{14} different molecules. Some modified SELEX methods were also developed for selecting high-affinity aptamers specific for a wide range of target molecules more efficiently (Kulbachinskiy 2007; Stoltenburg et al. 2007; Aquino-Jarquin and Toscano-Garibay 2011).

Natural toxins, such as mycotoxins and botulinum neurotoxins, which are produced by a number of fungal species, plants, and other living organisms, pose potential health problems even at very low doses (Roll et al. 1990). Because of their toxic properties, the presence of natural toxins in foods has been of great concern in many countries. Accordingly, the International Agency for Research on Cancer (IARC) has listed mycotoxins as a human carcinogen, and the World Health Organization has recommended upper limits of their concentration in food. For example, ochratoxin A, which is known to have nephrotoxic, teratogenic, and immunotoxic effects, has been listed as a possible carcinogen by IARC. Aflatoxin B1, a toxic food contaminant, has also been listed as a category 1 human carcinogen (Xu et al. 2016). Similarly, patulin, which mainly contaminates apples and apple products, has been classified as a category 3 human carcinogen. These natural toxins are also found as low-level contaminants in various grains and fruits. Therefore, reliable and easy-to-use methods for detecting these natural toxins are required in order to ensure food safety (Hu et al. 2013).

Currently, natural toxins are detected by chromatographic and chromatography-mass spectrometry-linked techniques, such as HPLC, GC, LC/MS, and GC/MS (Crosby 1984; Watanabe and Shimizu 2005; Zollner and Mayer-Helm 2006; Turner et al. 2009; Andrade and Lancas 2017; Breidbach 2017), all of which involve complicated sample preparation steps and require expensive instruments run by well-trained researchers for proper sample analysis. Because of these reasons, currently available detection techniques are not suitable for analyzing large number of toxin samples. To circumvent this problem, various biosensors were developed as an alternative to the aforementioned methods. Thus, the biosensors that were developed based on specific high-affinity interactions between antigens and antibodies, especially the ones that involved enzyme-linked immunosorbent assay (ELISA), have been extensively used for the detection of mycotoxins, such as ochratoxin A, deoxynivalenol, and fumonisin. However, the enzymes and antibodies used for developing these biosensors are generally expensive and have short shelf life due to their instability under various experimental and physiological conditions, such as temperature, pressure, pH, and exposure to UV, among others.

Here, we described methods for generating aptamers and aptamer-based methods for sensing natural toxins. As some small target molecules, including natural toxins, might not have an appropriate functionality for immobilization without altering the critical structural elements, we have discussed the selection strategies for generating aptamers without target immobilization in more detail.

5.2 Strategies for Aptamer Generation

5.2.1 Classic SELEX

The general selection process of the classic SELEX (Sassanfar and Szostak 1993; Klug and Famulok 1994) method involved four steps (Fig. 5.1). The process begins with the design of a RNA or ssDNA library, which consists of a 20–80 base-long randomized region, flanked by constant regions. This randomized region, which usually contains 10^{14}–10^{15} different molecules, provides diversity to the library for target binding. The constant regions include primer-binding site for reverse transcription, PCR amplification, and in vitro transcription. For selection, a RNA pool is produced by in vitro transcription with T7 RNA polymerase (as a T7 promoter is usually incorporated in preparing the library). The initial nucleic acid pool is incubated with target molecules under controlled binding conditions. One of the most crucial steps in the SELEX process is the separation of target-bound nucleic acid sequences from the unbound sequences. The separation process involves generating an immobilized

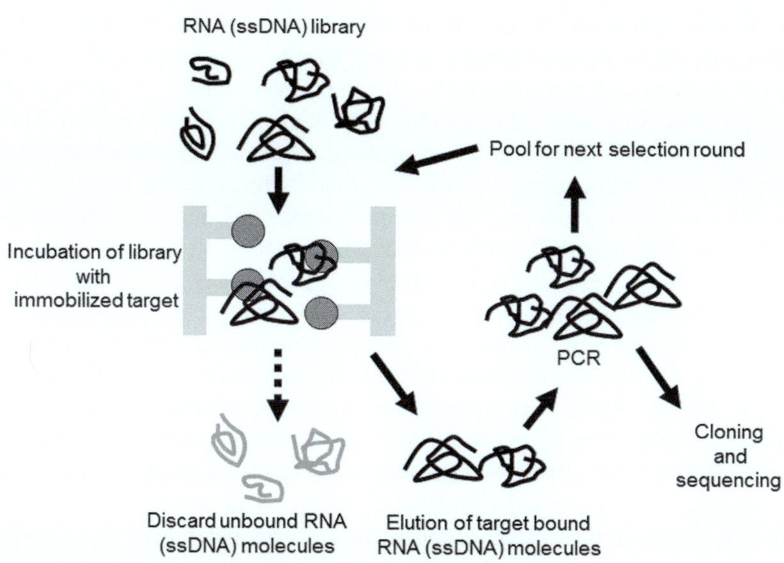

Fig. 5.1 Schematic representation of the classic SELEX technology

target on magnetic or Sepharose beads. For protein targets, immobilization options include nitrocellulose membrane and protein tags (Ogihara et al. 2015). At each selection round, a negative selection step is performed in the absence of the target to eliminate nonspecific adsorption of nucleic acids to immobilized substrate. To avoid concentrating unbound molecules, the reaction time for the negative selection step is often increased, and the negative selection step is repeated several times as the selection rounds progress. The resulting molecules are then subjected to a positive selection step to elute off the bound molecules from the immobilized target. The eluted molecules are subsequently amplified by PCR to produce the next-generation nucleic acid pool. This selection cycle continues up to as many as 15 rounds until the affinity of the nucleic acid population is maximized. After the last selection round, molecules eluted in the final pool are cloned and sequenced. Aptamers composed of natural nucleotides, especially RNA aptamers, are often not suitable for sensing applications as they are prone to enzymatic degradation in biological fluids. To circumvent this problem, a large number of chemically modified nucleotide analogs have been developed for synthesizing nuclease-resistant aptamers. In most cases, this was achieved by modifying the sugar residues, base, or the phosphate residue on the nucleotide. For this purpose, a variety of chemically modified nucleotides, such as 2′-amino modified, 2′-fluoro modified, 2′-O-methyl modified, 4′-thio modified, phosphorothioate modified, or uridine modified, have been synthesized to improve aptamer stability (Lipi et al. 2016).

5.2.2 Allosteric Selection

Allosteric selection is an alternative selection method, which is carried out using allosteric ribozymes or DNAzymes, created by fusing a random RNA or DNA domain to a known ribozyme or DNAzyme so that the binding of the target would trigger initiation of enzyme catalysis (Koizumi et al. 1999; Nutiu and Li 2005a, b; Chiuman and Li 2007). This strategy offers great potential for isolating aptamers for small molecules lacking an appropriate functionality for immobilization.

Breaker's group has pioneered the method for selecting allosteric ribozymes that respond to specific ligands from a pool of random RNA sequences (Soukup and Breaker 1999; Soukup et al. 2000; Zivarts et al. 2005; Gu et al. 2012; Furukawa et al. 2014). The allosteric selection procedure (Fig. 5.2) is almost the same as the classic SELEX procedure except for the pool design. An allosteric ribozyme consists of three domains, namely, hammerhead ribozyme (Forster and Symons 1987; Fedor and Uhlenbeck 1992), aptamer, and connecting domains. A suitable connecting domain is required for exhibiting ligand-dependent self-cleaving activity of each allosteric ribozyme (Soukup et al. 2000; Gu et al. 2012; Furukawa et al. 2014). Two types of RNA constructs have been used to isolate target-dependent allosteric ribozymes. The connecting domain in one of these constructs was previously used as a connecting domain for a flavin mononucleotide-sensing allosteric ribozyme (Soukup and Breaker 1999). The same connecting domain was also shown to function as a connecting

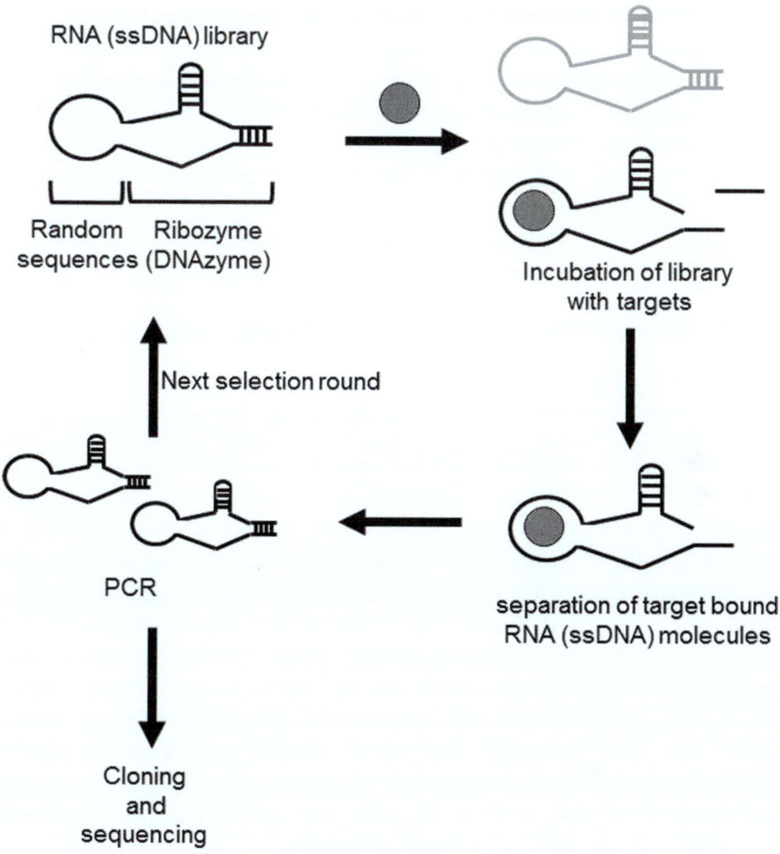

Fig. 5.2 Schematic representation of the allosteric selection

domain in a cyclic monophosphate compound sensing allosteric ribozyme (Koizumi et al. 1999), suggesting that this connecting domain might also function in a wide variety of target-dependent allosteric ribozymes. The other construct was composed of a hammerhead self-cleaving ribozyme and a random sequence region that included both aptamer and connecting domains. This construct was expected to optimize both aptamer and connecting domains simultaneously. Using the later construct, Zivarts et al. (2005) generated a divalent metal ion-sensing allosteric ribozyme.

5.2.3 CE-SELEX

Capillary electrophoresis selection (CE-SELEX) method was developed for the separation of aptamer-target complexes. Mendonsa and Bowser (2004) used capillary electrophoresis selection method for the first time to isolate human IgE binding aptamers based on mobility shifts due to differences in sizes and charges of target-bound and unbound molecules. They obtained aptamers with high-affinity and high selectivity in only four rounds of selection, which was less than that for the classic SELEX (as many as 15 rounds of selection). As the CE-SELEX method can also be used for isolating target-specific aptamers without target immobilization, this method is useful for selecting aptamers against small molecules. Accordingly, Yang and Bowser (2013) performed CE-SELEX using N-methyl mesoporphyrin as a small molecule target and obtained high-affinity aptamers (high nanomolar to low micromolar dissociation constants) in only three selection rounds.

5.2.4 GO-SELEX

GO-SELEX method is a modified SELEX method that requires no target immobilization and was developed based on π-π stacking interaction between graphene oxide (GO) and ssDNA (Park et al. 2012, 2014). Here, the ssDNA gets absorbed on the surface of the graphene oxide, and the complex protects the ssDNA from nuclease cleavage. In this method, a library composed of ssDNA and graphene oxide is incubated with the free target. In the presence of the target, only the aptamer that recognizes the target is released from the grapheme oxide and binds the target molecule, while the other ssDNAs remain absorbed on the graphene oxide. This immobilization-free selection method has been used successfully to identify aptamers specific for small molecules, such as okadaic acid (Gu et al. 2016) and T2-toxin (Chen et al. 2014). A modified GO-SELEX method for high-throughput aptamer screening was also developed. Using this method, Nguyen et al. (2014) obtained flexible multi-target aptamers, which can bind to three pesticides.

5.2.5 In Vitro Selection for Structure-Switching Signaling Aptamers

Nutiu and Li (2003, 2005a, b) developed a strategy for a target immobilization-free SELEX method. In this method, the ssDNA library consists of a constant region flanked by two random regions. The constant region is designed to hybridize with a biotinylated complementary sequence, which allows immobilizing the library to streptavidin-coated beads. Upon binding to the target molecule, ssDNA-target complex is released from the biotinylated complementary sequence. As the aptamers, selected by this method, are potentially useful as a signaling probe, this selection method was used to isolate signaling aptamers specific for pesticides (He et al. 2011; Wang et al. 2012).

5.2.6 *Other Types of Modified SELEX*

5.2.6.1 Modified SELEX Using Improved Libraries

Ever since the classic SELEX method was first reported in 1990, several methods, besides those mentioned above, modified to improve various aspects of the original SELEX method, have come into existence. Several technologies, such as photo-SELEX (Jensen et al. 1995; Yang et al. 2007), cDNA-SELEX (Dobbelstein and Shenk 1995), and Spiegelmer (Klussmann et al. 1996; Vater and Klussmann 2003), were developed to generate improved aptamer libraries. For example, photo-SELEX allows increasing the binding affinity between the aptamers and target molecules via UV-induced cross-linking. Such modifications of libraries can improve either the binding affinities or the stabilities of the aptamer-target complexes in vivo. For example, the Spiegelmer technology, which uses chiral nucleotide molecules to generate the library, was created to solve the stability problem in biological fluids. Using this technology, one can obtain a nuclease-resistant chiral oligonucleotide aptamer that can bind to the natural target molecule. The aptamers isolated from this pool of chiral nucleotides are called Spiegelmers. Advantages like this make these technologies suitable for in vivo applications.

5.2.6.2 Modified SELEX with Improved Aptamer Separation Process

In order to separate the aptamer-target complexes more efficiently and monitor the binding of the library at each selection round, Stoltenburg et al. (2005) developed FluMag-SELEX that used fluorescently labeled library and immobilized targets on the magnetic beads (Stoltenburg et al. 2005). Several other studies reported using microfluidic chips to efficiently separate aptamer-target complexes from the libraries (Oh et al. 2009; Cho et al. 2010; Xu et al. 2010). As the selection processes, such as reaction, separation, and detection, can be integrated within the chip's micro-channels, high-affinity aptamers against a target can be isolated efficiently using this selection method. Accordingly, high-affinity aptamers, specific for botulinum neurotoxin type A (Lou et al. 2009), C-reactive protein (Huang et al. 2010), and influenza A/H1N1 virus (Lai et al. 2014), were selected using this microfluidics SELEX method.

5.2.6.3 Modified SELEX for Cell Surface or Whole-Cell Targets

It is rather difficult to isolate aptamers specific for complex targets, such as intracellular proteins and transmembrane proteins, due to difficulties involved in expression and folding of such proteins with high purity. To solve this problem, Morris et al. (1998) were first to carry out a cell-based SELEX (cell-SELEX) procedure using human red blood cell membranes. In this method, expression and folding of cell surface proteins in their native states allow aptamer binding to the whole cell. Consequently, aptamers against a wide range of live cells, including tumor cells (Daniels et al. 2003; Cerchia et al. 2005; Shangguan et al. 2006; Tang et al. 2007; Fang and Tan 2010) and pathogenic bacteria (Marton et al. 2016), have been isolated. Recently, fluorescence-activated cell sorting (FACS) was used to separate aptamer-bound target cells (Raddatz et al. 2008; Mayer et al. 2010). In this selection method, named as FACS-SELEX, a fluorescently labeled nucleic acid library is incubated with the target cells, following which aptamer-bound cells are separated from the rest of the cells based on their fluorescence intensity.

5.3 Biosensor Application of Aptamers for Natural Toxins

5.3.1 Sensor Types

A number of aptamer-based biosensors, called as aptasensors, have been developed for the detection of small molecules, including natural toxins, using colorimetric, fluorescence, and electrochemical detection techniques. In many cases, the conformational change that occurs when an aptamer binds to a target molecule has been the basis for developing the sensing process. Target-induced conformational changes can be converted into measurable signals (such as fluorescence, colorimetric, or electrochemical) by using a variety of techniques including target-induced strand displacement (Feng et al. 2014), changes in the aggregation state of gold nanoparticles (AuNPs) (Liu and Lu 2006; Wu et al. 2010), self-cleavage by ribozymes or DNAzymes (Chiuman and Li 2007; Xiang et al. 2010), and activation of peroxidase activity by horseradish peroxidase (HRP)-mimicking DNAzyme (Travascio et al. 1998; Pelossof et al. 2010).

The strand displacement strategy is based on the assembly or disassembly of aptamer-complementary strand complex. In the absence of the target molecule, the aptamer binds to its complementary strand, forming a duplex. However, when the target molecule binds to the aptamer, the complementary strand is released, generating a detectable signal, such as fluorescence. This strategy is illustrated in Fig. 5.3 by using a fluorescence-labeled strand.

AuNPs have been widely used as a colorimetric probe in aptasensor development. The principle behind the AuNP-based aptasensors involves controlling the aggregation and dispersion states of AuNPs (Zhao et al. 2007, 2008). Interestingly, aggregation and dispersion of AuNPs leads to a visible change in solution color, blue color upon aggregation and red color upon dispersion, due to surface plasmon coupling. While ssDNAs can electrostatically adsorb onto the surface of dispersed AuNPs and stabilize them against aggregation at certain salt concentrations, dsDNAs are not able to interact with AuNPs, causing them to aggregate. Accordingly, several aptasensors have been developed based on this red to blue color transition property of AuNPs and the strand displacement strategy (Yang et al. 2011; Luan

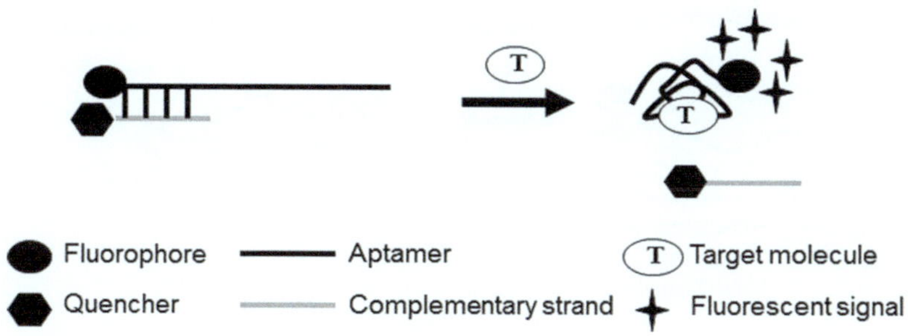

Fig. 5.3 A fluorescence-based assay for the detection of aptamers by target-induced strand separation

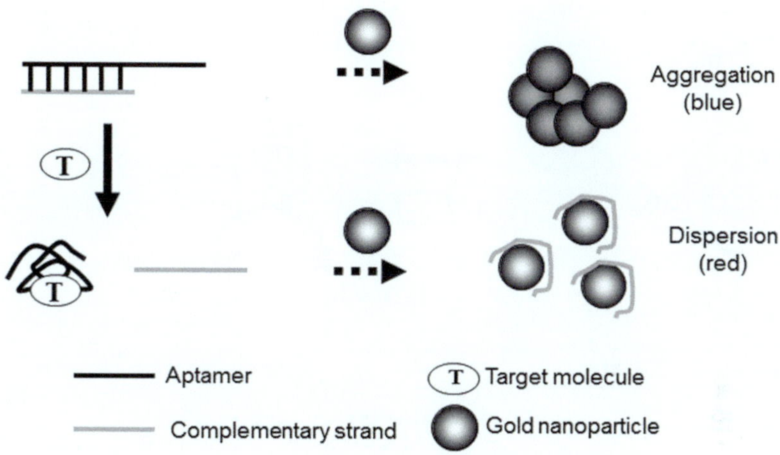

Fig. 5.4 Working principle of a colorimetric assay for an AuNP-based aptasensor

et al. 2015a; Soh et al. 2015). As illustrated in Fig. 5.4, in the presence of aptamer-complementary strand duplexes (dsDNA), AuNPs are in aggregated form (hence blue color); however, in the presence of target molecules, the aptamers change conformation in order to bind to the target molecules, as a result of which the complementary strands are released from the duplex. The free complementary strands then interact with AuNPs and stabilize them in dispersed form, leading to a change in solution color from blue to red. Thus, this target-induced conformational change in the aptamer plays a key role in altering the aggregation state of the AuNPs, and this change in the aggregation state is visible by the naked eye.

Fluorescence resonance energy transfer (FRET) technology, illustrated in Fig. 5.5, was often used to detect the self-cleaving activity of ribozymes or DNAzymes. Thus, FRET has been used to examine the cleavage kinetics of catalytic RNAs (Singh et al. 1999; Jenne et al. 2001). Several FRET-based methods were also developed for the detection of ligand-induced self-cleavage by a hammerhead ribozyme or a DNAzyme (Frauendorf and Jaschke 2001; Kim et al. 2007; de Silva and Walter 2009).

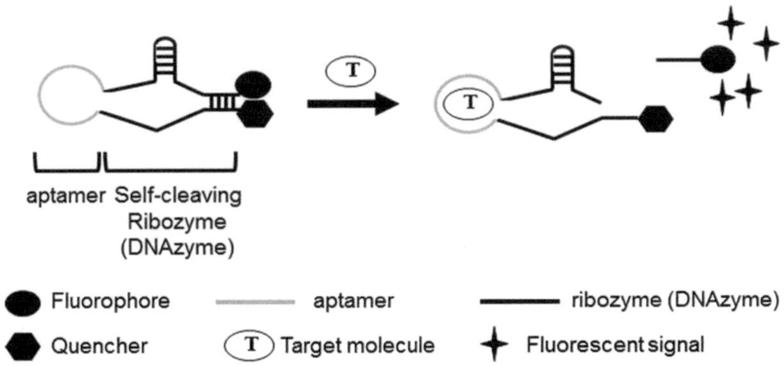

Fig. 5.5 Fluorescence resonance energy transfer (FRET)-based assay by using self-cleaving ribozyme

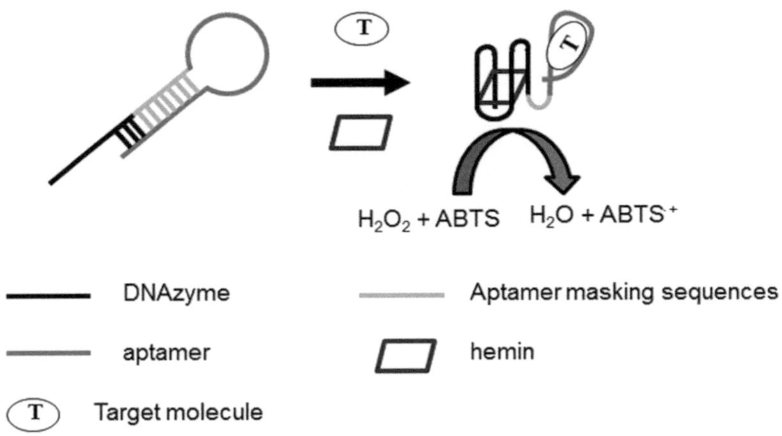

Fig. 5.6 Colorimetric detection scheme by using horseradish peroxidase (HRP)-mimicking DNAzyme

Horseradish peroxidase (HRP)-mimicking DNAzymes have a G-quadruplex structure that can be folded and stabilized by metal ions and hemin. HRP-mimicking DNAzyme-based aptasensors were constructed by connecting the aptamer with the DNAzyme. As illustrated in Fig. 5.6, in the absence of the target, a portion of the aptamer sequence masks the DNAzyme sequence, blocking the HRP activity of the DNAzyme. The formation of a target-aptamer complex leads to the release of the masking sequence, restoring the HRP activity of the HRP-mimicking DNAzyme. Although this type of aptasensors often shows limited sensitivity, in this case too, the signal can be detected by the naked eyes.

5.3.2 Aptasensors for Natural Toxins

Natural toxins are produced by a variety of fungi and plants. Mycotoxins, secondary metabolites derived from fungal species, are one of the major classes of natural toxins that not only contaminate grains and fruits but also contaminate products derived from them. As most of these mycotoxins are known to be carcinogenic (Richard 2007), availability of sensitive, reliable, and less laborious detection methods is needed by the food industry to ensure food safety. To meet their requirements, a large number of aptamer-based detection methods have been developed in the past two decades.

Among the mycotoxins, ochratoxin A (OTA) is one of the most dangerous food grain contaminants. Ever since Cruz-Aguado and Penner (2008) identified the first OTA aptamer, a large number of aptasensor have been proposed for OTA. There are some excellent reviews focusing on various aptamer-based detection methods (namely, colorimetric, fluorometric, electrochemical, and rolling circle amplification) for OTA (Rhouati et al. 2013; Ha 2015).

Aflatoxins, which are toxic metabolites, are known to contaminate agricultural products, such as peanuts, corn, and figs. Among the six types of aflatoxins (B1, B2, G1, G2, M1, and M2) that are known today, aflatoxin B1 is the most well known and most toxic subtype (Luan et al. 2015b). Aptamers for aflatoxin B1, B2, and M1 were isolated (Malhotra et al. 2014) and used for developing various types of aptamer-based detection methods, such as a HRP-mimicking DNAzyme-based colorimetric aptasensor (Seok et al. 2015), an AuNP-based fluorometric aptasensor (Wang et al. 2016), an electrochemical aptasensor (El-Moghazy et al. 2016), and a real-time quantitative polymerase chain reaction (RT-qPCR)-based detection method using the aptamer as a signal enhancer (Guo et al. 2014, 2016). There are also reports on aptamer-based detection of aflatoxins in real food samples, including milk (Sharma et al. 2016) and corn (Seok et al. 2015).

Aptasensors have also been developed for other natural toxins, such as fumonisin, patulin, and botulinum neurotoxins. For example, by using a fumonisin aptamer isolated by Mckeague et al. (2010), several fumonisin aptasensors, including one based on fluorescence resonance energy transfer (FRET), were developed (Wu et al. 2013). Patulin aptasensors, on the other hand, were isolated by using a DNA module platform for developing aptasensors, a process based on a microarray format combined with in silico secondary structure prediction (Tomita et al. 2016). An RNA aptasensor for the detection of type A botulinum neurotoxin was also developed (Chang et al. 2016; Janardhanan et al. 2013).

5.4 Conclusions and Future Prospects

Significant progress has been made in generating aptamers and developing aptasensors. Because of the availability of various SELEX methods, it is now possible to obtain specific aptamers against any target, including proteins, small molecules, and whole cells. Despite the existence of these tools, aptamers against all types of natural toxins are not available, thus warranting the need for the development of such aptamers. Although a number of aptasensor strategies are available for the detection of natural toxins, options for the on-site detection of natural toxins using aptasensors are still limited, mainly due to cost- and sensitivity-related problems. Recently, a novel detection method was developed using an aptamer and a glucose meter. Further studies on developing such effective aptamer-based signal amplification techniques would be needed in the future.

Acknowledgments We thank Dr. Yasuyuki Tomita for helpful discussions.

References

Andrade MA, Lancas FM (2017) Determination of Ochratoxin A in wine by packed in-tube solid phase microextraction followed by high performance liquid chromatography coupled to tandem mass spectrometry. J Chromatogr A 1493:41–48

Aquino-Jarquin G, Toscano-Garibay JD (2011) RNA aptamer evolution: two decades of SELEction. Int J Mol Sci 12:9155–9171

Breidbach A (2017) A greener, quick and comprehensive extraction approach for LC-MS of multiple mycotoxins. Toxins 9:E91

Bunka DH, Stockley PG (2006) Aptamers come of age – at last. Nat Rev Microbiol 4:588–596

Cerchia L, Duconge F, Pestourie C, Boulay J, Aissouni Y, Gombert K, Tavitian B, de Franciscis V, Libri D (2005) Neutralizing aptamers from whole-cell SELEX inhibit the RET receptor tyrosine kinase. PLoS Biol 3:e123

Chang TW, Janardhanan P, Mello CM, Singh BR, Cai S (2016) Selection of RNA aptamers against botulinum neurotoxin type A light chain through a non-radioactive approach. Appl Biochem Biotechnol 180:10–25

Chen X, Huang Y, Duan N, Wu S, Xia Y, Ma X, Zhu C, Jiang Y, Wang Z (2014) Screening and identification of DNA aptamers against T-2 toxin assisted by graphene oxide. J Agric Food Chem 62:10368–10374

Chiuman W, Li Y (2007) Simple fluorescent sensors engineered with catalytic DNA 'MgZ' based on a non-classic allosteric design. PLoS One 2:e1224

Cho M, Xiao Y, Nie J, Stewart R, Csordas AT, Oh SS, Thomson JA, Soh HT (2010) Quantitative selection of DNA aptamers through microfluidic selection and high-throughput sequencing. Proc Natl Acad Sci USA 107:15373–15378

Crosby NT (1984) Review of current and future analytical methods for the determination of mycotoxins. Food Addit Contam 1:39–44

Cruz-Aguado JA, Penner G (2008) Determination of ochratoxin a with a DNA aptamer. J Agric Food Chem 56:10456–10461

Daniels DA, Chen H, Hicke BJ, Swiderek KM, Gold L (2003) A tenascin-C aptamer identified by tumor cell SELEX: systematic evolution of ligands by exponential enrichment. Proc Natl Acad Sci USA 100:15416–15421

de Silva C, Walter NG (2009) Leakage and slow allostery limit performance of single drug-sensing aptazyme molecules based on the hammerhead ribozyme. RNA 15:76–84

Dobbelstein M, Shenk T (1995) In vitro selection of RNA ligands for the ribosomal L22 protein associated with Epstein-Barr virus-expressed RNA by using randomized and cDNA-derived RNA libraries. J Virol 69:8027–8034

Ellington AD, Szostak JW (1990) In vitro selection of RNA molecules that bind specific ligands. Nature 346:818–822

El-Moghazy AY, Soliman EA, Ibrahim HZ, Noguer T, Marty JL, Istamboulie G (2016) Ultra-sensitive biosensor based on genetically engineered acetylcholinesterase immobilized in poly (vinyl alcohol)/Fe-Ni alloy nanocomposite for phosmet detection in olive oil. Food Chem 203:73–78

Fang X, Tan W (2010) Aptamers generated from cell-SELEX for molecular medicine: a chemical biology approach. Acc Chem Res 43:48–57

Fedor MJ, Uhlenbeck OC (1992) Kinetics of intermolecular cleavage by hammerhead ribozymes. Biochemistry 31:12042–12054

Feng C, Dai S, Wang L (2014) Optical aptasensors for quantitative detection of small biomolecules: a review. Biosens Bioelectron 59:64–74

Forster AC, Symons RH (1987) Self-cleavage of virusoid RNA is performed by the proposed 55-nucleotide active site. Cell 50:9–16

Frauendorf C, Jaschke A (2001) Detection of small organic analytes by fluorescing molecular switches. Bioorg Med Chem 9:2521–2524

Furukawa K, Gu H, Breaker RR (2014) In vitro selection of allosteric ribozymes that sense the bacterial second messenger c-di-GMP. Methods Mol Biol 1111:209–220

Gu H, Furukawa K, Breaker RR (2012) Engineered allosteric ribozymes that sense the bacterial second messenger cyclic diguanosyl 5′-monophosphate. Anal Chem 84:4935–4941

Gu H, Duan N, Wu S, Hao L, Xia Y, Ma X, Wang Z (2016) Graphene oxide-assisted non-immobilized SELEX of okdaic acid aptamer and the analytical application of aptasensor. Sci Rep 6:21665

Guo X, Wen F, Zheng N, Luo Q, Wang H, Wang H, Li S, Wang J (2014) Development of an ultra-sensitive aptasensor for the detection of aflatoxin B1. Biosens Bioelectron 56:340–344

Guo X, Wen F, Zheng N, Li S, Fauconnier ML, Wang J (2016) A qPCR aptasensor for sensitive detection of aflatoxin M1. Anal Bioanal Chem 408:5577–5584

Ha TH (2015) Recent advances for the detection of Ochratoxin A. Toxins 7:5276–5300

He J, Liu Y, Fan M, Liu X (2011) Isolation and identification of the DNA aptamer target to acetamiprid. J Agric Food Chem 59:1582–1586

Hu W, Li X, He G, Zhang Z, Zheng X, Li P, Li CM (2013) Sensitive competitive immunoassay of multiple mycotoxins with non-fouling antigen microarray. Biosens Bioelectron 50:338–344

Huang CJ, Lin HI, Shiesh SC, Lee GB (2010) Integrated microfluidic system for rapid screening of CRP aptamers utilizing systematic evolution of ligands by exponential enrichment (SELEX). Biosens Bioelectron 25:1761–1766

Janardhanan P, Mello CM, Singh BR, Lou J, Marks JD, Cai S (2013) RNA aptasensor for rapid detection of natively folded type A botulinum neurotoxin. Talanta 117:273–280

Jenison RD, Gill SC, Pardi A, Polisky B (1994) High-resolution molecular discrimination by RNA. Science 263:1425–1429

Jenne A, Hartig JS, Piganeau N, Tauer A, Samarsky DA, Green MR, Davies J, Famulok M (2001) Rapid identification and characterization of hammerhead-ribozyme inhibitors using fluorescence-based technology. Nat Biotechnol 19:56–61

Jensen KB, Atkinson BL, Willis MC, Koch TH, Gold L (1995) Using in vitro selection to direct the covalent attachment of human immunodeficiency virus type 1 Rev protein to high-affinity RNA ligands. Proc Natl Acad Sci USA 92:12220–12224

Kim HK, Liu J, Li J, Nagraj N, Li M, Pavot CM, Lu Y (2007) Metal-dependent global folding and activity of the 8-17 DNAzyme studied by fluorescence resonance energy transfer. J Am Chem Soc 129:6896–6902

Klug SJ, Famulok M (1994) All you wanted to know about SELEX. Mol Biol Rep 20:97–107

Klussmann S, Nolte A, Bald R, Erdmann VA, Furste JP (1996) Mirror-image RNA that binds D-adenosine. Nat Biotechnol 14:1112–1115

Koizumi M, Soukup GA, Kerr JN, Breaker RR (1999) Allosteric selection of ribozymes that respond to the second messengers cGMP and cAMP. Nat Struct Biol 6:1062–1071

Kulbachinskiy AV (2007) Methods for selection of aptamers to protein targets. Biochem Biokhim 72:1505–1518

Lai HC, Wang CH, Liou TM, Lee GB (2014) Influenza A virus-specific aptamers screened by using an integrated microfluidic system. Lab Chip 14:2002–2013

Lipi F, Chen S, Chakravarthy M, Rakesh S, Veedu RN (2016) In vitro evolution of chemically-modified nucleic acid aptamers: pros and cons, and comprehensive selection strategies. RNA Biol 13:1232–1245

Liu J, Lu Y (2006) Preparation of aptamer-linked gold nanoparticle purple aggregates for colorimetric sensing of analytes. Nat Protoc 1:246–252

Lou X, Qian J, Xiao Y, Viel L, Gerdon AE, Lagally ET, Atzberger P, Tarasow TM, Heeger AJ, Soh HT (2009) Micromagnetic selection of aptamers in microfluidic channels. Proc Natl Acad Sci USA 106:2989–2994

Luan Y, Chen J, Li C, Xie G, Fu H, Ma Z, Lu A (2015a) Highly sensitive colorimetric detection of Ochratoxin A by a label-free aptamer and gold nanoparticles. Toxins 7:5377–5385

Luan Y, Chen Z, Xie G, Chen J, Lu A, Li C, Fu H, Ma Z, Wang J (2015b) Rapid visual detection of Aflatoxin B1 by label-free aptasensor using unmodified gold nanoparticles. J Nanosci Nanotechnol 15:1357–1361

Malhotra S, Pandey AK, Rajput YS, Sharma R (2014) Selection of aptamers for aflatoxin M1 and their characterization. J Mol Recognit 27:493–500

Marton S, Cleto F, Krieger MA, Cardoso J (2016) Isolation of an aptamer that binds specifically to *E. coli*. PLoS One 11:e0153637

Mayer G, Ahmed MS, Dolf A, Endl E, Knolle PA, Famulok M (2010) Fluorescence-activated cell sorting for aptamer SELEX with cell mixtures. Nat Protoc 5:1993–2004

McKeague M, Bradley CR, De Girolamo A, Visconti A, Miller JD, Derosa MC (2010) Screening and initial binding assessment of fumonisin b(1) aptamers. Int J Mol Sci 11:4864–4881

Mendonsa SD, Bowser MT (2004) In vitro selection of high-affinity DNA ligands for human IgE using capillary electrophoresis. Anal Chem 76:5387–5392

Morris KN, Jensen KB, Julin CM, Weil M, Gold L (1998) High affinity ligands from in vitro selection: complex targets. Proc Natl Acad Sci USA 95:2902–2907

Nguyen VT, Kwon YS, Kim JH, Gu MB (2014) Multiple GO-SELEX for efficient screening of flexible aptamers. Chem Commun 50:10513–10516

Nutiu R, Li Y (2003) Structure-switching signaling aptamers. J Am Chem Soc 125:4771–4778

Nutiu R, Li Y (2005a) Aptamers with fluorescence-signaling properties. Methods 37:16–25

Nutiu R, Li Y (2005b) In vitro selection of structure-switching signaling aptamers. Angew Chem Int Ed 44:1061–1065

Ogihara K, Savory N, Abe K, Yoshida W, Asahi M, Kamohara S, Ikebukuro K (2015) DNA aptamers against the Cry j 2 allergen of Japanese cedar pollen for biosensing applications. Biosens Bioelectron 63:159–165

Oh SS, Qian J, Lou X, Zhang Y, Xiao Y, Soh HT (2009) Generation of highly specific aptamers via micromagnetic selection. Anal Chem 81:5490–5495

Park JW, Tatavarty R, Kim DW, Jung HT, Gu MB (2012) Immobilization-free screening of aptamers assisted by graphene oxide. Chem Commun 48:2071–2073

Park JW, Jin Lee S, Choi EJ, Kim J, Song JY, Bock Gu M (2014) An ultra-sensitive detection of a whole virus using dual aptamers developed by immobilization-free screening. Biosens Bioelectron 51:324–329

Pelossof G, Tel-Vered R, Elbaz J, Willner I (2010) Amplified biosensing using the horseradish peroxidase-mimicking DNAzyme as an electrocatalyst. Anal Chem 82:4396–4402

Raddatz MS, Dolf A, Endl E, Knolle P, Famulok M, Mayer G (2008) Enrichment of cell-targeting and population-specific aptamers by fluorescence-activated cell sorting. Angew Chem Int Ed 47:5190–5193

Rhouati A, Yang C, Hayat A, Marty JL (2013) Aptamers: a promising tool for ochratoxin A detection in food analysis. Toxins 5:1988–2008

Richard JL (2007) Some major mycotoxins and their mycotoxicoses – an overview. Int J Food Microbiol 119:3–10

Robertson DL, Joyce GF (1990) Selection in vitro of an RNA enzyme that specifically cleaves single-stranded DNA. Nature 344:467–468

Roll R, Matthiaschk G, Korte A (1990) Embryotoxicity and mutagenicity of mycotoxins. J Environ Pathol Toxicol Oncol 10:1–7

Sassanfar M, Szostak JW (1993) An RNA motif that binds ATP. Nature 364:550–553

Seok Y, Byun JY, Shim WB, Kim MG (2015) A structure-switchable aptasensor for aflatoxin B1 detection based on assembly of an aptamer/split DNAzyme. Anal Chim Acta 886:182–187

Shangguan D, Li Y, Tang Z, Cao ZC, Chen HW, Mallikaratchy P, Sefah K, Yang CJ, Tan W (2006) Aptamers evolved from live cells as effective molecular probes for cancer study. Proc Natl Acad Sci USA 103:11838–11843

Sharma A, Catanante G, Hayat A, Istamboulie G, Ben Rejeb I, Bhand S, Marty JL (2016) Development of structure switching aptamer assay for detection of aflatoxin M1 in milk sample. Talanta 158:35–41

Singh KK, Parwaresch R, Krupp G (1999) Rapid kinetic characterization of hammerhead ribozymes by real-time monitoring of fluorescence resonance energy transfer (FRET). RNA 5:1348–1356

Soh JH, Lin Y, Rana S, Ying JY, Stevens MM (2015) Colorimetric detection of small molecules in complex matrixes via target-mediated growth of aptamer-functionalized gold nanoparticles. Anal Chem 87:7644–7652

Soukup GA, Breaker RR (1999) Engineering precision RNA molecular switches. Proc Natl Acad Sci USA 96:3584–3589

Soukup GA, Emilsson GA, Breaker RR (2000) Altering molecular recognition of RNA aptamers by allosteric selection. J Mol Biol 298:623–632

Stoltenburg R, Reinemann C, Strehlitz B (2005) FluMag-SELEX as an advantageous method for DNA aptamer selection. Anal Bioanal Chem 383:83–91

Stoltenburg R, Reinemann C, Strehlitz B (2007) SELEX – a (r)evolutionary method to generate high-affinity nucleic acid ligands. Biomol Eng 24:381–403

Tang Z, Shangguan D, Wang K, Shi H, Sefah K, Mallikratchy P, Chen HW, Li Y, Tan W (2007) Selection of aptamers for molecular recognition and characterization of cancer cells. Anal Chem 79:4900–4907

Tomita Y, Morita Y, Suga H, Fujiwara D (2016) DNA module platform for developing colorimetric aptamer sensors. BioTechniques 60:285–292

Travascio P, Li Y, Sen D (1998) DNA-enhanced peroxidase activity of a DNA-aptamer-hemin complex. Chem Biol 5:505–517

Tuerk C, Gold L (1990) Systematic evolution of ligands by exponential enrichment: RNA ligands to bacteriophage T4 DNA polymerase. Science 249:505–510

Turner NW, Subrahmanyam S, Piletsky SA (2009) Analytical methods for determination of mycotoxins: a review. Anal Chim Acta 632:168–180

Vater A, Klussmann S (2003) Toward third-generation aptamers: Spiegelmers and their therapeutic prospects. Curr Opin Drug Discov Dev 6:253–261

Wang L, Liu X, Zhang Q, Zhang C, Liu Y, Tu K, Tu J (2012) Selection of DNA aptamers that bind to four organophosphorus pesticides. Biotechnol Lett 34:869–874

Wang B, Chen Y, Wu Y, Weng B, Liu Y, Lu Z, Li CM, Yu C (2016) Aptamer induced assembly of fluorescent nitrogen-doped carbon dots on gold nanoparticles for sensitive detection of AFB1. Biosens Bioelectron 78:23–30

Watanabe M, Shimizu H (2005) Detection of patulin in apple juices marketed in the Tohoku district, Japan. J Food Protect 68:610–612

Wu ZS, Lu H, Liu X, Hu R, Zhou H, Shen G, Yu RQ (2010) Inhibitory effect of target binding on hairpin aptamer sticky-end pairing-induced gold nanoparticle assembly for light-up colorimetric protein assay. Anal Chem 82:3890–3898

Wu S, Duan N, Li X, Tan G, Ma X, Xia Y, Wang Z, Wang H (2013) Homogenous detection of fumonisin B(1) with a molecular beacon based on fluorescence resonance energy transfer between NaYF4: Yb, Ho upconversion nanoparticles and gold nanoparticles. Talanta 116:611–618

Xiang Y, Wang Z, Xing H, Wong NY, Lu Y (2010) Label-free fluorescent functional DNA sensors using unmodified DNA: a vacant site approach. Anal Chem 82:4122–4129

Xu Y, Yang X, Wang E (2010) Review: aptamers in microfluidic chips. Anal Chim Acta 683:12–20

Xu L, Zhang Z, Zhang Q, Li P (2016) Mycotoxin determination in foods using advanced sensors based on antibodies or aptamers. Toxins 8:E239

Yang J, Bowser MT (2013) Capillary electrophoresis-SELEX selection of catalytic DNA aptamers for a small-molecule porphyrin target. Anal Chem 85:1525–1530

Yang Y, Yang D, Schluesener HJ, Zhang Z (2007) Advances in SELEX and application of aptamers in the central nervous system. Biomol Eng 24:583–592

Yang C, Wang Y, Marty JL, Yang X (2011) Aptamer-based colorimetric biosensing of Ochratoxin A using unmodified gold nanoparticles indicator. Biosens Bioelectron 26:2724–2727

Zhao W, Chiuman W, Brook MA, Li Y (2007) Simple and rapid colorimetric biosensors based on DNA aptamer and noncrosslinking gold nanoparticle aggregation. Chembiochem 8:727–731

Zhao W, Brook MA, Li Y (2008) Design of gold nanoparticle-based colorimetric biosensing assays. Chembiochem 9:2363–2371

Zivarts M, Liu Y, Breaker RR (2005) Engineered allosteric ribozymes that respond to specific divalent metal ions. Nucleic Acids Res 33:622–631

Zollner P, Mayer-Helm B (2006) Trace mycotoxin analysis in complex biological and food matrices by liquid chromatography-atmospheric pressure ionisation mass spectrometry. J Chromatogr A 1136:123–169

Chapter 6
Rational Design of Artificial Riboswitches

Atsushi Ogawa

Abstract A riboswitch is a cis-regulatory RNA element that controls gene expression in response to a specific ligand. This functional RNA is composed of two domains, an aptamer and an expression platform. A ligand binds to the former to induce the latter's conformational change or alternative folding, which turns on or off the expression of the downstream (or upstream in some cases) gene. Although natural riboswitches are limited in terms of the variation of their ligands, an in vitro-selected aptamer enables us to construct an artificial riboswitch responsive to a user-defined ligand molecule. However, it is difficult to functionally couple such an in vitro-selected aptamer with an expression platform for their efficient communication, which generally requires ligand-dependent hybridization switches of RNA duplexes over a wide range of mRNA. Nonetheless, we have thus far developed several rational methods for designing artificial riboswitches that function in bacterial or eukaryotic translation systems. The methods are described herein in historical order.

Keywords Riboswitch · Aptamer · Ribozyme · Aptazyme · IRES · Shunting

6.1 Introduction

Systems for regulating gene expression in response to environmental conditions are expected to be widely applicable in various fields, including synthetic biology, metabolic engineering, systems biology, biological analyses, medical science, and so on. Molecule-responsive systems would have particularly broad utility, because additional external stimulus responsiveness (e.g., to light) could be added by modifying the molecule with an appropriate functional moiety (Hayashi and Nakatani 2014; Walsh et al. 2014). To design such a gene regulation system with responsiveness to a user-defined molecule, it would be best to consult natural gene regulation systems. One of the most promising candidates to reference would be a *riboswitch*

A. Ogawa (✉)
Proteo-Science Center, Ehime University, Matsuyama, Ehime, Japan
e-mail: ogawa.atsushi.mf@ehime-u.ac.jp

S. Masuda, S. Izawa (eds.), *Applied RNA Bioscience*,
https://doi.org/10.1007/978-981-10-8372-3_6

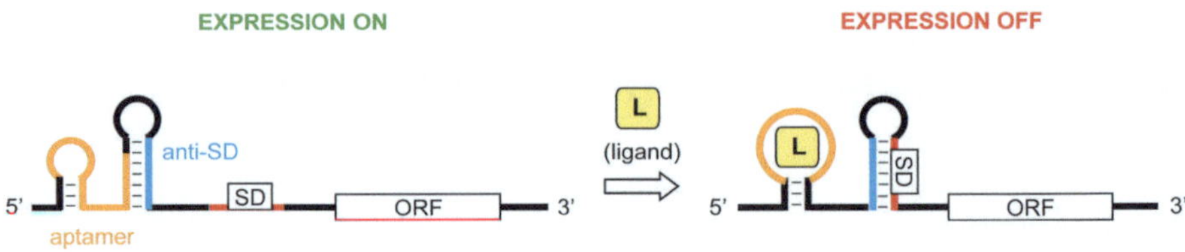

Fig. 6.1 A typical natural OFF-riboswitch modulating translation initiation in bacteria. SD and ORF represent a Shine-Dalgarno sequence and an open reading frame, respectively

(Fig. 6.1) (Roth and Breaker 2009). A riboswitch is a ligand-responsive, cis-regulatory RNA element that is composed of two domains, a sensory domain (called an *aptamer*) and a regulatory domain (called an *expression platform*). A specific ligand molecule bound to the former domain guides the latter's structural change or alternative folding, which switches the gene expression on or off. Natural riboswitches are widespread in the 5′ untranslated region (UTR) of bacterial transcripts (and rare in eukaryotic ones) and are principally used for sensing endogenous metabolites to regulate the relevant gene expression on the same mRNA (Breaker 2011). In contrast, an artificial riboswitch responsive to an arbitrary ligand molecule can be generated using the corresponding in vitro-selected aptamer, which can be obtained relatively easily using well-established methods (Murata and Sato 2014; Ohuchi 2014). In fact, several types of artificial riboswitches have already been obtained by introducing an in vitro-selected aptamer into mRNA, mainly through selection or screening methods (Berens and Suess 2015; Chang et al. 2012; Topp and Gallivan 2010). However, it is very hard to rationally construct artificial riboswitches, even by harnessing an existing riboswitch sequence as a foundation. This is because the two constituent domains generally communicate with each other by several major hybridization switches of duplexes, and thus merely replacing the aptamer domain is not sufficient to alter the ligand specificity. Nonetheless, our research group has developed several versatile methods for rationally and easily designing artificial riboswitches that function in either bacterial or eukaryotic translation systems. In this chapter, I introduce these methods in historical order.

6.2 Research Initiation

We began conducting research on artificial riboswitches at RIKEN in 2006. At that time, several types of artificial riboswitches had already been reported. The main type was eukaryotic, downregulating riboswitches (OFF-riboswitches) that were constructed simply by inserting one or more aptamers into a 5′ UTR (Grate and Wilson 2001; Harvey et al. 2002; Suess et al. 2003; Werstuck and Green 1998). Because the eukaryotic ribosome is loaded onto the 5′ terminus in the canonical translation mechanism, a ligand-aptamer complex in a 5′ UTR prevents the ribosome from being loaded onto or scanning the mRNA, meaning that hybridization switches (between sequences in the aptamer domain and other regions) are not principally required for regulation. This is why these eukaryotic OFF-riboswitches can be artificially constructed with ease. However, this regulation mechanism was not readily applicable in the development of more technologically useful, upregulating riboswitches (ON-riboswitches), of course including bacterial ones.[1] Although a few artificial ON-riboswitches that function in bacteria had also been obtained (Desai and Gallivan 2004; Suess et al. 2004), there was no versatile rational design method that was shown to be applicable to more than two different ligand-aptamer pairs. Thus, we sought to establish a versatile method for rationally constructing bacterial ON-riboswitches.

6.3 Bacterial ON-Riboswitches Modulating mRNA Cleavage

When designing artificial riboswitches, the first step is to decide which expression mechanism should be harnessed for the gene regulation. Although there are several candidates (transcription termination, translation initiation, splicing, and mRNA degradation), we focused on translation initiation, because it can be easily repressed in bacterial expression systems by sequestering a relatively short expression platform (i.e., a Shine-Dalgarno (SD) sequence) with the complementary sequence (anti-SD) inserted just upstream thereof. Since any aptamer sequence can also be inserted with the anti-SD into the 5′ UTR, it was seemingly straightforward to artificially design a bacterial ON-riboswitch. However, the problem was how to dehybridize the anti-SD/SD duplex in response to aptamer-ligand complex formation. We tried to refer to hybridization switches in known riboswitches, but we quickly realized they are not easily applicable due to their complexity. This meant that we needed to rationally design (from scratch) ligand-dependent hybridization switches specific for each aptamer to be implanted.

However, since we had only just launched our research into artificial riboswitches, this hurdle seemed difficult to overcome. We therefore considered several different approaches to somehow lower this hurdle – specifically, to circumvent the design of hybridization switches. Our analysis pointed toward the use of an *aptazyme*, a conjugate of two functional RNAs: an aptamer and a ribozyme (i.e., RNA enzyme) (Soukup and Breaker 1999). In this relatively short RNA, only when a specific ligand binds to the former part, the latter is activated (or deactivated) by changing its conformation via hybridization switches. Briefly, an aptazyme intrinsically has a ligand-responsive hybridization switch ability. More importantly, it can also be obtained from a random RNA library with an in vitro selection method (Furukawa et al. 2014), meaning that there is no need to design the hybridization switches.

[1] Recently, a versatile cis-acting inverter module has been developed for eukaryotic ON-riboswitches (Endo et al. 2013).

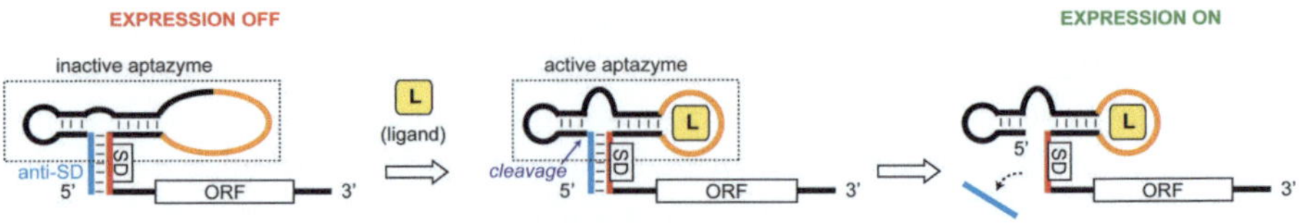

Fig. 6.2 An artificial ON-riboswitch modulating mRNA cleavage in bacterial expression systems

The aptazymes that we chose to construct our artificial ON-riboswitches were self-cleaving ribozyme-based ones (Soukup and Breaker 1999). We designed our ON-riboswitches to dissociate the anti-SD from the SD sequence through the ligand-responsive self-cleavage of mRNA (Fig. 6.2) (Ogawa and Maeda 2007). In this design, when a ligand binds to the corresponding aptazyme inserted between the anti-SD/SD duplex, the activated aptazyme cleaves the upstream anti-SD out. Because an interduplex is less stable than an intraduplex, the anti-SD should be spontaneously released if its length (and sequence) is appropriately adjusted. In keeping with this design strategy, we first constructed a bacterial ON-riboswitch that is responsive to a small molecule, theophylline, by using the theophylline-dependent self-cleaving aptazyme that had been selected in vitro (Soukup et al. 2000). Although optimization of the anti-SD was required to maximize the switching efficiency (the ON/OFF induction ratio in the optimized riboswitch was 7.2 at 1 mM theophylline at 37 °C in a bacterial cell-free translation system), the optimized anti-SD (or more precisely, a slightly different one) was found to be effective for another ligand (cGMP)-responsive aptazyme (Soukup et al. 2001). More specifically, we could readily construct a cGMP-dependent riboswitch just by replacing the aptazyme segment in the theophylline-dependent one. This is how we succeeded in the development of a versatile method (that was proved to be available for at least two different ligands) for rationally constructing bacterial ON-riboswitches, through the use of a self-cleaving aptazyme to circumvent the design of hybridization switches.[2]

Although the foregoing theophylline-responsive riboswitch was optimized in a reconstituted cell-free translation system (Shimizu et al. 2001), we next confirmed its activity in living cells (*E. coli*) (Ogawa and Maeda 2008a). As a result, while it was much less active at 37 °C, it showed a moderate switching efficiency at 23 °C (ON/OFF = 3.2 at 5 mM theophylline in culture solution). In addition, the switching efficiency was increased about fourfold by inserting the same riboswitch upstream of the RNA polymerase gene to construct a cascading system (ON/OFF = 12.9 at 5 mM theophylline in culture solution).[3]

[2] We also succeeded in the rational design of a trans-acting gene regulator using an aptazyme/suppressor-tRNA conjugate (Ogawa and Maeda 2008b). Although this regulator is suited for a cell-free translation system, another group improved it so as to function in living cells (*E. coli*) (Berschneider et al. 2009).

[3] This cascade strategy would be available for enhancing the overall switching efficiency in any riboswitch, regardless of the type of organisms.

It was at that time unclear why the riboswitch activity at 37 °C was much lower in cells than in vitro. However, there were several possible explanations: the cleaved mRNA may have been more susceptible to degradation; the relative speed of the ribosome binding to the mRNA folding may have been much higher; and/or the anti-SD/SD duplex may have been less stable. We therefore decreased the cultivation temperature to reduce these effects, which led to an increase in the switching efficiency.[4] Nonetheless, the efficiency was still moderate. The soundest remaining reason for the lower activity was that the concentration of magnesium ions in cells was much lower than that in the reconstituted cell-free translation system (Shimizu et al. 2001). The self-cleaving hammerhead ribozyme in the aptazyme that we used lacks tertiary interactions between loops I and II and thus requires a relatively high magnesium ion concentration for folding into the active structure (Khvorova et al. 2003; Martick and Scott 2006). Therefore, we exogenously added magnesium ions to the culture solution to adjust the concentration to that in the cell-free system. However, the riboswitch activity was not improved, probably because it was hard to sufficiently increase the magnesium ion concentration inside living cells.[5]

After we reported the aptazyme-based riboswitches described above, another group constructed a similar, theophylline-dependent, bacterial ON-riboswitch by utilizing a highly active self-cleaving ribozyme with loop I–II interactions, which functions well even at a low magnesium ion concentration (Wieland and Hartig 2008). In fact, the ON/OFF ratio of this aptazyme-based riboswitch in *E. coli* cells was approximately three times higher than that of ours in stand-alone (i.e., non-cascading) mode (about 10 at 4 mM theophylline in culture solution). However, this riboswitch was obtained through in vivo screening, perhaps because it was difficult to rationally design the hybridization switches in the implanted aptazyme, which was not in vitro-selected. Incidentally, although an aptazyme with a highly active self-cleaving ribozyme can also be selected in vitro, it has been reported that an in vitro-selected one did not function in vivo for some reason (Link et al. 2007).

6.4 Eukaryotic ON-Riboswitches Modulating mRNA Cleavage

At the end of 2008, I started a new lab at Ehime University, where cell-free science based on wheat germ extract (WGE) had become an active area of research (Endo and Sawasaki 2006). I thus sought to aim to establish a versatile method for rationally designing *eukaryotic* ON-riboswitches by harnessing WGE at the new lab. I first tried to construct aptazyme-based riboswitches again, though this time by

[4] The low cultivation temperature probably contributed also to the active ribozyme folding.

[5] This is my current view, though at that time we thought that the magnesium ion concentration was sufficiently increased but the efficiency remained low for other reasons (Ogawa and Maeda 2008a).

exploiting the characteristics of cell extracts, whose conditions are much closer to cell environments than those of a reconstituted system are.[6]

However, at that time, eukaryotic ON-riboswitches that were artificially constructed by the use of aptazymes derived from a highly active self-cleaving ribozyme (with loop-loop interactions different from those described above) had already been reported (Win and Smolke 2007). These aptazyme-based ON-riboswitches had been designed to modulate mRNA degradation, which is believed to occur rapidly in eukaryotic cells when mRNA is cleaved at a certain site. They are composed of OFF-type self-cleaving aptazymes, whose cleavage is inhibited only in the presence of their specific ligands, to turn translation on through the signal inversion (OFF to ON).[7] More specifically, mRNA with the aptazyme in the 3′ UTR is cleaved without the ligand to suppress the expression, though it is unclear how fast the cleaved mRNA undergoes degradation (the shortened 3′ UTR may make the main contribution to the translation suppression), while the translation occurs in the presence of the ligand by inhibiting the self-cleavage (i.e., by suppressing the translation suppression). It should be noted that, although the implanted aptazyme was not in vitro-selected, the hybridization switches thereof were very well designed.[8]

Nonetheless, I had doubts as to the versatility of aptazymes derived from high-activity ribozymes, which are highly sensitive to a narrow range of magnesium ion concentrations and are thus likely to be highly dependent on slight differences in conditions. I thus tried to construct another type of aptazyme-based eukaryotic ON-riboswitches by utilizing in vitro-selected aptazymes derived from the low-activity ribozyme that we had previously used for creating bacterial riboswitches. Under the assumption that cleaved mRNA would also be available for translation, I chose ON-type self-cleaving aptazymes to design signal inversion-free ON-riboswitches in which a translational suppressor sequence (specifically, an implanted aptazyme with a 5′ terminal rigid stem-loop structure) was removed by self-cleavage, as in the case of our previous bacterial riboswitches. Unfortunately, however, the constructed riboswitches showed no activity at a low magnesium concentration in WGE even at low temperature, unlike in *E. coli* cells. Eventually, to maximize the aptazyme activity, I needed to carry out an aptazyme reaction at a higher magnesium ion concentration and then mask extra magnesium ions by EDTA before a translation reaction in WGE (Ogawa 2009). As a result of separating the two reactions, a very high ON/OFF induction ratio was observed (approximately 50 at 500 μM ligand), indicating that the only remaining problem was in the difference

[6] In particular, a high-quality WGE provides more cell-like conditions because it has almost no contamination of the suicide system directed against ribosomes, which usually does not exist in cells but invades when a cell wall is damaged (Madin et al. 2000).

[7] OFF-riboswitches had also been constructed by using ON-type self-cleaving aptazymes and the opposite signal inversion (ON–OFF) (Win and Smolke 2007).

[8] The hybridization switches had been designed for responding to theophylline by using its aptamer. Although the paper reported that these switches could also be used for tetracycline, tetracycline is known to inhibit the hammerhead ribozyme itself (Murray and Arnold 1996).

in the optimal magnesium ion concentration between the self-cleavage reaction and the following ribosomal translation.

6.5 Eukaryotic ON-Riboswitches Modulating Internal Ribosome Entry

Having established that self-cleaving aptazymes (ribozymes) are highly dependent on magnesium ions, I next attempted to construct eukaryotic ON-riboswitches *without any ribozyme*. At the time of this attempt, this meant that the ligand-dependent hybridization switches had to be rationally designed for each aptamer to be implanted.[9] In addition, there was another issue to be solved to create eukaryotic ON-riboswitches modulating translation initiation. In the canonical eukaryotic translation system, the 40S (or 43S) ribosome enters from the 5′ terminus, so that an aptamer-ligand complex in a 5′ UTR inevitably prevents the ribosome from being loaded onto and/or scanning mRNA, as described above (Werstuck and Green 1998). This suggests that it is physically impossible to construct eukaryotic ON-riboswitches that modulate the canonical translation without mRNA cleavage (including splicing). In fact, to my knowledge, mRNA cleavage-free eukaryotic ON-riboswitches had still not been obtained at the time of these investigations, even by the use of selection or screening methods.

I attempted to address the second issue first. For this purpose, I focused on an internal ribosome entry site (IRES). An IRES is a unique sequence that literally enables ribosome loading in the middle of an mRNA, as a bacterial SD does, meaning that an aptamer-ligand complex upstream of an IRES should not interfere with the ribosome progression. I thus thought that it would be possible to create mRNA cleavage-free eukaryotic ON-riboswitches that modulate IRES-mediated translation. Nonetheless, it was at first doubtful whether deactivation of an IRES could be realized as easily as that of the bacterial SD merely by inserting a short antisense sequence before the IRES, because the IRES generally has a high-order tertiary structure composed of hundreds of bases, while the bacterial SD has only several bases. However, I found with surprising ease that an 8-mer antisense sequence (named aIRES) could be used for binding to and sufficiently blocking the *Plautia stali* intestine virus (PSIV)-IRES, which functions well in various eukaryotic translation systems (including WGE) (Ogawa 2011). In addition, an anti-aIRES sequence (named aaIRES) that was inserted upstream of the aIRES inhibited aIRES/IRES duplex formation by hybridizing to the aIRES, which restored IRES-mediated translation considerably. These results proved that IRES-mediated translation can be regulated by the hybridization switches of these duplexes.

Having resolved the second issue, I next addressed the first one. Although I devised several patterns of ligand-dependent hybridization switches that could be

[9] Afterward, I succeeded in constructing hybridization switch-free eukaryotic ON-riboswitches (Ogawa 2013).

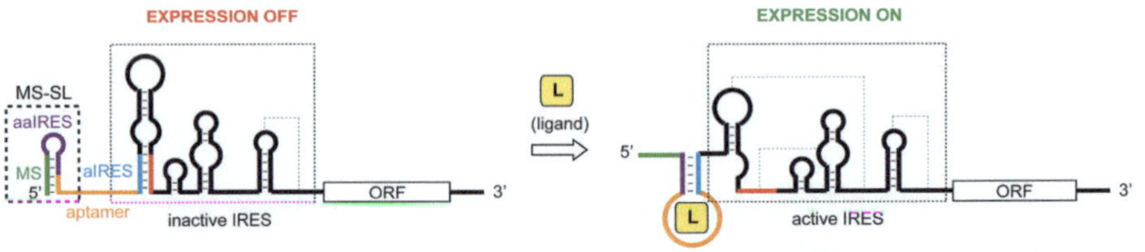

Fig. 6.3 An artificial ON-riboswitch modulating internal ribosome entry in eukaryotic expression systems. The thinnest dotted lines represent pseudoknots

rationally designed, the most successful one had a relatively straightforward design. Namely, (1) a well-minimized aptamer for recognizing its ligand was inserted between the aaIRES and the aIRES; (2) a modulator sequence (MS) was inserted before the aaIRES to ligand-dependently modulate formation of the aaIRES/aIRES duplex (and thus also that of the aIRES/IRES duplex) (Fig. 6.3) (Ogawa 2011). In the OFF state, the MS forms a stem-loop structure (named MS-SL) with the aaIRES and a part of the aptamer to facilitate aIRES/IRES duplex formation, thereby repressing translation. In contrast, in the ON state, ligand-aptamer complex formation causes the aaIRES to be released from the MS and to hybridize to the aIRES, thereby promoting IRES-mediated translation. In this simple design strategy, one only has to optimize the length of MS to maximize the ON/OFF induction ratio. In fact, I readily succeeded in constructing an IRES-based eukaryotic ON-riboswitch that was highly responsive to theophylline, by using its aptamer (Jenison et al. 1994). The ON/OFF induction ratio of this riboswitch with the optimized MS length was 9.6 at 1 mM theophylline in WGE.

Although the versatility of this design method would have been confirmed by applying it to other aptamer-ligand pairs, I next reviewed the results of MS optimization in the theophylline-responsive riboswitch to circumvent the optimization for other pairs (i.e., to simplify the method). Of course, the longer the MS was, the less the translation efficiency both in the OFF and ON state became, due to the longer MS-SL stem. However, the decrease was saturated only in the OFF state at a certain length of MS, meaning that this length is sufficient for inhibiting IRES-mediated translation. Therefore, this length of MS (i.e., MS-SL stability) should give (and actually gave) the maximum ON/OFF induction ratio. I designed several IRES-based ON-riboswitches with other ligand-binding aptamers such that their MS-SL stability was comparable to that in the optimized theophylline-responsive one, by reference only to the aptamer sequences.[10] As a result, all riboswitches brilliantly exhibited high riboswitch activities in response to their corresponding ligands, depending on the aptamer-ligand affinity (e.g., the ON/OFF induction ratio of the tetracycline-responsive riboswitch with a stronger tetracycline aptamer (K_D is <1 nM) was 29 at 300 μM tetracycline). In this way, I successfully established a

[10]A ΔG value was used as an index of the MS-SL stability. This value can be calculated using only the sequence information of the implanted aptamer and a standard personal computer (Mathews et al. 2004).

versatile method for rationally designing eukaryotic ON-riboswitches using only aptamer sequence information [see elsewhere for details on the design method (Ogawa 2014a)]. In addition, I developed a similar design method for eukaryotic OFF-riboswitches that downregulate IRES-mediated translation (Ogawa 2012).

Because these IRES-based riboswitches are ribozyme-free, they are expected to be much less sensitive to magnesium ions than aptazyme-based ones are (Muranaka et al. 2009; Ogawa 2011). Although I confirmed their function only in WGE, they are likely to also function well (with minor adjustments such as the length optimization of each duplex) in other eukaryotic expression systems including living cells. Especially in eukaryotic cells, transcription and the subsequent translation proceed relatively slowly in different places, so that thermodynamically controlled riboswitches such as these IRES-based riboswitches are expected to be usable in vivo. Incidentally, the design strategy for IRES-based riboswitches may also be available for constructing bacterial ON-riboswitches (by considering an aIRES as an anti-SD).[11]

6.6 Eukaryotic ON-Riboswitches Modulating Ribosomal Shunt

Although IRES-based riboswitches can be relatively easily designed and exert relatively high switching efficiencies with strong aptamers, I was eager to create an even more efficient type of eukaryotic ON-riboswitches. For this purpose, I focused on the fact that IRES-based riboswitches essentially require thermodynamic energy to change their conformations via hybridization switches, which are likely to reasonably lose the switching efficiency. Thus, my next objective was to construct *hybridization switch-free* eukaryotic ON-riboswitches, which were expected not only to accomplish the conformational changes using less energy and thus exhibit higher efficiency but also to have a much more straightforward design due to the lack of complicated switches. In fact, hybridization switch-free eukaryotic OFF-riboswitches had already been created very easily just by inserting an aptamer into a 5′ UTR, as described above. In this type of OFF-riboswitch, an aptamer-ligand complex itself inhibits the ribosomal loading or scanning without (major) hybridization switches. However, it is likely impossible to design hybridization switch-free ON-riboswitches that upregulate the canonical translation, because a rigid structure on mRNA generally prevents the ribosomal progression. I thus searched for a report on a noncanonical translation mechanism in which a rigid structure conversely promotes a ribosomal movement. This search suggested that the *ribosomal shunt* might be a good candidate (Pooggin et al. 2000). It had been reported that some viruses employ this atypical mechanism in eukaryotes and in particular

[11] In bacterial expression systems, translation rapidly proceeds in synchrony with transcription, so that thermodynamically controlled riboswitches may function less efficiently than kinetically controlled ones (Mishler and Gallivan 2014).

that it efficiently takes place in the 35S RNA leader of the cauliflower mosaic virus (CaMV), which has a short open reading frame (sORF) and the following rigid stem structure at a short distance. Specifically, the 40S ribosome that has finished translation of sORF can surprisingly shunt over the rigid stem to reach the downstream ORF (dORF). Because it had also been shown that a rigid stem is required for this phenomenon (Pooggin et al. 2000), I could well imagine that a rigid aptamer-ligand complex could be substituted for the stem to promote the ribosomal shunt and the subsequent dORF expression.

However, the key to constructing ON-riboswitches that modulate the ribosomal shunt was determining how to suppress the dORF expression in the absence of their ligands. Because an aptamer is generally flexible without its ligand, if the rigid stem was replaced with an aptamer, the ribosome that has finished sORF translation would not *shunt* but rather *scan* the downstream region and then reinitiate dORF translation.[12] In fact, it was found that the reinitiation efficiency without the rigid stem was twice higher than that with it in WGE (Ogawa 2013). This meant that it was not possible to construct ON-riboswitches merely by replacing the whole of the stem with an aptamer.

I solved this issue by utilizing the difference between two modes of ribosomal movement after sORF translation, *shunting* and *scanning*. I first found that the choice of mode was determined by the distance between sORF and the stem (Dss). Whereas the original 6-nt Dss gave the most efficient expression of dORF, I found that the longer the Dss, the lower the efficiency, and a Dss of more than approximately 26 nt completely inhibited the dORF expression, indicating the mode gradually changed from shunting to scanning as the Dss was lengthened. The reason the dORF expression was suppressed with longer Dss is that the rigid stem that is required as a promoter in the shunting mode conversely behaved as an inhibitor in the scanning mode. To harness this feature, I replaced only the lower half segment of the stem with a split aptamer (or two split aptamers in tandem) to successfully construct hybridization switch-free eukaryotic ON-riboswitches that functioned well in WGE (Fig. 6.4) (Ogawa 2013). In these riboswitches, the remaining upper

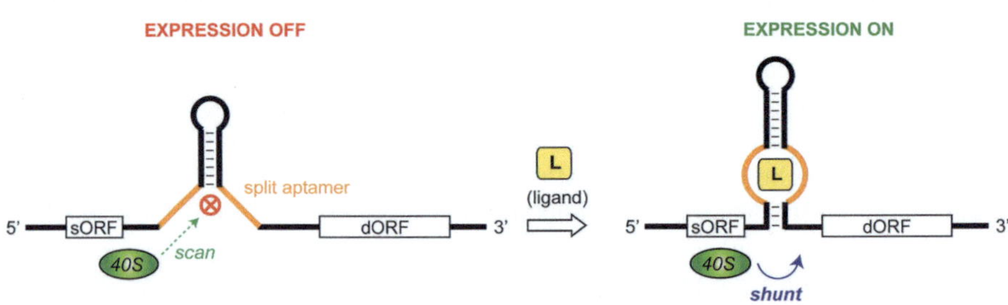

Fig. 6.4 An artificial ON-riboswitch modulating the ribosomal shunt in eukaryotic expression systems. sORF and dORF represent short and downstream open reading frames, respectively

[12] It is known that a large population of eukaryotic 40S ribosomal subunits remains on mRNA even after the translation of a relatively short ORF and scans the downstream region (Kozak 1987).

half stem at a sufficiently long Dss blocks the ribosome in the scanning mode in the OFF state, while the ligand forms the rigid stem-like structure with its split aptamer(s) at a proper (i.e., 6-nt) Dss to activate the ribosomal shunt and the subsequent reinitiation of dORF translation in the ON state. Incidentally, these riboswitches were designed to trap, with a mimic gene(s), the 40S ribosome that somehow unwound or shunted over the upper half stem, to further suppress dORF translation in the OFF state [see elsewhere for details on the design method (Ogawa 2015)].

Although there are some limitations to the design and/or use of these eukaryotic ON-riboswitches modulating the ribosomal shunt (e.g., only splittable aptamers are available; it is unclear whether the ribosomal shunt will occur efficiently in other eukaryotic expression systems), they have two main advantages over hybridization switch-based riboswitches. First, they are easier to design. In fact, the ligand specificity was able to be changed just by replacing the aptamer domain without any calculation. Secondly, they require much less energy for the conformational changes and thus exhibit higher switching efficiencies: the ON/OFF induction ratio of the optimized theophylline-responsive riboswitch was 14.4 at 1 mM theophylline in WGE, which is 1.5-fold higher than that of the IRES-based one.

6.7 Conclusions and Future Perspectives

In this chapter, I briefly introduced several methods for rationally designing artificial ON-riboswitches that function in bacterial or eukaryotic translation systems. Although details were withheld due to space limitations, they can be seen in our relevant papers or book chapters.[13] These methods would enable one to readily design artificial ON-riboswitches responsive to a ligand of interest, using only the sequence information of the corresponding well-minimized aptamer (or aptazyme).[14] Because all the various types of riboswitches described herein function independently from the genes they regulate, they are expected to upregulate any gene. Moreover, by virtue of the rationality of these designs, one can further optimize riboswitches with relative ease, depending on the use conditions. To enhance the switching efficiency and the ligand specificity, a cascading system and/or combined use with other gene regulators (e.g., another regulation-type riboswitch) would be useful (Ogawa and Maeda 2008a). When the rational design strategies fail, selection or screening strategies that other research groups have reported may be a good choice, especially for a bacterial riboswitch (Lynch and Gallivan 2009; Nomura and Yokobayashi 2014). These approaches generally require time and effort but could yield a desirable riboswitch.

[13] Recently, I edited a book that covers various detailed methods for constructing artificial riboswitches (including IRES-based one) and ligand-responsive gene regulators (Ogawa 2014b).

[14] In the case that there has been no report on an appropriate aptamer (or aptazyme), it is first necessary to obtain one using an in vitro selection method (under conditions suited to the expression system to be used, if possible). In addition, the selected aptamer should be minimized by eliminating extra sequences.

If the objective is to simply turn on a gene expression in bacteria, the most powerful candidates are currently the theophylline-responsive ON-riboswitches obtained through in vivo screening (Topp et al. 2010). However, these riboswitches may not function satisfactorily if they are blindly inserted before the gene to be regulated. This is because they must be in a position very close to the gene, and thus their functions are highly susceptible to influence by the 5′ terminal sequence of the gene. Even in such a case, however, the knowledge and experience acquired in researches on the rational design of riboswitches provide us with appropriate guidelines. In fact, we successfully utilized these riboswitches in cyanobacteria to regulate various genes whose 5′ terminal sequences were rationally silently mutated toward efficient hybridization switches (Nakahira et al. 2013): the highest ON/OFF induction ratio was as high as 190 at 2 mM theophylline in culture solution. In contrast, to my knowledge, there has been no report on equally efficient artificial ON-riboswitches that function in eukaryotic expression systems. We are now trying to rationally design new regulation types of eukaryotic riboswitches both in order to discover a more efficient riboswitch and to expand the riboswitch repertoire. In terms of the latter goal, a wide variety of riboswitches based on different regulation mechanisms would not only provide more choices for application under different conditions but also permit us to couple them together on the same mRNA to achieve sharper induction or more complex circuits.[15]

As described in the Introduction section, environment-responsive gene regulation systems are expected to be used in various fields. In particular, riboswitches are one of the most promising systems because they (and their ligands) can now be artificially customized with relative ease. If an efficient riboswitch could be developed, it would be possible to detect its ligand molecules in vitro and/or in vivo with the expression of a reporter gene regulated by the riboswitch (Muranaka et al. 2009; Ogawa 2011). It would also be possible to ligand-dose-dependently control the cell behavior by regulating the expression of an endogenous gene with an artificial riboswitch (Chen et al. 2010; Culler et al. 2010; Topp and Gallivan 2007). Incidentally, position-specific gene regulation can also be achieved with photo-responsive caged ligands (Walsh et al. 2014). Further, it would be possible to construct new gene circuits by using several artificial riboswitches. The genome-editing technologies that are now widely used should facilitate their implantation. In addition, much attention has recently been focused on vaccines based on mRNAs with artificial riboswitches (Andries et al. 2015). Last but not least, artificial riboswitches are also useful in cell-free synthetic biology (Hodgman and Jewett 2012; Lentini et al. 2016). Collectively, these studies with artificial riboswitches hold great promise, and I eagerly await their advances.

Acknowledgment This work was partially supported by JSPS KAKENHI Grant Number 16K05846.

[15] We have recently constructed a novel regulation type of eukaryotic ON-riboswitch that ligand-dose-dependently upregulates translation initiation mediated by a 3′ cap-independent translation element (3′ CITE) with no major hybridization switches (Ogawa et al. 2017).

References

Andries O, Kitada T, Bodner K, Sanders NN, Weiss R (2015) Synthetic biology devices and circuits for RNA-based 'smart vaccines': a propositional review. Expert Rev Vaccines 14:313–331

Berens C, Suess B (2015) Riboswitch engineering – making the all-important second and third steps. Curr Opin Biotechnol 31:10–15

Berschneider B, Wieland M, Rubini M, Hartig JS (2009) Small-molecule-dependent regulation of transfer RNA in bacteria. Angew Chem Int Ed 48:7564–7567

Breaker RR (2011) Prospects for riboswitch discovery and analysis. Mol Cell 43:867–879

Chang AL, Wolf JJ, Smolke CD (2012) Synthetic RNA switches as a tool for temporal and spatial control over gene expression. Curr Opin Biotechnol 23:679–688

Chen YY, Jensen MC, Smolke CD (2010) Genetic control of mammalian T-cell proliferation with synthetic RNA regulatory systems. Proc Natl Acad Sci USA 107:8531–8536

Culler SJ, Hoff KG, Smolke CD (2010) Reprogramming cellular behavior with RNA controllers responsive to endogenous proteins. Science 330:1251–1255

Desai SK, Gallivan JP (2004) Genetic screens and selections for small molecules based on a synthetic riboswitch that activates protein translation. J Am Chem Soc 126:13247–13254

Endo Y, Sawasaki T (2006) Cell-free expression systems for eukaryotic protein production. Curr Opin Biotechnol 17:373–380

Endo K, Hayashi K, Inoue T, Saito H (2013) A versatile cis-acting inverter module for synthetic translational switches. Nat Commun 4:2393

Furukawa K, Gu H, Breaker RR (2014) In vitro selection of allosteric ribozymes that sense the bacterial second messenger c-di-GMP. In: Ogawa A (ed) Artificial riboswitches. . Methods in molecular biology, vol 1111. Springer, New York, pp 209–220

Grate D, Wilson C (2001) Inducible regulation of the *S. cerevisiae* cell cycle mediated by an RNA aptamer-ligand complex. Bioorg Med Chem 9:2565–2570

Harvey I, Garneau P, Pelletier J (2002) Inhibition of translation by RNA-small molecule interactions. RNA 8:452–463

Hayashi G, Nakatani K (2014) Development of photoswitchable RNA aptamer-ligand complexes. In: Ogawa A (ed) Artificial riboswitches. Methods in molecular biology, vol 1111. Springer, New York, pp 29–40

Hodgman CE, Jewett MC (2012) Cell-free synthetic biology: thinking outside the cell. Metab Eng 14:261–269

Jenison RD, Gill SC, Pardi A, Polisky B (1994) High-resolution molecular discrimination by RNA. Science 263:1425–1429

Khvorova A, Lescoute A, Westhof E, Jayasena SD (2003) Sequence elements outside the hammerhead ribozyme catalytic core enable intracellular activity. Nat Struct Biol 10:708–712

Kozak M (1987) Effects of intercistronic length on the efficiency of reinitiation by eucaryotic ribosomes. Mol Cell Biol 7:3438–3445

Lentini R, Martín NY, Mansy SS (2016) Communicating artificial cells. Curr Opin Chem Biol 34:53–61

Link KH, Guo L, Ames TD, Yen L, Mulligan RC, Breaker RR (2007) Engineering high-speed allosteric hammerhead ribozymes. Biol Chem 388:779–786

Lynch SA, Gallivan JP (2009) A flow cytometry-based screen for synthetic riboswitches. Nucleic Acids Res 37:184–192

Madin K, Sawasaki T, Ogasawara T, Endo Y (2000) A highly efficient and robust cell-free protein synthesis system prepared from wheat embryos: plants apparently contain a suicide system directed at ribosomes. Proc Natl Acad Sci USA 97:559–564

Martick M, Scott WG (2006) Tertiary contacts distant from the active site prime a ribozyme for catalysis. Cell 126:309–320

Mathews DH, Disney MD, Childs JL, Schroeder SJ, Zuker M, Turner DH (2004) Incorporating chemical modification constraints into a dynamic programming algorithm for prediction of RNA secondary structure. Proc Natl Acad Sci USA 101:7287–7292

Mishler DM, Gallivan JP (2014) A family of synthetic riboswitches adopts a kinetic trapping mechanism. Nucleic Acids Res 42:6753–6761

Muranaka N, Sharma V, Nomura Y, Yokobayashi Y (2009) Efficient design strategy for whole-cell and cell-free biosensors based on engineered riboswitches. Anal Lett 42:108–122

Murata A, Sato S (2014) In vitro selection of RNA aptamers for a small-molecule dye. In: Ogawa A (ed) Artificial riboswitches. Methods in molecular biology, vol 1111. Springer, New York, pp 17–28

Murray JB, Arnold JRP (1996) Antibiotic interactions with the hammerhead ribozyme: tetracyclines as a new class of hammerhead inhibitor. Biochem J 317:855–860

Nakahira Y, Ogawa A, Asano H, Oyama T, Tozawa Y (2013) Theophylline-dependent riboswitch as a novel genetic tool for strict regulation of protein expression in cyanobacterium *Synechococcus elongatus* PCC 7942. Plant Cell Physiol 54:1724–1735

Nomura Y, Yokobayashi Y (2014) Dual genetic selection of synthetic riboswitches in *Escherichia coli*. In: Ogawa A (ed) Artificial riboswitches. Methods in molecular biology, vol 1111. Springer, New York, pp 131–140

Ogawa A (2009) Biofunction-assisted sensors based on a new method for converting aptazyme activity into reporter protein expression with high efficiency in wheat germ extract. Chembiochem 10:2465–2468

Ogawa A (2011) Rational design of artificial riboswitches based on ligand-dependent modulation of internal ribosome entry in wheat germ extract and their applications as label-free biosensors. RNA 17:478–488

Ogawa A (2012) Rational construction of eukaryotic OFF-riboswitches that downregulate internal ribosome entry site-mediated translation in response to their ligands. Bioorg Med Chem Lett 22:1639–1642

Ogawa A (2013) Ligand-dependent upregulation of ribosomal shunting. Chembiochem 14:1539–1543

Ogawa A (2014a) Rational design of artificial ON-riboswitches. In: Ogawa A (ed) Artificial riboswitches. . Methods in molecular biology, vol 1111. Springer, New York, pp 165–181

Ogawa A (ed) (2014b) Artificial riboswitches. Springer, New York

Ogawa A (2015) Engineering of ribosomal shunt-modulating eukaryotic ON riboswitches by using a cell-free translation system. In: Burke-Aguero DH (ed) Riboswitches as targets and tools. Methods in enzymology, vol 550. Springer, New York, pp 109–128

Ogawa A, Maeda M (2007) Aptazyme-based riboswitches as label-free and detector-free sensors for cofactors. Bioorg Med Chem Lett 17:3156–3160

Ogawa A, Maeda M (2008a) An artificial aptazyme-based riboswitch and its cascading system in *E. coli*. Chembiochem 9:206–209

Ogawa A, Maeda M (2008b) A novel label-free biosensor using an aptazyme-suppressor-tRNA conjugate and an amber-mutated reporter gene. Chembiochem 9:2204–2208

Ogawa A, Murashige Y, Tabuchi J, Omatsu T (2017) Ligand-responsive upregulation of 3' CITE-mediated translation in a wheat germ cell-free expression system. Mol BioSyst 13:314–319

Ohuchi S (2014) Identification of RNA aptamers against recombinant proteins with a hexahistidine tag. In: Ogawa A (ed) Artificial riboswitches. Methods in molecular biology, vol 1111. Springer, New York, pp 41–56

Pooggin MM, Hohn T, Fütterer J (2000) Role of a short open reading frame in ribosome shunt on the *Cauliflower mosaic virus* RNA leader. J Biol Chem 275:17288–17296

Roth A, Breaker RR (2009) The structural and functional diversity of metabolite-binding riboswitches. Annu Rev Biochem 78:305–334

Shimizu Y, Inoue A, Tomari Y, Suzuki T, Yokogawa T, Nishikawa K, Ueda T (2001) Cell-free translation reconstituted with purified components. Nat Biotechnol 19:751–755

Soukup GA, Breaker RR (1999) Engineering precision RNA molecular switches. Proc Natl Acad Sci USA 96:3584–3589

Soukup GA, Emilsson GAM, Breaker RR (2000) Altering molecular recognition of RNA aptamers by allosteric selection. J Mol Biol 298:623–632

Soukup GA, DeRose EC, Koizumi M, Breaker RR (2001) Generating new ligand-binding RNAs by affinity maturation and disintegration of allosteric ribozymes. RNA 7:524–536
Suess B, Hanson S, Berens C, Fink B, Schroeder R, Hillen W (2003) Conditional gene expression by controlling translation with tetracycline-binding aptamers. Nucleic Acids Res 31:1853–1858
Suess B, Fink B, Berens C, Stentz R, Hillen W (2004) A theophylline responsive riboswitch based on helix slipping controls gene expression in vivo. Nucleic Acids Res 32:1610–1614
Topp S, Gallivan JP (2007) Guiding bacteria with small molecules and RNA. J Am Chem Soc 129:6807–6811
Topp S, Gallivan JP (2010) Emerging applications of riboswitches in chemical biology. ACS Chem Biol 5:139–148
Topp S, Reynoso CM, Seeliger JC, Goldlust IS, Desai SK, Murat D, Shen A, Puri AW, Komeili A, Bertozzi CR, Scott JR, Gallivan JP (2010) Synthetic riboswitches that induce gene expression in diverse bacterial species. Appl Environ Microbiol 76:7881–7884
Walsh S, Gardner L, Deiters A, Williams GJ (2014) Intracellular light-activation of riboswitch activity. Chembiochem 15:1346–1351
Werstuck G, Green MR (1998) Controlling gene expression in living cells through small molecule-RNA interactions. Science 282:296–298
Wieland M, Hartig JS (2008) Improved aptazyme design and in vivo screening enable riboswitching in bacteria. Angew Chem Int Ed 47:2604–2607
Win MN, Smolke CD (2007) A modular and extensible RNA-based gene-regulatory platform for engineering cellular function. Proc Natl Acad Sci USA 104:14283–14288

Chapter 7
mRNA Engineering for the Control of Mammalian Cells in Medical Applications

Kei Endo and Hirohide Saito

Abstract Messenger RNA (mRNA) is an important carrier of genetic information and shows increasing medical application potential. The transfer of in vitro synthesized mRNA molecules into cells enables the expression of genes of interest without unexpected damage to the genomic DNA that risks cellular defects or tumorigenesis. Along with forcing the expression of external genes, engineered mRNAs can detect intracellular information for the artificial regulation of gene expressions. These features indicate the potential of mRNAs as central devices to engineer and control cells both in vitro and in vivo. Moreover, such devices can act as components of complex and sophisticated cellular systems or networks. In this article, we summarize recent advances in mRNA engineering and their application in the biomedical field and discuss future perspectives concerning mRNA-based biotechnology.

Keywords Translation · mRNA engineering · Cell engineering · Riboswitch · microRNA · Gene network · Synthetic biology

K. Endo (✉)
Department of Computational Biology and Medical Sciences, Graduate School of Frontier Sciences, The University of Tokyo, Kashiwa, Japan
e-mail: kei-endo@k.u-tokyo.ac.jp

H. Saito
Department of Life Science Frontiers, Center for iPS Cell Research and Application, Kyoto University, Kyoto, Japan
e-mail: hirohide.saito@cira.kyoto-u.ac.jp

S. Masuda, S. Izawa (eds.), *Applied RNA Bioscience*,
https://doi.org/10.1007/978-981-10-8372-3_7

7.1 Introduction

7.1.1 Engineering Living Mammalian Cells by RNA-Only Delivery

Cell engineering describes manipulation of the cell state and is most effective through controlling the expression of transgenes, such as the transdifferentiation of somatic cells (Davis et al. 1987), and reprogramming to the induced pluripotent stem (iPS) cells (Takahashi and Yamanaka 2006). Besides changing the cell state, transgenes can impart new functions to the cells. For example, the introduction of antigen-coding mRNAs allows dendritic cells to present artificial antigens that induce the production of specific antibodies in vivo (Gilboa and Vieweg 2004). With its continuing development, cell engineering has increasing potential for medical applications.

Cell engineering often tries to mimic natural processes. During differentiation, the cell state is autonomously and dynamically driven by the collective behaviors of networks of gene expression or protein activities. Transgenes in cell engineering regulate these networks. Although the expression of endogenous regulatory proteins is within the domain of regulatory networks, the expression of transgenes is without and is therefore unregulated. A major challenge in cell engineering is having the transgenes express during only a desired time period. To determine the period dynamically and autonomously, cell engineering devices should express transgenes in accordance with the cell state (Fig. 7.1).

The delivery method of the device plays a crucial role in the transgene expression. Normally, transgenes are delivered by treating cells with DNA molecules, such as plasmid DNA, the genomic DNA of viral vectors, or short single-stranded DNA. However, DNA molecules may integrate into the genome of the host cell where they can cause unexpected genomic alterations, which risks harmful phenotypes, compromising the cells for medical applications. RNA delivery, on the other hand, does not lead to genome integration and thus circumvents these concerns.

Because transgenes were previously delivered using DNA, many cell engineering devices control transgene expression by transcriptional regulation. However, devices delivered with RNA are incompatible with transcriptional regulation and instead depend on posttranscriptional regulation. Posttranscriptional regulation often involves the recruitment of regulatory proteins on the untranslated region (UTR) of a mRNA to control the stability or rate of translation initiation (Sonenberg and Hinnebusch 2009). These regulatory proteins specifically interfere with general machineries involved in mRNA processing, translation, or mRNA metabolism and are recruited in posttranscriptional processes that involve RNA interference (RNAi) or microRNA (miRNA). Alternatively, some mRNAs directly bind to a metabolite and consequently change their secondary structures to regulate translation. These metabolite-binding mRNAs are called riboswitches (Tucker and Breaker 2005). Riboswitches convert the input metabolite (or another specific molecule) into an output protein without the need for regulatory proteins and thus are suitable as a

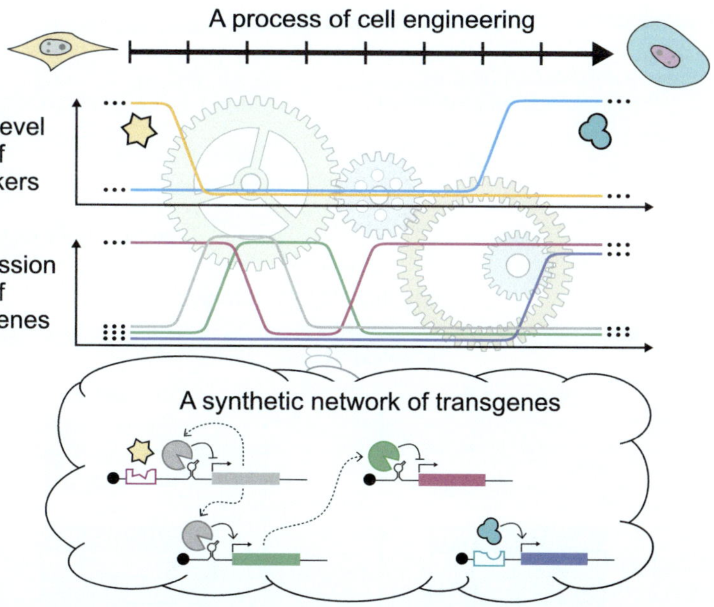

Fig. 7.1 Dynamic gene expression networks during a process of cell engineering. A cell engineering device outputs transgenes that control the cellular state. The device responds to the marker molecules, whose level changes with the cell state

component of artificially RNA-delivered engineered devices that modulate the cell state.

Riboswitches in principle are composed of two domains: an aptamer domain, which binds to the input molecule, and a protein-coding domain, which acts as a template for the output protein (Fig. 7.2). As illustrated in Fig. 7.2, appropriate connection of the aptamer and coding region leads to switch-like behavior of the mRNA. OFF switches output their coding protein only in the absence of the input molecule. On the contrary, ON switches only produce an output when the input is present. The sequences and functions of the aptamer and coding region are independent and thus interchangeable, which enhances the modularity of the device. Due to these properties, artificially designed riboswitches have been actively developed in recent years (Groher and Suess 2014).

In this chapter, the means for engineering living mammalian cells using in vitro synthesized RNA molecules, in particular artificial riboswitches, are described. First, briefly introduced are methods for delivering RNA molecules into cultured cells and for synthesizing designed mRNAs. Next, molecular devices that function like a genetic switch and their compatibility with RNA delivery are discussed. Then, the application of devices that detect miRNA activity in a cell is presented. Finally, the connection of the devices to form artificial networks is discussed for future application.

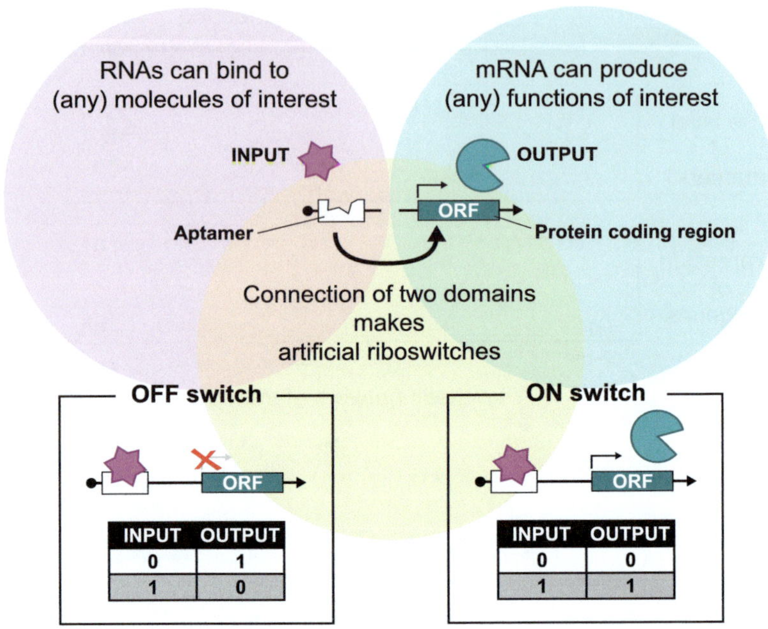

Fig. 7.2 Design principle of artificial riboswitches

7.1.2 Transferring mRNA Molecules into Cells

In contrast to DNAs, which function in the nucleus, mRNAs function in the cytoplasm. This feature allows mRNAs to be applied to nondividing cells and attains extremely high efficiency of the delivery such that almost all cells in a dish take in the transgenes. The delivery methods of mRNAs are quite similar to or almost the same as those of DNAs and short-interfering RNAs (siRNAs), which induce RNAi. The two major methods, electroporation and lipofection, are briefly introduced in the following paragraphs. In addition to these methods, gene gun (Qiu et al. 1996), empty viral particles (Kiyohara et al. 2012), and synthetic nanoscale particles (Uchida et al. 2013) have been reported for RNA delivery.

7.1.2.1 Electroporation

The application of an electric field to cultured mammalian cells allows RNA as well as DNA and other chemicals to enter a cell. This method requires a specific apparatus, an electroporator, and cuvettes with electrodes. Most electroporators require cells to be suspended in a specific solution, which might affect the cell fate. To avoid this effect, the authors have often used NEPA21 (Nepa Gene, Japan), because it is applicable to cells in any media of interest.

7.1.2.2 Lipofection

Cultured cells take in a complex of nucleotides and carrier reagents, many of which are made of cationic lipids. Commercially available reagents for both DNA and siRNA transfection are in most cases applicable to mRNA transfection with no or few modifications. More recently, reagents specific to mRNA transfection have been released. Similar to DNA and siRNA transfection, the efficiency and cytotoxicity of the method depend on both the reagents and the cell type. Thus, compatibility of the reagents needs to be screened for each cell type of interest. Some reagents require serum-free medium, which might affect the cell fate. To avoid this effect, the authors often use Stemfect RNA transfection kit (Stemgent, USA), TransIT-mRNA transfection kit (Mirus Bio, USA), or Lipofectamine MessengerMAX reagent (Thermo Fisher Scientific, USA) at conditions optimized for RNA transfection and compatible with media containing serum.

7.1.3 Synthesizing mRNAs that Function in Cells

Eukaryotic mRNAs consist of a 5′ terminal cap structure, 5′ UTR, protein-coding region, 3′ UTR, and 3′ terminal poly(A) tail. Because the protein-coding region typically stretches thousands of nucleotides (nt), making it too long to be synthesized chemically, mRNAs are synthesized via in vitro transcription by T7 RNA polymerase from a template DNA. For the template, both plasmid DNAs and PCR products are used. The poly(A) tail is synthesized two ways: transcription from a template containing poly(T) tract or enzymatic polyadenylation after the transcription. In the former, the poly(T) tract is designed in a reverse PCR primer, and the resulting PCR products are subjected to transcription. The reason is because a long poly(T) tract is unstable in plasmid. Thus, the length of the tail is limited to 100–150 nt depending on the maximum length of the synthetic primer. In the latter, bacterial poly(A) polymerase adds approximately 1000 adenosines to the 3′ end of the transcripts. The length of the tail, which affects the translation efficiency and the mRNA stability, is variable among mRNA molecules in a tube. This variation may be inadequate to precisely control the behavior of delivered mRNAs. The 5′ cap is also attached to the mRNA both during and after transcription. The addition of cap analogs to the reaction leads to a one-step synthesis of capped mRNA. Thus, capping during transcription requires fewer experimental procedures, but the efficiency of modification is limited to no more than 80–90%. We usually synthesize mRNAs from PCR products containing a 120 nt poly(T) tract with capping during transcription.

Researchers have modified transcripts to avoid cytotoxicity of the delivered RNAs. The chemical properties of RNA molecules are known to induce the innate immune system, which functions to counter viral infection. A triphosphate at the 5′ terminal of the in vitro synthesized RNAs triggers the interferon response. This triphosphate is removed in capped mRNAs; however, uncapped transcripts remain

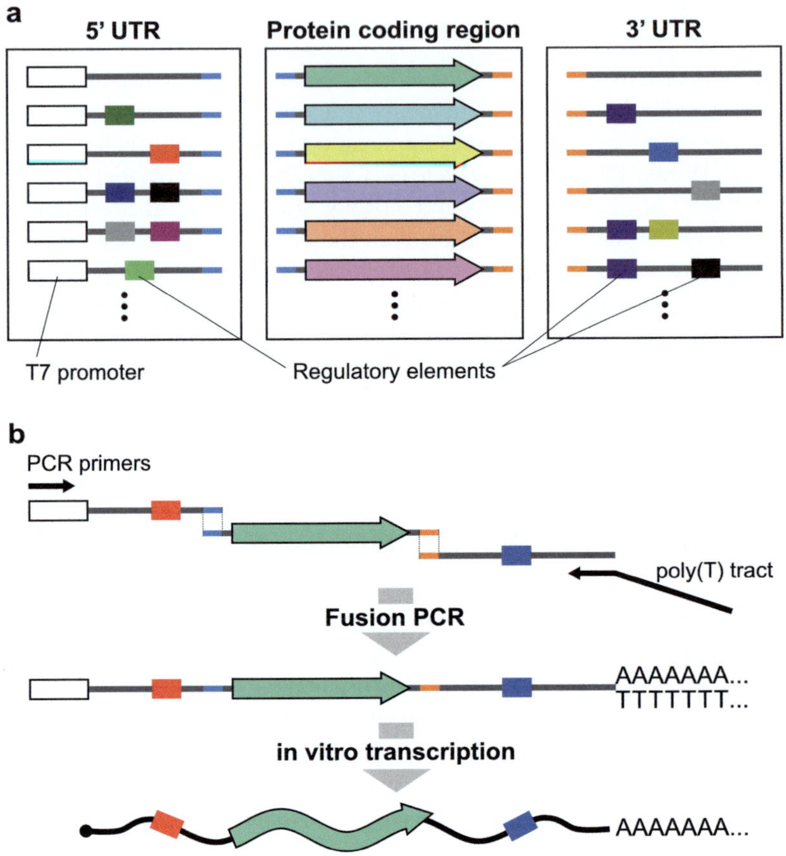

Fig. 7.3 (**a**) Libraries of mRNA domains for the systematic construction of synthetic mRNAs. (**b**) Procedures for mRNA production

in the product. Thus, synthesized mRNAs need to be treated with a phosphatase prior to delivery. The modification of bases is also known to reduce the immunogenicity of RNA (Durbin et al. 2016). For the RNA delivery of transgenes, modified nucleotides should be compatible with transcription in vitro and translation in cells. In many protocols, some uridine and cytosine residues are modified by pseudouridine and 5-methyl cytosine, respectively. In our protocol, all pyrimidines are replaced with these modified bases (Miki et al. 2015; Endo et al. 2016).

The sequence of mRNAs is divided into three domains of separate functions: the 5′ UTR, the protein-coding region, and the 3′ UTR. The 5′ and 3′ UTRs can be designed to contain regulatory elements for the expression of an encoded protein. Multiple elements in a single UTR are acceptable in some cases. The combination of the three domains independently prepared as DNA fragments enables systematic construction of designer mRNAs (Fig. 7.3). To concatenate the fragments by PCR (i.e., fusion PCR), the sequences of each domain in a library are flanked by consensus sequences. The 5′ end of the 5′ UTR and the 3′ end of the 3′ UTR bind to PCR primers. The 3′ end of the 5′ UTR and the coding region share the consensus sequence with the 5′ end of the coding region and the 3′ UTR, respectively. Finally, the PCR products are used as DNA templates for in vitro transcription, which produces a designer mRNA that expresses a protein of interest.

7.2 Devices Compatible with RNA Delivery

7.2.1 Protein Devices and (m)RNA Devices

To modulate the cell state, a device will detect specific molecules in a cell as input and regulate specific cellular functions, such as the expression of genes or the activity of proteins, as output. Depending on the complexity of the engineering, multiple molecular devices that behave as molecular switches will be used. Additionally, some devices use the output of other devices as their own input. Various proteins fulfill these conditions. RNA devices like riboswitches can be delivered into a cell as mRNAs that code for the proteins.

7.2.1.1 Protein Devices

Inducible transcription factors (TFs; Fig. 7.4a) are well-studied molecular switches that change their activity in response to a ligand to control the expression of downstream genes. One of the best-known examples is *E. coli* lactose repressor, *lacI*. Ligand binding to the repressor induces protein conformation changes and switches off the repressor activity. In a similar manner to bacterial inducible repressors, nuclear receptors function in animal cells, although they are typically transcriptional activators. Some bacterial TFs have been engineered for applications in eukaryotic cells. For example, tetracycline transactivator and its variants have been used in cells from a wide variety of species. The ligand-binding domains of some inducible TFs are reported to function as fusion proteins with the DNA-binding domain from another TF, which indicates the modularity and reusability of the domains (Keeley et al. 2005 and references therein). Ligand-binding domains can also be engineered to develop an affinity for other chemical compounds (Taylor et al. 2016).

Yeast two-hybrid systems that screen protein-protein interactions are applicable to sensing specific molecules in a cell. As proof of concept, one yeast TF, GAL4, was divided into two fragments: a DNA-binding domain and a transcription activation domain. These two fragments were fused with a pair of proteins of interest, the interaction of which brings the two TF domains close together to recover the TF activity. Taking advantage, protein devices can use two fused proteins that interact in a ligand-dependent manner to detect specific ligands (Fig. 7.4b). For example, a yeast three-hybrid system that screens protein-RNA interactions (Martin

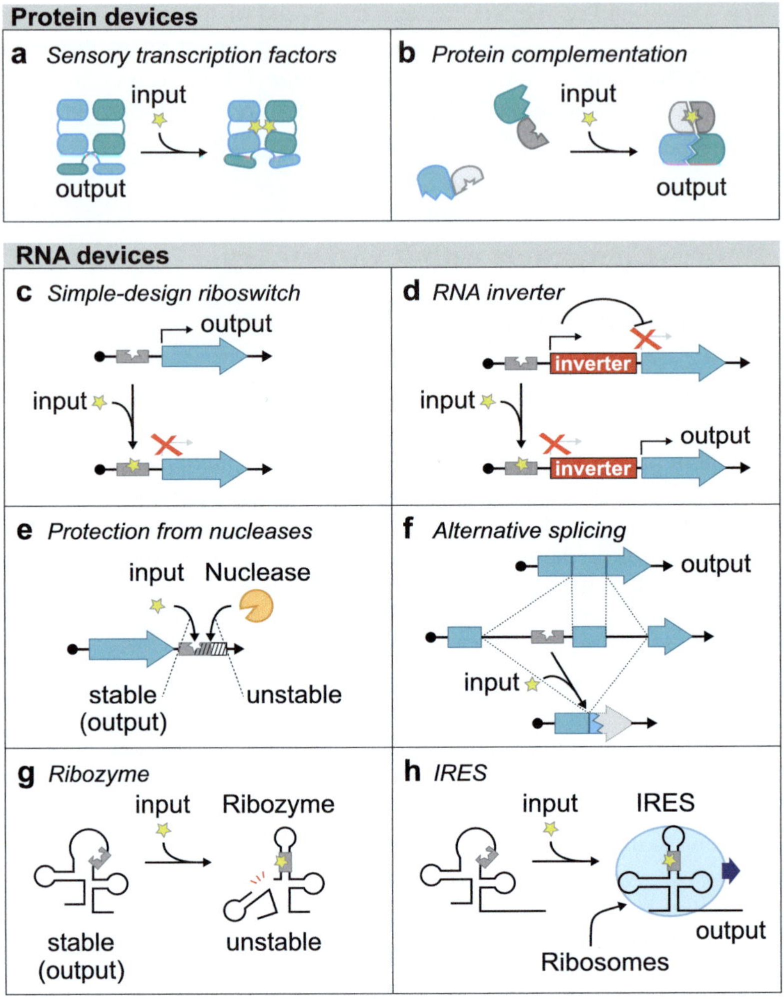

Fig. 7.4 (**a, b**) Examples of protein devices. (**c–h**) Examples of RNA devices

2012) can detect a specific RNA fragment. Notably, these systems are not limited to the recovery of TF activity. A more generalized system, called the protein-fragment complementation assay (Piehler 2005), has been used to recover the activity of enzymes, such as luciferase (Ozawa et al. 2001; Paulmurugan and Gambhir 2003) and β-lactamase (Galarneau et al. 2002), in a manner dependent on specific protein-protein interactions. Problems with this approach remain, however. Proper selection of the split site when developing protein devices based on protein-fragment complementation is difficult, and the range of proteins that can be used in protein-fragment complementation assays is limited.

One can also engineer TF DNA-binding domains to construct a desired TF. To design domains that bind to a DNA sequence of interest, small protein modules that specifically and independently recognize a single or few nucleotides have been developed (Gaj et al. 2013). More recently, CRISPR/Cas, a RNA-protein complex that specifically recognizes DNA sequences with a complementary RNA fragment, was discovered in various bacteria and has since been engineered to expand its recognition ability (Wang and Qi 2016). Although these designer DNA-binding domains were originally made to cut a single site from a huge genomic sequence, they have since been applied to TF engineering. For example, a defective Cas protein that loses DNA cleavage activity has been fused with transcription activation domains, histone modification enzymes, or DNA modification enzymes to construct designer TFs (Wang and Qi 2016). Furthermore, replacement of GAL4 with designer TFs expands the ligand detection system described above to construct a designer inducible TF that binds to a DNA sequence of interest to control the expression of specific genes in a ligand-dependent manner (Gao et al. 2016). However, the output of exogenous proteins under transcriptional regulation requires the introduction of DNA fragments into the target cells. This requirement is particularly problematic in engineering systems that depend on networking devices with RNA delivery.

7.2.1.2 RNA Devices

As described above, RNAs, in particular mRNAs, are suitable molecular switches that function in cells. Notably, the development of artificial riboswitches (Werstuck and Green 1998) preceded the discovery of natural riboswitches. The original artificial riboswitch was designed to bind to a derivative of Hoechst dye in the 5′ UTR and functioned as an OFF switch (Fig. 7.4c). The regulatory mechanism of the switch is similar to that of ferritin mRNA controlled by iron-responsive protein (IRP) in erythrocytes (Wilkinson and Pantopoulos 2014). IRP binds to the 5′ UTR of ferritin mRNA and blocks the initiation of cap-dependent translation by steric hindrance instead of recruiting regulatory proteins. These studies suggest that the binding of small molecules or proteins to the 5′ UTR generally blocks translation initiation. In other words, designed mRNAs that contain an aptamer domain in the 5′ UTR could behave as an OFF riboswitch (Fig. 7.2).

To expand the modularity and reusability of simple design riboswitches (Fig. 7.4c), a module for inverting an OFF switch into an ON switch was developed (Fig. 7.4d) (Endo et al. 2013a). This inverter module contains a bait open reading frame (ORF) and internal ribosome entry site (IRES). The coding region of the bait ORF is designed to evoke a response by surveillance machinery that recognizes aberrant mRNA. Thus, translation of the bait ORF triggers degradation of the switch mRNA. Conversely, binding of an input to the aptamer domain inhibits translation of the bait ORF and degradation of the switch. Consequently, output proteins are produced from the IRES of the stabilized switch.

To date, artificial riboswitches have been developed based on a variety of molecular mechanisms. Functional RNA molecules, such as aptamers, often fold into stem-loop structures. The modular structure of RNAs enables replacing one region with another in a cut-and-paste manner. Moreover, the two domains of the switch function independently. These features are why a variety of RNA fragments are integrated into mRNA modules when designing artificial riboswitches. Similar to natural transferrin receptor 1 (TfR1) mRNA, which is another mRNA regulated by IRP, an artificial mRNA that contains an aptamer domain overlapped with a destabilized motif recognized by specific RNase has been reported (Fig. 7.4e) (Babiskin and Smolke 2011). Binding of an input to the aptamer blocks the attack of RNase and stabilizes the switch. Input binding to a transcript can control the selection of the splice site, thus permitting artificial alternative splicing (Fig. 7.4f) (Kim et al. 2005; Culler et al. 2010). Furthermore, functional RNA molecules such as ribozymes (Fig. 7.4g) (Win and Smolke 2007; Ausländer et al. 2014) and IRES (Fig. 7.4h) (Ogawa 2011) can be fused with an aptamer to form molecular switches, which are then integrated into mRNAs as a module of riboswitches.

In addition to the cut-and-paste design, the activity of the switch can be tuned rationally. Some switches depend on changes in RNA folding in response to the input. Thus, alterations in the thermostability of the stems and the loops can tune the performance of a switch (Beisel et al. 2008; Ogawa 2011). As for simple design riboswitches, their responses can also be tuned by the number and position of aptamers in the 5′ UTR (Endo et al. 2013b).

Noncoding RNAs have also been engineered into other types of RNA-based switches. For example, short hairpin RNAs (shRNAs) that induce RNAi were fused with aptamer domains that bind to small molecules or proteins (An et al. 2006; Saito et al. 2011). In the absence of input molecules, shRNA switches reduce the expression level of their target mRNAs in a cell. In the presence of the input, the input aptamer complex blocks shRNAs from cleavage by Dicer protein. Consequently, shRNAs remain intact and fail to induce RNAi, which allows the mRNA expression level to recover. shRNA switches normally stretch less than 100 nt and form simple and stable stem-loop structures. These features enable rational design of the switch according to the 3D structure of the aptamer domain (Kashida et al. 2012).

7.2.2 Compatibility of RNA Devices with RNA Delivery

The compatibility of RNA devices with RNA delivery has also been explored. Simple design riboswitches (Fig. 7.4c) have been synthesized in vitro and delivered into cells with the same methods used to deliver mRNAs (Wroblewska et al. 2015; Kawasaki et al. 2017). Similarly, TfR1-type riboswitches (Fig. 7.4e) are compatible with RNA delivery. On the contrary, the RNA inverter (Fig. 7.4d) and alternative splicing-based switches (Fig. 7.4f) require mRNA processing in the nucleus, which complicates RNA delivery. Ribozymes and IRES that form specific structures to function might lose their activity in the presence of modified bases that are introduced to avoid cytotoxicity. To overcome this problem, ribozymes and IRESs that are compatible with 5-methylcytidine, pseudouridine, or other types of modified bases are required. In addition, the activity of ribozymes must be blocked during in vitro transcription and prior to delivery. RNA delivery is also problematic in switches that are based on noncoding RNAs that form higher-order structures to function. As well as the activity of functional RNAs, modified bases also affect aptamer affinity to the ligand. The incorporation of modified bases prevents the use of known aptamers for switches compatible with RNA delivery and thus requires de novo aptamer creation.

Despite the above issues, modified bases in the aptamer domain can be advantageous for RNA devices. The modified bases often improve the affinity of aptamers (Keefe and Cload 2008). In addition, the use of engineered unnatural bases possibly imparts further activity, e.g., photoreactivity. Thus, it may be beneficial to incorporate more modified bases into the aptamer domain without compromising function, because the domain is synthesized in vitro and distinct of the mRNA sequence that encodes amino acids.

7.2.3 Intracellular Information Detected by RNA Devices

One way in which RNA devices control the cell state is by using aptamers that bind to biomolecules. However, natural RNA sequences capture only few biomolecules. Systematic evolution of ligands by exponential enrichment (SELEX) can be used to build artificial aptamers that expand the library of biomolecules (Wu and Kwon 2016). However, these RNA devices do not reliably function inside cells. Below we describe successful artificial aptamers built by SELEX.

7.2.3.1 Small Molecules

The first artificial riboswitch includes an aptamer that responds to a derivative of Hoechst dye (Werstuck and Green 1998). Since then, aptamers that respond to other small molecules have been used. Among them, a theophylline aptamer (Jenison et al. 1994) and a tetracycline aptamer (Berens et al. 2001) are well characterized and used in riboswitches compatible with RNA delivery (Harvey et al. 2002; Suess et al. 2003; Babiskin and Smolke 2011). Similarly, synthetic shRNA switches can respond to small molecules (An et al. 2006; Beisel et al. 2008). In general, aptamers with high affinity require candidates to be highly concentrated in cells. A two-step screening is used to design such RNA devices. The first screening is for an input molecule with high intracellular concentration during the specific period of the biological process of interest, and the second screening is for its specific aptamer via SELEX.

7.2.3.2 Proteins

In contrast to aptamers against small molecules, aptamers against proteins usually show higher affinity. However, unlike small molecules that easily and quickly diffuse inside the cell, protein localization in the cell is more strictly controlled. For example, TFs are key marker proteins that represent the state of the cell but localize in the nucleus where RNAs are harder to deliver. Even in the cytoplasm, where many macromolecules are crowded, proteins gather according to their function. The localization constraint limits the types of proteins that can be captured by delivered RNAs. RNA-binding proteins (RBPs) make good candidates as input proteins because they localize close to and have binding affinity for intracellular RNA. Their general affinity to RNA, however, might require the careful selection of specific aptamers.

7.2.3.3 RNAs

Intracellular RNA molecules also can be used as biomarkers. Among them, miRNA would be the first choice. Since most miRNAs are expressed by RNA polymerase II, the profile of their expression is expected to indicate the state of the cell as a subset of the transcriptome. miRNAs are known to bind to and inhibit the expression of specific mRNAs. Thus, they can control the expression of synthesized mRNAs that are transferred into the cell. Moreover, mRNAs responding to a miRNA can be simply designed by inserting the antisense sequence of the miRNA. This insertion removes the need for aptamer selection. Conveniently, modified bases for RNA delivery are compatible with interactions through base pairings. As with simple design riboswitches, some mRNAs contain a miRNA target sequence in their 5′ UTR. Interestingly, transfected mRNAs that carry a single miRNA target in the 5′ UTR more sensitively detect miRNA activity in the cell than those that carry four targets in the 3′ UTR (Miki et al. 2015).

7.2.4 Identification of Living Cells by RNA Devices According to miRNA Activity

The delivery of synthetic mRNAs that respond to a specifically expressed miRNA enables cell purification. For example, the expressions of miR-1, miR-208a, and miR-499a are specifically upregulated after the differentiation of cardiomyocytes from human iPS cells (Miki et al. 2015). To perform live cell purification via fluorescence-activated cell sorting, fluorescent proteins need to be encoded in the RNA devices as outputs (Fig. 7.5a). The delivery of blue fluorescent protein mRNAs that respond to one of these miRNAs together with green fluorescent protein mRNAs (control mRNAs) efficiently isolates cells that express the cardiac isoform of troponin T, which is a cardiomyocyte-specific marker (Miki et al. 2015). Similarly, endothelial cells, hepatocytes, and insulin-producing cells differentiated from iPS

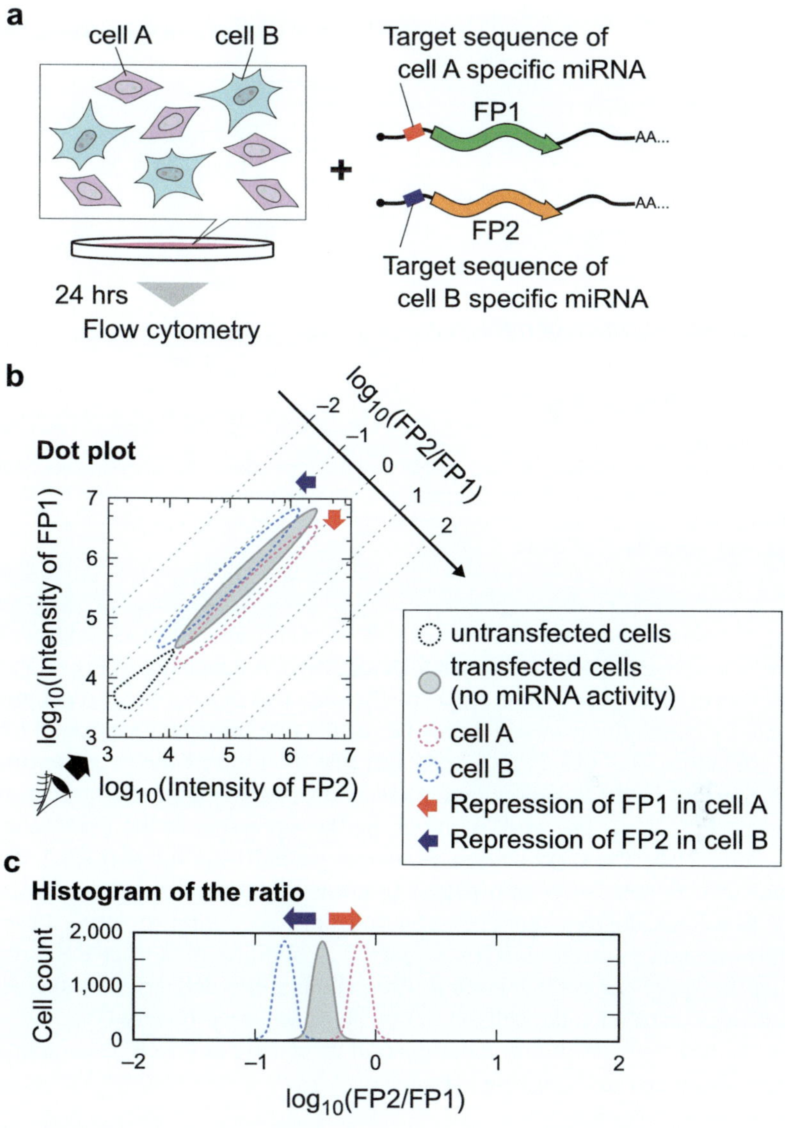

Fig. 7.5 (**a**) Scheme for detecting miRNA activity in living cells by miRNA-responsive mRNAs. (**b**) Schematic dot plots of mRNA-delivered cells analyzed by flow cytometry. (**c**) Schematic histograms of the cell population in HRIC

cells can be purified using RNA devices that detect the intracellular activity of miR-126, miR-122, and miR-375, respectively. (Miki et al. 2015) In some cases, undifferentiated cells need to be removed from a cell culture selectively. Pluripotent stem cells and partially differentiated cells can be removed using a RNA device that targets miR-302, which is specifically and highly expressed in human iPS cells (Parr et al. 2016).

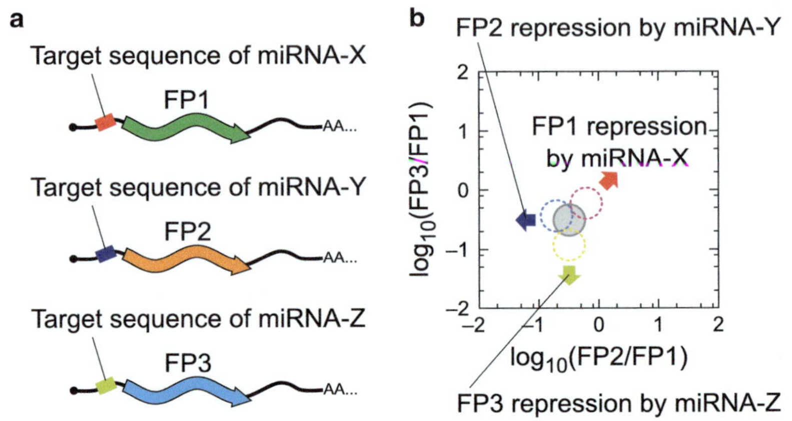

Fig. 7.6 (**a**) The design of three distinct mRNAs for 2D separation. The mRNAs detect different miRNAs and output different fluorescent proteins. (**b**) Schematic dot plots of mRNA-delivered cells analyzed by a flow cytometer. Arrows indicate movement of the cell population in a 2D plane by the activity of the three miRNAs

Simultaneous detection of the activity of multiple miRNAs improves the resolution of live cell purification (Endo et al. 2016). As depicted in Fig. 7.5b, after the delivery of two mRNAs that, respectively, code two distinct fluorescent proteins (FP1 and FP2), the fluorescence intensities of the two proteins were found to have high correlation. This observation indicates that the ratio of the two intensities is almost constant in the cell population and is distributed like a sharp peak in the histogram (Fig. 7.5c). The peak is shifted by the repression of the fluorescence of one protein. Thus, the combination of two target miRNAs for two RNA devices separates two fluorescence ratio peaks. In proof-of-concept experiments, specific sets of two RNA devices separated two cell lines according to subtle (less than twofold) differences in the miRNA activity. The addition of further RNA devices expands the space for cell separation. For example, the delivery of three or four RNA devices distributes the cells in 2D or 3D space, respectively (Fig. 7.6). This system, named high-resolution identification of cell types (HRIC), is compatible with flow cytometry and imaging cytometry.

7.3 Connecting RNA Devices to Construct Higher Networks

7.3.1 Network Motifs Composed of RNA Devices

An artificial gene network that implements complex functions must connect multiple devices. As in natural metabolic pathways, connections could be established between devices via metabolites. That is, one device outputs an enzyme that produces metabolites, which in turn behave as inputs for other devices. In some

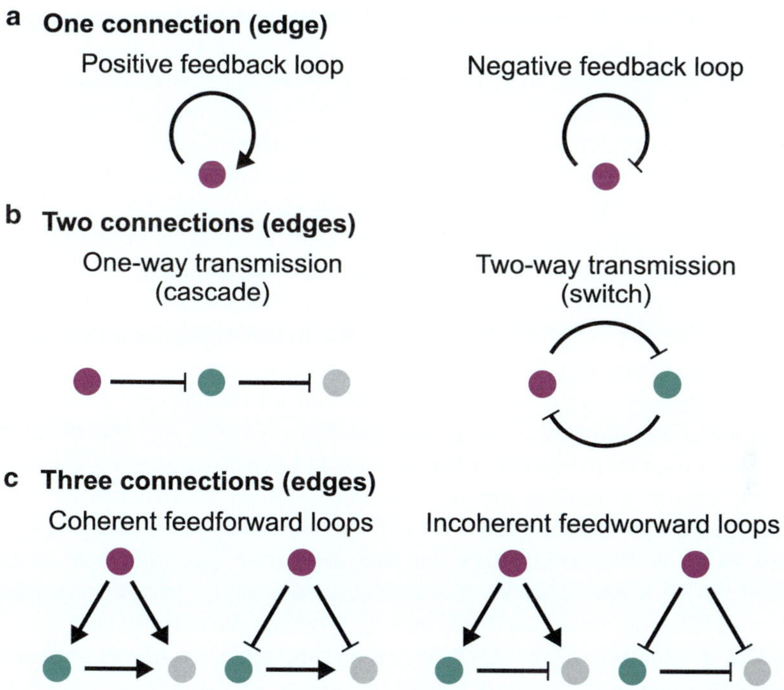

Fig. 7.7 Network motifs in gene expression networks composed of one (**a**), two (**b**), and three (**c**) factors

designs, outputted proteins such as RBPs act as the input for other devices. Additionally, RBPs can be fused with an effector to construct a designer regulator that controls the expression from mRNA. For example, MS2 coat protein, which is a well-known RBP of bacteriophage, has been fused with CCN4, which is a component of the deadenylase complex, to trigger the rapid degradation of target mRNAs (Wroblewska et al. 2015).

Feedback loops are a motif of gene regulatory networks composed of a single device (Fig. 7.7a). In many natural feedback loops, a regulatory protein controls the expression of other genes as well as the expression of itself. A negative feedback loop based on translational control can be constructed by inserting a binding motif for a RBP into the 5′ UTR of the RBP mRNA (Stapleton et al. 2012), resulting in a simple design riboswitch (Fig. 7.4c), which, as mentioned above, is compatible with RNA delivery. On the contrary, the construction of positive feedback loops needs the inclusion of ON switches. Among the ON switches developed, the TfR1-type switch (Fig. 7.4e) is the only candidate for devices compatible with RNA delivery. TfR1-type switches are designed to rapidly degrade after cell entry unless bound to an input.

One-directional connections between devices lead to signaling cascades (left panel, Fig. 7.7b). Here, artificial protein-responsive riboswitches respond to a RBP so that the RNA device can output a signaling protein. When two OFF switches are connected, the network output is inverse the output of the first OFF switch (the sensory device). Adding a third OFF switch creates a three-step cascade that inverts the network behavior again so that the network output and sensory device output are the same. Compared to the cascade of RNA devices that are constitutively transcribed in cells, cascades of RNA-delivered RNA devices show signal attenuation as the number of cascade steps is increased (Wroblewska et al. 2015). The bidirectional connection of two OFF switches makes a network that acts like a toggle switch (right panel, Fig. 7.7b). Although such a network behaves as expected when its components are constitutively transcribed in cells (Wroblewska et al. 2015), the network fails when the components are delivered with RNAs. These RNA-delivered devices are designed to rapidly degrade in cells. To achieve optimal function, RNA devices that do not rapidly degrade are recommended.

Connecting three devices can also make a feed-forward loop (Fig. 7.7c), which is one of the best-studied motifs in gene regulatory networks. As far as we know, no feed-forward loop composed of RNA devices and delivered as RNA exists. Some incoherent feed-forward loops can be composed of three OFF switches, but coherent feed-forward loops require both ON and OFF switches. The behavior of feed-forward loops is characterized by its dynamics over time; however, networks delivered by RNA have only been evaluated in snapshots. Means for tracing the behavior of RNA devices after RNA delivery remain to be developed.

Besides the lifetime of the RNA device and the inclusion of ON switches, the nonlinear behavior of the network should be considered in the design. Nonlinearity could cause variation in the RNA uptake leading to inconsistent effects on different cell subpopulations. Coupled RNA devices need to suppress this variation.

7.3.2 *RNA Replicons as Vectors for RNA-Deliverable Devices*

RNA replicons are expected to maintain RNA devices in a cell without DNA delivery (Fig. 7.8a). They are derived from the genome of RNA viruses that replicate in the cytoplasm of host cells. Inside cells, a replicon produces its own RNA-dependent RNA polymerases, which transcribe the antisense strand of the replicon and then the sense strand from the antisense strand. The polymerases also produce mRNAs from the sense or antisense strands of the replicon. Thus, the replicon keeps supplying the RNA devices to the cells. Because it maintains itself even after the division of the host cells, the level of RNA devices in a cell remains high. On the other hand, in the case of direct delivery of RNA devices, the devices in a cell are halved after cell division (Fig. 7.8b). This loss is particularly problematic in devices that regulate cell differentiation, which often takes several days or weeks and/or requires cell divisions. Thus RNA replicons are a promising platform for RNA-deliverable devices that engineer mammalian cells in medical applications.

Among RNA replicons, those derived from an alphavirus, Venezuelan equine encephalitis virus, have been applied to supply RNA devices constitutively in a cell (Wroblewska et al. 2015). Alphaviruses have a linear positive- and single-stranded

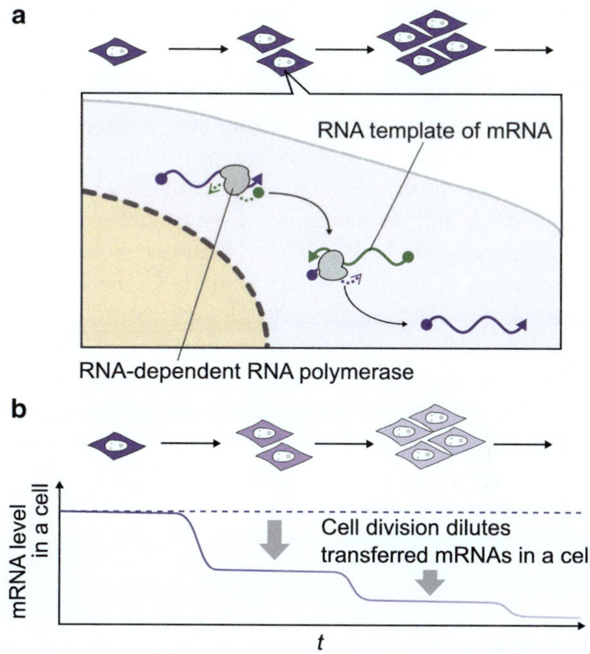

Fig. 7.8 (**a**) Maintenance of mRNA by RNA replicons. (**b**) The level of mRNAs delivered in a cell

RNA genome, which directly codes viral proteins as the positive sense of mRNA. Their replicon contains genes of interest with one huge viral nonstructural protein that acts as a replicase. The alphaviral replicon constitutively supplies RNA devices from its antisense strand in the cytoplasm; however, it strongly stimulates the innate immune system and shows high immunotoxicity. Thus, the application of this replicon is extremely limited. A replicon from Sendai virus is another potential vector for RNA-deliverable devices. Sendai virus has a linear negative- and single-stranded RNA genome and has been used as a viral vector. Negative-stranded replicons are harder to reconstitute in cells compared with positive-stranded replicons such as the alphaviral replicon. Thus, the development of a useful RNA replicon needs to begin with the reconstruction of Sendai viral replicons via RNA delivery.

Once the replicon has completed its function, the replicon needs to be removed from the cells. For the Sendai virus vector, an inducible replicon that was maintained by a temperature-sensitive replicase was developed (Ban et al. 2011). Here, the temperature shift is non-autonomous and affects all cells in a dish simultaneously regardless of the cell state. To achieve autonomous removal of the replicons, the expression of the replicase could be controlled by a riboswitch that responds to a marker molecule at the final state of the target cells.

7.4 Summary

The delivery of RNAs into living cells is a promising means for introducing genes of interest without the risks of genomic damage. Molecular devices made of RNA have the potential to process the information of any molecule of interest into the output of any protein of interest. RNA devices can be designed to communicate with each other and to compose a gene network that behaves dynamically and autonomously in a cell. These features make RNA attractive material for the engineering of living cells, particularly in medical applications. Already, a variety of RNA devices have been developed, including more recently those compatible with RNA delivery. Yet the development of RNA-deliverable devices is still in its infancy. As the field progresses steadily, RNA-deliverable devices should benefit medicine by improving the quality and efficiency of engineering and producing useful living cells.

Acknowledgments The authors thank Dr. Peter Karagiannis for his critical comments and language editing on the original version of this manuscript.

References

An CI, Trinh VB, Yokobayashi Y (2006) Artificial control of gene expression in mammalian cells by modulating RNA interference through aptamer-small molecule interaction. RNA 12:710–716

Ausländer S, Stücheli P, Rehm C, Ausländer D, Hartig JS, Fussenegger M (2014) A general design strategy for protein-responsive riboswitches in mammalian cells. Nat Methods 11:1154–1160

Babiskin AH, Smolke CD (2011) Engineering ligand-responsive RNA controllers in yeast through the assembly of RNase III tuning modules. Nucleic Acids Res 39:5299–5311

Ban H, Nishishita N, Fusaki N, Tabata T, Saeki K, Shikamura M, Takada N, Inoue M, Hasegawa M, Kawamata S, Nishikawa S (2011) Efficient generation of transgene-free human induced pluripotent stem cells (iPSCs) by temperature-sensitive Sendai virus vectors. Proc Natl Acad Sci U S A 108:14234–14239

Beisel CL, Bayer TS, Hoff KG, Smolke CD (2008) Model-guided design of ligand-regulated RNAi for programmable control of gene expression. Mol Syst Biol 4:224

Berens C, Thain A, Schroeder R (2001) A tetracycline-binding RNA aptamer. Bioorg Med Chem 9:2549–2556

Culler SJ, Hoff KG, Smolke CD (2010) Reprogramming cellular behavior with RNA controllers responsive to endogenous proteins. Science 330:1251–1255

Davis RL, Weintraub H, Lassar AB (1987) Expression of a single transfected cDNA converts fibroblasts to myoblasts. Cell 51:987–1000

Durbin AF, Wang C, Marcotrigiano J, Gehrke L (2016) RNAs containing modified nucleotides fail to trigger RIG-I conformational changes for innate immune signaling. MBio 7:e00833–e00816

Endo K, Hayashi K, Inoue T, Saito H (2013a) A versatile cis-acting inverter module for synthetic translational switches. Nat Commun 4:2393

Endo K, Stapleton JA, Hayashi K, Saito H, Inoue T (2013b) Quantitative and simultaneous translational control of distinct mammalian mRNAs. Nucleic Acids Res 41:e135

Endo K, Hayashi K, Saito H (2016) High-resolution identification and separation of living cell types by multiple microRNA-responsive synthetic mRNAs. Sci Rep 6:21991

Gaj T, Gersbach CA, Barbas CF (2013) ZFN, TALEN, and CRISPR/Cas-based methods for genome engineering. Trends Biotechnol 31:397–405

Galarneau A, Primeau M, Trudeau LE, Michnick SW (2002) β-lactamase protein fragment complementation assays as in vivo and in vitro sensors of protein protein interactions. Nat Biotechnol 20:619–622

Gao Y, Xiong X, Wong S, Charles EJ, Lim WA, Qi LS (2016) Complex transcriptional modulation with orthogonal and inducible dCas9 regulators. Nat Methods 13:1043–1049

Gilboa E, Vieweg J (2004) Cancer immunotherapy with mRNA-transfected dendritic cells. Immunol Rev 199:251–263

Groher F, Suess B (2014) Synthetic riboswitches – a tool comes of age. Biochim Biophys Acta 839:964–973

Harvey I, Garneau P, Pelletier J (2002) Inhibition of translation by RNA-small molecule interactions. RNA 8:452–463

Jenison RD, Gill SC, Pardi A, Polisky B (1994) High-resolution molecular discrimination by RNA. Science 263:1425–1429

Kashida S, Inoue T, Saito H (2012) Three-dimensionally designed protein-responsive RNA devices for cell signaling regulation. Nucleic Acids Res 40:9369–9378

Kawasaki S, Fujita Y, Nagaike T, Tomita K, Saito H (2017) Synthetic mRNA devices that detect endogenous proteins and distinguish mammalian cells. Nucl Acid Res 45(12):e117–e117

Keefe AD, Cload ST (2008) SELEX with modified nucleotides. Curr Opin Chem Biol 12:448–456

Keeley MB, Busch J, Singh R, Abel T (2005) TetR hybrid transcription factors report cell signaling and are inhibited by doxycycline. BioTechniques 39:529–536

Kim DS, Gusti V, Pillai SG, Gaur RK (2005) An artificial riboswitch for controlling pre-mRNA splicing. RNA 11:1667–1677

Kiyohara E, Tamai K, Katayama I, Kaneda Y (2012) The combination of chemotherapy with HVJ-E containing Rad51 siRNA elicited diverse anti-tumor effects and synergistically suppressed melanoma. Gene Ther 19:734–741

Martin F (2012) Fifteen years of the yeast three-hybrid system: RNA-protein interactions under investigation. Methods 58:367–375

Miki K, Endo K, Takahashi S, Funakoshi S, Takei I, Katayama S, Toyoda T, Kotaka M, Takaki T, Umeda M, Okubo C, Nishikawa M, Oishi A, Narita M, Miyashita I, Asano K, Hayashi K, Osafune K, Yamanaka S, Saito H, Yoshida Y (2015) Efficient detection and purification of cell populations using synthetic microRNA switches. Cell Stem Cell 16:699–711

Ogawa A (2011) Rational design of artificial riboswitches based on ligand-dependent modulation of internal ribosome entry in wheat germ extract and their applications as label-free biosensors. RNA 17:478–488

Ozawa T, Kaihara A, Sato M, Tachihara K, Umezawa Y (2001) Split luciferase as an optical probe for detecting protein-protein interactions in mammalian cells based on protein splicing. Anal Chem 73:2516–2521

Parr CJ, Katayama S, Miki K, Kuang Y, Yoshida Y, Morizane A, Takahashi J, Yamanaka S, Saito H (2016) MicroRNA-302 switch to identify and eliminate undifferentiated human pluripotent stem cells. Sci Rep 6:32532

Paulmurugan R, Gambhir SS (2003) Monitoring protein-protein interactions using split synthetic renilla luciferase protein-fragment-assisted complementation. Anal Chem 75:1584–1589

Piehler J (2005) New methodologies for measuring protein interactions in vivo and in vitro. Curr Opin Struct Biol 15:4–14

Qiu P, Ziegelhoffer P, Sun J, Yang NS (1996) Gene gun delivery of mRNA in situ results in efficient transgene expression and genetic immunization. Gene Ther 3:262–268

Saito H, Fujita Y, Kashida S, Hayashi K, Inoue T (2011) Synthetic human cell fate regulation by protein-driven RNA switches. Nat Commun 2:160

Sonenberg N, Hinnebusch AG (2009) Regulation of translation initiation in eukaryotes: mechanisms and biological targets. Cell 136:731–745

Stapleton JA, Endo K, Fujita Y, Hayashi K, Takinoue M, Saito H, Inoue T (2012) Feedback control of protein expression in mammalian cells by tunable synthetic translational inhibition. ACS Synth Biol 1:83–88

Suess B, Hanson S, Berens C, Fink B, Schroeder R, Hillen W (2003) Conditional gene expression by controlling translation with tetracycline-binding aptamers. Nucleic Acids Res 31:1853–1858

Takahashi K, Yamanaka S (2006) Induction of pluripotent stem cells from mouse embryonic and adult fibroblast cultures by defined factors. Cell 126:663–676

Taylor ND, Garruss AS, Moretti R, Chan S, Arbing MA, Cascio D, Rogers JK, Isaacs FJ, Kosuri S, Baker D, Fields S, Church GM, Raman S (2016) Engineering an allosteric transcription factor to respond to new ligands. Nat Methods 13:177–183

Tucker BJ, Breaker RR (2005) Riboswitches as versatile gene control elements. Curr Opin Struct Biol 15:342–348

Uchida S, Itaka K, Uchida H, Hayakawa K, Ogata T, Ishii T, Fukushima S, Osada K, Kataoka K (2013) In vivo messenger RNA introduction into the central nervous system using polyplex nanomicelle. PLoS One 8:e56220

Wang F, Qi LS (2016) Applications of CRISPR genome engineering in cell biology. Trends Cell Biol 26:875–888

Werstuck G, Green MR (1998) Controlling gene expression in living cells through small molecule-RNA interactions. Science 282:296–298

Wilkinson N, Pantopoulos K (2014) The IRP/IRE system in vivo: insights from mouse models. Front Pharmacol 5:176

Win MN, Smolke CD (2007) A modular and extensible RNA-based gene-regulatory platform for engineering cellular function. Proc Natl Acad Sci U S A 104:14283–14288

Wroblewska L, Kitada T, Endo K, Siciliano V, Stillo B, Saito H, Weiss R (2015) Mammalian synthetic circuits with RNA binding proteins for RNA-only delivery. Nat Biotechnol 33:839–841

Wu YX, Kwon YJ (2016) Aptamers: the "evolution" of SELEX. Methods 106:21–28

Chapter 8
Modulation of Abnormal Splicing of RNA Diseases by Small Chemical Compounds

Naoyuki Kataoka

Abstract Pre-mRNA splicing is a critical step for protein gene expression in higher eukaryotes. Another mode of splicing, termed alternative splicing, contributes to diversity of the expressed proteins from the limited number of genes in chromosomes. Those steps are highly regulated and must be accurate. Therefore, disruption of splicing regulation often results in hereditary and sporadic diseases called as "RNA diseases" in human. Modulation of splicing by small chemical compounds and nucleic acids has been targeting aberrant splicing in those diseases. In this chapter, I will introduce several RNA diseases and splicing-target therapeutic approaches with chemical compounds. Unveiling molecular mechanism and correction of aberrant splicing by small chemical compounds are important for both RNA biologists and clinicians who desire therapies for those diseases.

Keywords RNA splicing · Chemical compound · RNA disease · Aberrant splicing

8.1 Introduction

Splicing of precursor of mRNA (pre-mRNA) is a critical step for protein gene expression in higher eukaryotes (Wahl et al. 2009). There are two modes for splicing, called as constitutive splicing and alternative splicing. For constitutive splicing, all exons in a pre-mRNA are ligated in order without any insertions and deletions of nucleotides, which cause frame shift. In contrast, alternative splicing (AS) utilizes several alternative exons for both inclusion and exclusion (Fu and Ares 2014; Lee and Rio 2015). There are several types of AS, such as exon skipping/inclusion, alternative 5′ splice sites, alternative 3′ splice sites, mutually exclusive exons, and intron retention (Fig. 8.1). AS can produce many variants of mRNA that are

N. Kataoka (✉)
Laboratory of Cell Regulation, Departments of Applied Animal Sciences and Applied Biological Chemistry Graduate School of Agriculture and Life Sciences, The University of Tokyo, Tokyo, Japan
e-mail: akataoka@mail.ecc.u-tokyo.ac.jp

S. Masuda, S. Izawa (eds.), *Applied RNA Bioscience*,
https://doi.org/10.1007/978-981-10-8372-3_8

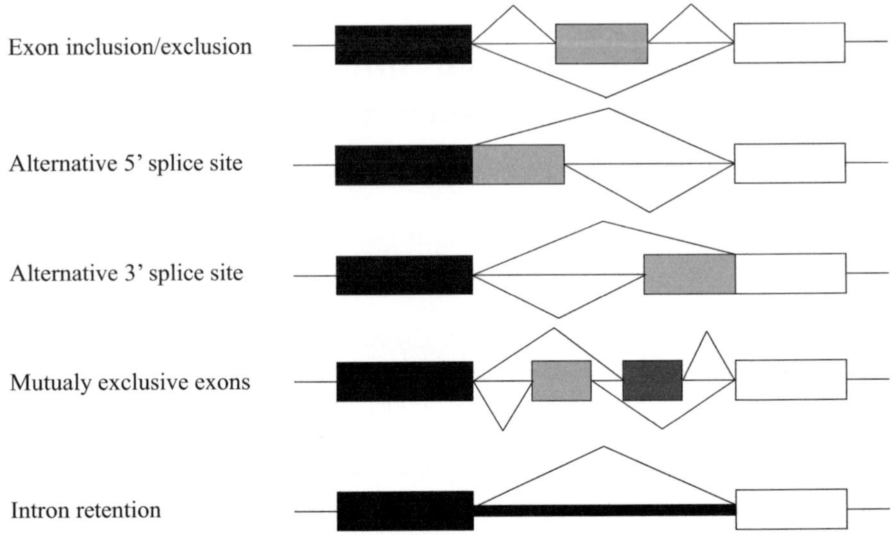

Fig. 8.1 Major alternative splicing patterns in higher eukaryotes. Major patterns of alternative splicing in higher eukaryotes are schematically illustrated. Boxes and lines in between show exons and introns, respectively. In intron retention pattern, intron shown in a thick line is recognized as an alternative exon

translated into proteins that have different structures and functions. Thus, AS likely contributes to generate diversity of the expressed proteins in higher eukaryotes from the limited number of genes (Nilsen and Graveley 2010). The Human Genome Project completed more than 10 years ago, and it turned out that over 95% of human genes undergo AS (Pan et al. 2008; Wang et al. 2008). Abnormalities of constitutive and alternative splicing in human may cause dysfunctions or absence of the encoded proteins, which often result in hereditary diseases (Krawczak et al. 1992).

The essential signals for splicing reaction mostly reside in introns (Fig. 8.2) (Wahl et al. 2009). At the 5′ end of intron, which participates in first-step reaction, consensus sequence of GURRGU (R stands for purine) can be found in most of human introns (Fig. 8.2). At the 3′ end of introns, CAG consensus sequence is often discovered (Fig. 8.2). In addition to this sequence, pyrimidine residue stretch precedes to 3′ splice site in order to support recognition of 3′ splice site (Fig. 8.2). A branch point sequence, in which lariat formation occurs by 2′-5′ phosphodiester formation with Guanine residue at 5′ splice site, resides to 20–30 nucleotide upstream of 3′ splice site (Fig. 8.2). Over 10 years have passed since the completion of the Human Genome Project, and it turned out that the average size of exons is 163 nucleotides, whereas the average length of introns is 5849 nucleotides (Zhu et al. 2009). Since most of human introns are much longer than exons, exon recognition is likely a major mode for splicing in human. Many regulatory elements were identified in exons to modulate exon recognition, such as exonic splicing enhancer (ESE) and exonic splicing silencer (ESS) (Fig. 8.2) (Cartegni et al. 2002). ESE is an element to enhance recognition of the exon in which it resides, whereas

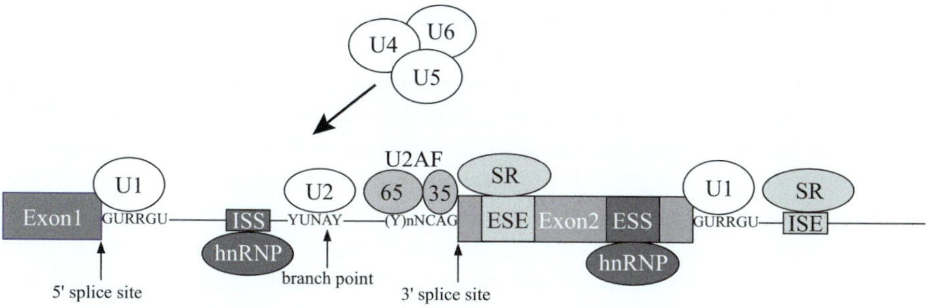

Fig. 8.2 Schematic representation of sequences and factors required for splicing reaction and regulation. Conserved sequence elements of metazoan pre-mRNAs are shown. R and Y stand for purine and pyrimidine residues, respectively. N indicates any nucleotides. Adenosine residue used for branch nucleotide is underlined. U1, U2, U4, U5, U6: U1 snRNP (U1 small nuclear ribonucleoprotein), U2 snRNP, U4 snRNP, U5 snRNP, U6 snRNP. U4, U5, and U6 snRNPs form a tri-snRNP complex. U2AF: U2 small nuclear ribonucleoprotein auxiliary factor. 65: U2AF 65 kD subunit, 35: U2AF 35 kD subunit. *SR* serine-arginine-rich protein, *hnRNP* heterogeneous nuclear ribonucleoprotein, *ESE* exonic splicing enhancer, *ESS* exonic splicing silencer, *ISE* intronic splicing enhancer, *ISS* intronic splicing silencer

ESS inhibits exon recognition that it exists. These elements are also found in introns, which, in turn, are called as intronic splicing enhancer (ISE) and intronic splicing silencer (ISS) (Fig. 8.2).

A large ribonucleoprotein complex, termed spliceosome, carries out splicing reaction (Wahl et al. 2009). The stepwise assembly of spliceosome on pre-mRNA takes place by association of the uridine (U)-rich small nuclear RNPs (snRNPs) (Fig. 8.2) and a multitude of non-snRNP splicing factors (Wahl et al. 2009). Among them, well-characterized splicing regulators are serine-arginine-rich (SR) proteins and heterogeneous nuclear ribonucleoproteins (hnRNPs) (Fig. 8.2). SR proteins are a family of proteins that have one or two RNA-binding domains (RBDs) and arginine-serine-rich (RS) domain at amino-terminus and carboxy-terminus, respectively (Howard and Sanford, 2015; Manley and Krainer 2010). To date, there are 12 SR proteins in human (Manley and Krainer 2010). The hnRNPs are nuclear abundant RNA-binding proteins that have many cellular functions, such as transcription, splicing, RNA transport/localization, translation, and RNA stability (Geuens et al. 2016). The hnRNP family consists of 20 proteins, which are named A1 to U (Geuens et al. 2016). SR proteins are known to bind ESEs and facilitate exon recognition by bridging both sides of the exon through interaction with U2AF heterodimer and U170K proteins via its RS domain (De Conti et al. 2013). In contrast, some of the hnRNP proteins have been demonstrated to bind to ESS (Geuens et al. 2016). For hnRNP A1, it was demonstrated that hnRNP A1 mediates silencing of exon recognition by initially binding to a high-affinity binding site in exon and then promotes further hnRNP A1 association with both upstream and downstream regions of the exon (Zhu et al. 2001). Thus, this causes inhibition of ESE binding of SR proteins. In addition to these regulators, several tissue-specific splicing factors have been identified in order to regulate splicing in a tissue-specific manner (Witten and Ule 2011).

Mutations in those *cis*-elements and *trans*-acting factors, which are required for correct splicing reaction and regulation, cause aberrant splicing that often results in diseases in human. These diseases that have defects in RNA metabolism steps are called as "RNA diseases," and it is estimated that up to 15% of all point mutations causing human genetic disease cause an RNA splicing defect (Krawczak et al. 1992). Continuous and numerous efforts have been made to find cure for those diseases. In this review, I will describe several RNA diseases that have mis-splicing and therapeutic approaches using chemical compounds and oligonucleotides for them.

8.2 Duchenne Muscular Atrophy (DMD)

Duchenne muscular atrophy is an inherited muscle disease. This disease is caused by a mutation in the dystrophin gene, the largest human gene on the X chromosome (Fairclough et al. 2013; Hoffman et al. 1987). Because of progressive muscle wasting, DMD patients usually succumb to cardiac or respiratory failure in their 20s. Becker muscular dystrophy (BMD) is a milder variant of DMD, usually affecting adult males. An open reading frame rule is applicable to explain the difference between DMD and BMD. Nonsense mutations or internal deletions causing frame shifts in the dystrophin mRNAs create premature termination codons (PTCs), which usually result in a severe DMD phenotype, because of a lack of the dystrophin protein (Fairclough et al. 2013; Hoffman et al. 1987). In contrast, mutations/deletions that maintain the original open reading frame in the mRNAs cause the milder BMD phenotype. The reason for this is that a mutated or internally deleted, but still functional, dystrophin protein can be expressed from the mRNA. However, in some mild BMD cases, the patients had nonsense mutations in exons but still produced novel in-frame dystrophin mRNAs by skipping the exons containing the nonsense codon. The current major therapeutic approach is to convert DMD to BMD phenotypes by restoring dystrophin protein open reading frame by inducing exon skipping with antisense oligonucleotides (AONs). Several different AONs have been designed against either splice sites or splicing enhancer elements to induce exon skipping in cells of DMD patients. For example, the AON designed for exon 51 skipping EXONDYS 51 (eteplirsen) had completed clinical trials, and it was approved by FDA in 2016 (Aartsma-Rus and Krieg 2017; Stein 2016). In addition to AONs, considering the therapeutic cost and convenience, small chemical compounds have also been highly awaited.

One chemical compound, TG003, was reported to be a potential candidate for such therapy for DMD by using splicing reporters (Nishida et al. 2011). In one Kobe dystrophin patient (KUCG797), there is a nonsense mutation in exon 31, which replaces GAG for glutamate with the TAG for a stop codon (c.4303G>T, p. Glu1435X) (Nishida et al. 2011). Although the patient was expected no dystrophin

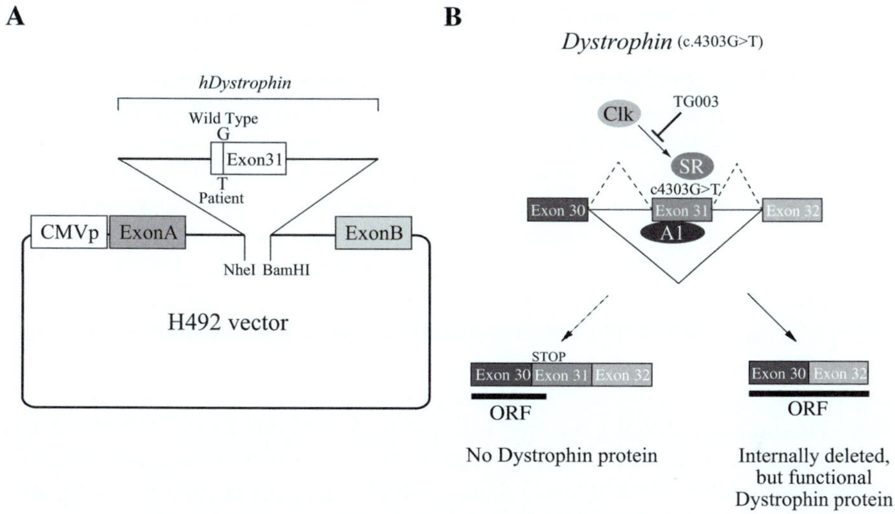

Fig. 8.3 Exon skipping strategy with a kinase inhibitor in human *dystrophin* gene. (**a**) A schematic representation of the minigene plasmids that harbor either the wild-type or mutated exon 31 of the dystrophin gene. The H492 vector encodes two cassette exons (A and B) and an intron sequence containing a multiple cloning site. The dystrophin gene region containing exon 31 with flanking introns was inserted between the NheI and BamHI sites. These plasmids were transfected into HeLa cells, and then the pre-mRNAs were transcribed from the cytomegalovirus (CMV) promoter (CMVp). (**b**) Splicing of c4303G>T mutated human dystrophin gene and modulation with TG003. This mutation disrupts ESE for SR protein and creates hnRNP A1 (A1) binding site. This causes exon 31 exclusion, and TG003 inhibits SR protein phosphorylation that promotes exon 31 skipping. With a stop codon in exon 31, the open reading frame (ORF) is disrupted with the mRNA that includes exon 31. In contrast, from exon 31-skipped mRNA, the internally deleted but functional dystrophin protein is produced

protein production, which results in severe DMD, the immunostaining of a biopsied skeletal muscle showed patchy and discontinuous dystrophin signals with antibodies, which are comparable to BMD (Nishida et al. 2011). In order to explain the discrepancy between the genotype and the dystrophin immunostaining pattern, it was presumed that this nonsense mutation in exon 31 causes skipping of the mutated exon in the patient's muscle cells. To test this possibility, the dystrophin mRNA in the patient skeletal muscle was analyzed by the reverse transcription polymerase chain reaction (RT-PCR) amplification of the region extending from exon 27 to exon 32. In the patient sample, there were two kinds of RT-PCR products. The longer one is the expected size with exons 27–32, and the other smaller product was lacking exon 31 (Nishida et al. 2011). To study this exon skipping, a minigene splicing vector (H492) that contained two exons (exons A and B) and one intron in between encoding a multicloning site was used (Fig. 8.3a) (Tran et al. 2007). Either the wild-type or mutant dystrophin gene fragment that retains exon 31 and the flanking introns was inserted between *Nhe*I and *Bam*HI sites of H492 vector (Fig. 8.3a). These reporter plasmids were transfected into HeLa cells, and the exon31-skipped product was detected with the mutant plasmid, but not with the wild-type plasmid (Nishida et al. 2011). This demonstrated that the chimera splicing reporter could recapitulate exon31 skipping splicing pattern in HeLa cells, so some small chemical compounds that were already demonstrated to affect splicing were adopted for in cell splicing assay with these splicing reporters. Among them, TG003, a specific inhibitor for cdc2-like kinases (CLKs) (Muraki et al. 2004), was shown to induce the mutated exon 31 skipping, but not with the wild-type exon. The c.4303G>T point mutation disrupts ESE element for SRSF9/SRp30c and concomitantly creates ESS element for hnRNP A1 (Nishida et al. 2011). TG003 was originally identified as an inhibitor of Clks, one of the SR protein-specific kinases, and it was shown that TG003 reduces phosphorylation of several SR proteins and causes SR protein subcellular localization from diffused nucleoplasm to speckle localization (Muraki et al. 2004). Although the precise molecular mechanism for TG003 to modulate splicing reaction is uncertain, it is highly likely TG003 reduces SRSF9 access to ESE and/or protein-protein interaction with other splicing factors through RS domain of SRSF9 to enhance exon31 recognition (Fig. 8.3b). Finally, TG003 promoted the expression of an internally deleted but functional dystrophin protein by inducing exon 31 skipping in cells derived carrying the c.4303G>T mutation (Nishida et al. 2011), suggesting that splicing can be a therapeutic target of DMD by small chemical compounds.

8.3 Familial Dysautonomia (FD)

As described above with DMD, a splicing pattern of typical splicing reporter is analyzed by RT-PCR. However, this type of reporter was not suitable for the large-scale screening method to test vast variety of small chemical compounds. For this purpose, splicing reporters using fluorescent proteins were developed, since splicing pattern change can be detected by checking color intensity under microscope. Initially, single-color fluorescence reporters were designed to monitor correct or aberrant splicing and were successfully used in some studies and screenings (Ellis et al. 2004; Sheives and Lynch 2002; Wang et al. 2004). However, expression of the fluorescent protein from monochromatic reporters can be affected by alterations in many steps, such as transcription, mRNA transport, mRNA stability, and translation (MacMorris et al. 1999). Then, multicolor fluorescence splicing reporters were developed to overcome this limitation of single-color reporters. For these reporters, expression of each fluorescent protein from reporter minigenes indicates a specific splicing event, and all reporters are subject to the same cellular influences. Splicing pattern alteration can be detected simply by color change in cells. Many color-splicing reporter systems were developed in a worm, mouse, and cultured cells, and they have been used for identification of several trans-acting factors of alternative splicing (Kuroyanagi et al. 2006; Kuroyanagi et al. 2010; Takeuchi et al. 2010). Recently, this system was applied to the disease gene in order to recapitulate aberrant splicing in familial dysautonomia (FD).

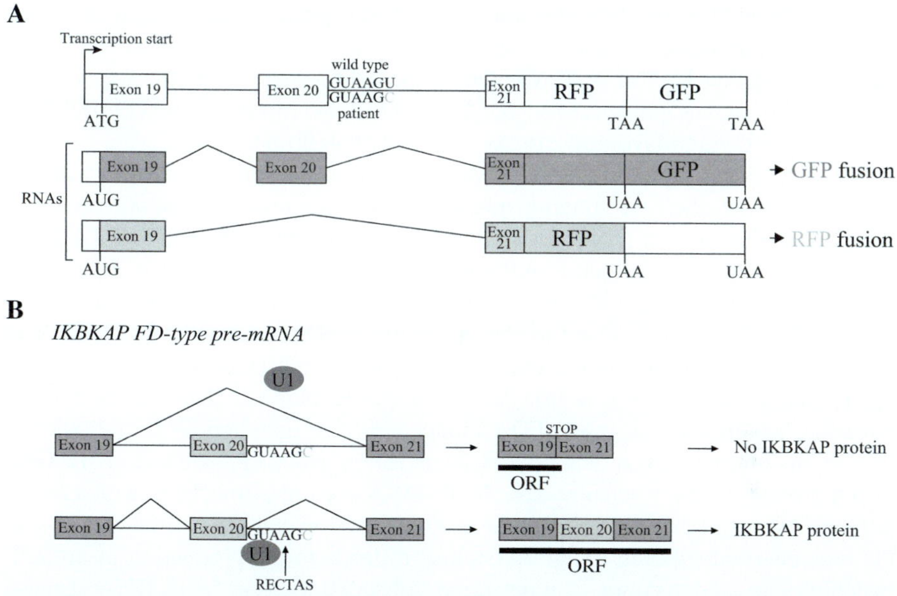

Fig. 8.4 SPREADD reporter and both normal and aberrant splicing patterns of human *IKBKAP* gene. (**a**) Schematic representation of the SPREADD reporter using a portion of human *IKBKAP* gene. The top indicates the reporter structure in SPREADD reporter. With the wild-type reporter, the sequence of the 5′ splice site is GTAAGT, while GTAAGC is a 5′ splice site sequence for FD-type mutant reporter. The middle and the bottom show the mRNAs derived from those reporters schematically. When the exon 20 is included, the ORF matches to that of GFP, resulting in GFP fusion protein production. In contrast, if the exon 20 is excluded, the ORF is continuous to RFP, but not to that of GFP, leading to RFP fusion protein expression. The sixth position nucleotide of 5′ splice site that is mutated in FD is highlighted in gray. (**b**) Normal- and FD-type aberrant splicing of human *IKBKAP* gene and correction by RECTAS. RECTAS promotes association of U1 snRNP to 5′ splice site in intron 20. The position mutated in FD is highlighted by gray. STOP indicates a premature termination codon produced by a frame shift

FD is a hereditary sensory and autonomic neuropathy with progressive degeneration. 99.5% of FD patients carry homozygous mutations at the sixth nucleotide of the 5′ splice site of intron 20 (IVS20 +6T>C) in the inhibitor of kappa light polypeptide gene enhancer in B cells, kinase complex-associated protein (*IKBKAP*) gene (Anderson et al. 2001; Slaugenhaupt et al. 2001). This mutation reduces base pairing with U1 snRNP that causes the skipping of exon 20 (Fig. 8.4b) (Ibrahim et al. 2007). The exon 20 skipping ratio varies in different tissues of FD patients (Cuajungco et al. 2003). The lowest production of exon 20 including *IKBKAP* mRNA is observed in neuronal tissues, and this is thought to be a cause of FD. Since FD mutation is in the intronic sequence, all exonic sequences are same as the wild type. The several attempts to correct *IKBKAP* gene splicing in FD patients have been made, and several chemical compounds were identified to improve exon 20 inclusion, such as tocotrienols (Anderson et al. 2003b), epigallocatechin gallate (EGCG) (Anderson et al. 2003a), phosphatidylserine (Bochner et al. 2013; Keren et al. 2010), and kinetin (Slaugenhaupt et al. 2004). Among them, kinetin was well studied for therapeutic research. Oral administration studies with animals have demonstrated that kinetin is distributed to the central nervous system as well as plasma, and oral administration to FD patients increased full-length *IKBKAP* transcripts without unfavorable effects (Axelrod et al. 2011).

The bichromatic color system was applied to prepare a splicing reporter with FD gene to analyze aberrant splicing mechanism and to search for small chemical compounds that correct splicing. This system was named as SPREADD (*sp*licing *re*porter *a*ssay for *d*isease genes with *d*ual color) (Yoshida et al. 2015). To prepare the wild-type reporter construct, human *IKBKAP* gene fragment containing from exon 19 to exon 21 was fused to the upstream of cDNAs of RFP and GFP. The cDNAs encoding those fluorescent proteins were fused tandemly with different open reading frames (Fig. 8.4a). Another reporter plasmid that carries a point mutation, which is found in the FD patients, was also prepared. With the reporter system, the normal-type splicing (exon 20 inclusion) results in expression of a GFP fusion protein. In this particular case, RFP cDNA portion is translated with a frame different from an original RFP protein without any stop codons. On the other hand, FD-type aberrant splicing (exon 20 exclusion) leads to the production of an RFP fusion protein, and the open reading frame is discontinuous to GFP cDNA because of the stop codon in RFP cDNA (Fig. 8.4a). These *hIKBKAP* splicing reporters were introduced into human neuroblastoma SH-SY5Y cells and human cervix epitheloid carcinoma HeLa cells. When the wild-type reporter was transfected, GFP signal was predominantly observed with both cell lines, whereas RFP signal was mostly detected with the FD-type splicing reporter as expected (Yoshida et al. 2015). The splicing patterns of the reporters were confirmed by RT-PCR, and exon 20 was predominantly included with the WT-type reporter RNA but almost entirely excluded with the FD-type RNA. These results indicated that both WT- and FD-type reporters reflect normal- and abnormal-type splicing, respectively, in culture cells. Screening with SPREADD reporter with mutant *IKBKAP* gene identified a novel chemical compound that fix aberrant splicing of *IKBKAP* gene, which was named as *rect*ifier of *a*berrant *s*plicing (RECTAS) (Fig. 8.4b) (Yoshida et al. 2015). Administration of RECTAS corrected aberrant splicing of *IKBKAP* gene that has a point mutation in intron 20 (Fig. 8.4b) and increased IKBKAP protein expression in primary fibroblasts established from several FD patients with lower concentration compared to the previous promising compound, kinetin (Yoshida et al. 2015). Furthermore, RECTAS recovered the hypomodified levels of uridine residues at the wobble position in several cytoplasmic tRNAs and results in improved cell growth of FD patient fibroblasts (Yoshida et al. 2015). After oral administration, RECTAS was rapidly absorbed and showed higher stability in the blood plasma than kinetin. Furthermore, RECTAS was transported into the brain at sufficient levels to reach the effective dose (Yoshida et al. 2015). These results suggest that RECTAS has therapeutic potential in FD.

8.4 Spinal Muscular Atrophy (SMA)

SMA is one of the common human genetic diseases, and it shows high infantile mortality (Lunn and Wang 2008). SMA is a motor neuron degenerative disease that is caused by deletions or mutations in the *survival of motor neuron* (*SMN*) gene, which is duplicated as an inverted repeat on human chromosome 5 at 5q13 (Lunn and Wang 2008). The telomeric copy is named as *SMN1*, whereas the centromeric one is called *SMN2* (Lefebvre et al. 1995). SMA patients have deletions or mutations in *SMN1* gene, and *SMN2* is nearly identical to *SMN1*, but it fails to compensate *SMN1* function due to one nucleotide difference in the exon 7 and the existence of intronic splicing silencer (ISS) in intron 7 (Fig. 8.5a), which results in exon 7 skipping and production of unstable SMN protein (Singh et al. 2006). The mechanism of *SMN2* exon 7 skipping has been extensively studied (Cartegni and Krainer 2002; Kashima and Manley 2003), and correction of aberrant splicing of

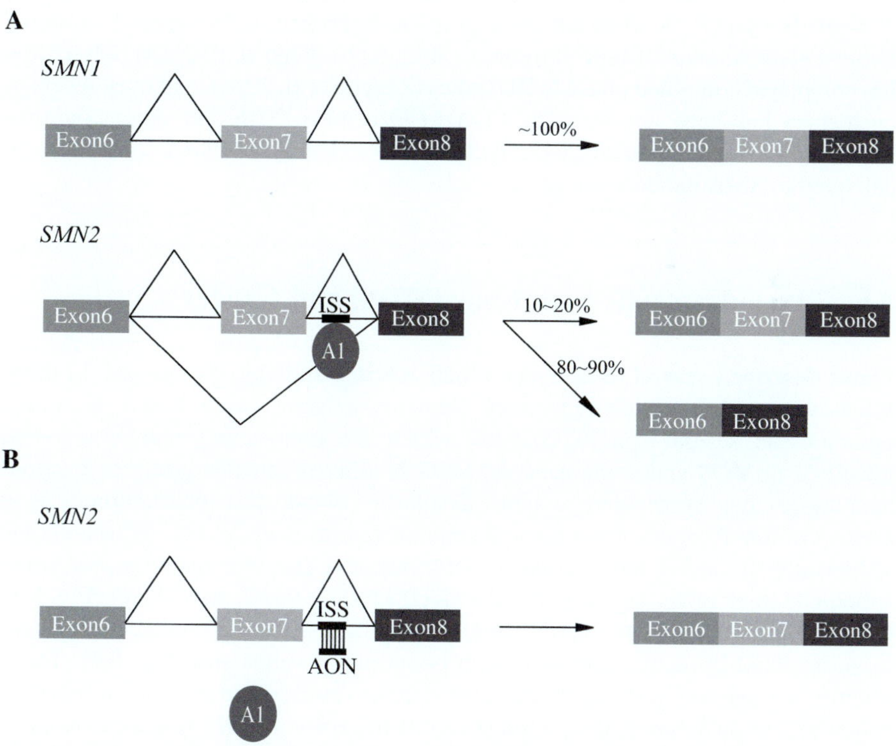

Fig. 8.5 Different splicing patterns of two copies of human *SMN* genes, *SMN1* and *SMN2*. (**a**) Two splicing patterns of human *SMN1* and *SMN2* genes. With *SMN1*, almost 100% mRNAs include exon 7. In contrast, exon 7 is mostly excluded (~80–90%) because of intronic splicing silencer (ISS) in intron 7, which is bound to hnRNP A1(A1). (**b**) A scheme for exon 7 inclusion with *SMN2* gene by using antisense-oligonucleotide (AON) that covers ISS to prevent hnRNP A1 from binding to pre-mRNA

SMN2 gene has been assumed to be a good target for compensating the defect of SMN1. Some of chemical compounds inducing exon 7 inclusion of the *SMN2* gene have been identified. First identified compound was sodium butyrate, known as an inhibitor of histone deacetylase (HDAC) (Sealy and Chalkley 1978), to induce *SMN2* exon 7 inclusion (Chang et al. 2001). It was also shown that sodium butyrate induces the expression of two kinds of SR proteins, resulting in correction of *SMN* mis-splicing (Chang et al. 2001). Aclarubicin (Andreassi et al. 2001) and a tetracycline derivative, PTK-SMA1 (Hastings et al. 2009), were also found to correct *SMN2* exon 7 splicing. Several compounds were also identified to correct aberrant splicing of *SMN2* exon 7 (Seo et al. 2013), and recently high-throughput screening identified NVS-SM1 that stabilizes a transient double-stranded RNA formation, in turn enhancing binding of the U1 snRNP to the 5′ splice site in intron 7 (Palacino et al. 2015).

Among many chemicals, olesoxime and nusinersen exhibited significant efficacy in clinical trials. Olesoxime has a strong neuroprotective compound with a cholesterol-like structure (Bordet et al. 2007). Nusinersen is a uniform 2′-O-methoxyethyl modified antisense drug that hybridizes to ISS in *SMN2* intron 7 to prevent association of hnRNP proteins (Fig. 8.5b) (Rigo et al. 2014). Olesoxime has completed combined phase II/III studies (Zanetta et al. 2014), and most recently, nusinersen has been approved by FDA in December 2016 and other countries including EU and Japan in 2017. Nusinersen, called as SPINRAZA™, will be delivered to SMA patients soon.

8.5 Myelodysplastic Syndrome (MDS) and Cancer

I have described several diseases in which aberrant splicing is observed. In those diseases, mutations in cis-regulatory elements in genes were found. Recently, mutations in several splicing factors, which are trans-acting regulators, were identified in MDS and some cancers. MDS is a heterogeneous group of myeloid neoplasms that show deregulated, dysplastic blood cell production and a predisposition to acute myeloid leukemia (Cazzola et al. 2013). Whole-exome sequencing of paired tumor/control DNA revealed frequent mutations of some splicing factors, including *SF3B1*, *U2AF1(U2AF35)*, *SRSF2*, and *ZRSR2* (Fig. 8.5) (Yoshida et al. 2011). Interestingly, the splicing factors mutated in MDS are likely to be involved in recognition of branch point and 3′ splice site (Fig. 8.6). These findings strongly suggest that deregulated splicing (both constitutive and alternative splicing) with mutated splicing factors results in cancer. SF3B1 is a component of functional U2 snRNP complex, which is required for branch point sequence recognition in an early step of splicing reaction (Will et al. 2002). SF3B1 inhibitors, such as spliceostatin A (SSA) (Kaida et al. 2007), pladienolide-B (Kotake et al. 2007), and GEX1A (Hasegawa et al. 2011), have been identified by a screening of anticancer drugs (Fig. 8.5). Since the SF3B1 mutations found in MDS patients are concentrated in some amino acid residues (i.e., K700E), it was thought that these

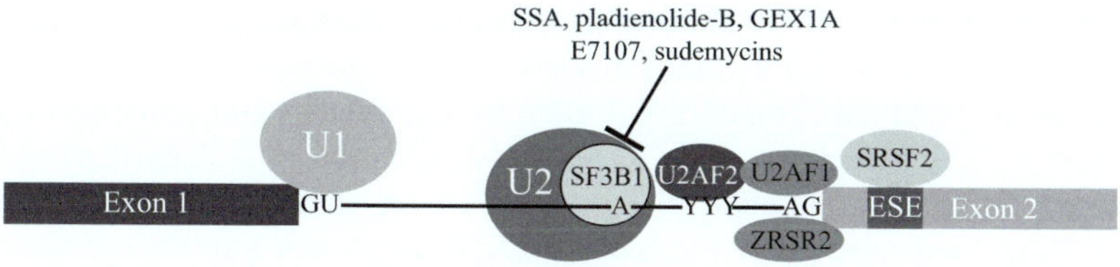

Fig. 8.6 Splicing factors involved in early splicing step are shown schematically with a pre-mRNA. The names of splicing genes whose mutations were observed in myelodysplastic syndrome (MDS) are shown in black letters. Interestingly, these splicing factors are involved in 3′ splice site recognition during splicing reaction. SSA, pladienolide-B, GEX1A E7107, and sudemycins target SF3B. ESE, exonic splicing enhancer; U1, U2, U1 small nuclear ribonucleoprotein, U2 small nuclear ribonucleoprotein

mutations are likely to be gain-of-function mutations (Yoshida et al. 2011). Therefore, it was expected that SF3B1 inhibitors have potentials in treating the myeloid disorders. SF3B1 is an essential splicing factor, thus it was assumed that SF3B1 inhibitors completely block splicing of all pre-mRNAs. However, inhibition of splicing by SSA was unexpectedly partial in cultured cells, and it generated shorter transcripts and carboxy-terminal truncated proteins in tumor cells (Corrionero et al. 2011; Kaida et al. 2007). Pladienolide B is an antitumor macrolide, and it was also reported to interact with SF3b (Kotake et al. 2007). E7107 is a urethane derivative of pladienolide D and exhibited strong antitumor activity (Kotake et al. 2007). It turned out that E7107 blocks spliceosome assembly by preventing stable binding of U2 snRNP to pre-mRNA (Folco et al. 2011). Most recently, it was demonstrated that treatment of isogenic mouse leukemias and patient-derived xenograft AMLs that carry srsf2 point mutation (P95H), also found in MDS, with E7107 resulted in substantial reductions in leukemic burden (Lee et al. 2016). This work provides evidence for both genetics and pharmacology that SRSF2-mutant leukemias are more sensitive to splicing modulation by a chemical compound than SRSF2-WT leukemias. It is highly expected that mutants of splicing factors in MDS (SF3B1, SRSF2, and U2AF35) have less activity to support gene expression patterns required for hematopoiesis compared to wild-type proteins, which contributes to sensitivity to E7107. Novel analogues of E7107 called sudemycins were also demonstrated to exhibit better chemical stability and half maximal inhibitory concentration (IC50) values in tumor cell lines (Fan et al. 2011). It was also recently shown that the cells expressing mutant U2AF1 (S34F), which was also identified in MDS patients (Yoshida et al. 2011), are sensitive to sudemycin and hematopoietic cells expressing this mutant protein show reduced survival and altered cell cycle by sudemycin (Shirai et al. 2017). These results indicate that modulation of spliceosome function by chemical compound may provide a new therapeutic approach with MDS and AML patients who harbor mutations in some splicing factors.

SF3B1 mutations are also detected in solid tumors, including uveal melanoma and breast cancer (Dvinge et al. 2016). It has been demonstrated that those cancer-causing SF3B1 mutations affect branch point selection that results in promotion of splice pattern changes, likely altering 3′ splice site selection (Darman et al. 2015; DeBoever et al. 2015). In this point of view, it was demonstrated that SSA causes recognition of upstream branch point sites by U2 snRNP and splicing of genes important for cell division, such as cyclin A2 and Aurora A kinase, is affected (Corrionero et al. 2011). This explains why SSA has anti-proliferative effects for tumors. E7107 was also shown to block an ATP-dependent remodeling event that results in expose of the branch point-binding region of U2 snRNA to pre-mRNA (Folco et al. 2011). Further comparison of molecular mechanisms for splicing pattern changes between SF3B1 mutations and SF3B1 inhibitor will lead to uncover how MDS and other cancers are caused by aberrant splicing by SF3B1 mutations. It also contributes to understand a precise molecular mechanism of splicing.

8.6 Conclusions

In this review, I introduced some of RNA diseases and chemicals that affect aberrant splicing of those diseases. In the era of high-throughput sequencing technique, many tissue- and disease-specific splicing patterns have been detected. This technique is a powerful tool to identify both disease-causing mis-splicing. Several lines of evidence designate that splicing is a possible target for therapy of inherited diseases and cancers. The other technique, high-throughput screening method using splicing reporters, including H492 and SPREADD, has been established and improved. To apply chemical compounds for clinical studies, it is important and required to select the best compound for each inherited disease or cancer by combining these two techniques. Once the best compound is found, it is required to know how it modulates splicing and what it targets. It will be, of course, essential to evaluate whether they have off-target effects or not with both culture cells and animals before human clinical trials.

Acknowledgments I would like to thank the people in Dr. Shin-Ichiro Takahashi lab (the University of Tokyo) for helpful discussion and comments. This work was supported by Grants-in-Aid for Scientific Research (Grant 23112706) from the Ministry of Education, Culture, Sports, Science and Technology (MEXT) of Japan. I apologize to all scientists whose research could not be properly discussed and cited in this review owing to space limitations.

Conflicts of Interest The author declares no conflict of interest.

References

Aartsma-Rus A, Krieg AM (2017) FDA approves eteplirsen for Duchenne muscular dystrophy: the next chapter in the eteplirsen Saga. Nucleic Acid Ther 27:1–3

Anderson SL, Coli R, Daly IW, Kichula EA, Rork MJ, Volpi SA, Ekstein J, Rubin BY (2001) Familial dysautonomia is caused by mutations of the IKAP gene. Am J Hum Genet 68:753–758

Anderson SL, Qiu J, Rubin BY (2003a) EGCG corrects aberrant splicing of IKAP mRNA in cells from patients with familial dysautonomia. Biochem Biophys Res Commun 310:627–633
Anderson SL, Qiu J, Rubin BY (2003b) Tocotrienols induce IKBKAP expression: a possible therapy for familial dysautonomia. Biochem Biophys Res Commun 306:303–309
Andreassi C, Jarecki J, Zhou J, Coovert DD, Monani UR, Chen X, Whitney M, Pollok B, Zhang M, Androphy E, Burghes AH (2001) Aclarubicin treatment restores SMN levels to cells derived from type I spinal muscular atrophy patients. Hum Mol Genet 10:2841–2849
Axelrod FB, Liebes L, Gold-Von Simson G, Mendoza S, Mull J, Leyne M, Norcliffe-Kaufmann L, Kaufmann H, Slaugenhaupt SA (2011) Kinetin improves IKBKAP mRNA splicing in patients with familial dysautonomia. Pediatr Res 70:480–483
Bochner R, Ziv Y, Zeevi D, Donyo M, Abraham L, Ashery-Padan R, Ast G (2013) Phosphatidylserine increases IKBKAP levels in a humanized knock-in IKBKAP mouse model. Hum Mol Genet 22:2785–2794
Bordet T, Buisson B, Michaud M, Drouot C, Galéa P, Delaage P, Akentieva NP, Evers AS, Covey DF, Ostuni MA, Lacapère JJ, Massaad C, Schumacher M, Steidl EM, Maux D, Delaage M, Henderson CE, Pruss RM (2007) Identification and characterization of cholest-4-en-3-one, oxime (TRO19622), a novel drug candidate for amyotrophic lateral sclerosis. J Pharmacol Exp Ther 322:709–720
Cartegni L, Krainer AR (2002) Disruption of an SF2/ASF-dependent exonic splicing enhancer in SMN2 causes spinal muscular atrophy in the absence of SMN1. Nat Genet 30:377–384
Cartegni L, Chew SL, Krainer AR (2002) Listening to silence and understanding nonsense: exonic mutations that affect splicing. Nat Rev Genet 3:285–298
Cazzola M, Della Porta MG, Malcovati L (2013) The genetic basis of myelodysplasia and its clinical relevance. Blood 122:4021–4034
Chang JG, Hsieh-Li HM, Jong YJ, Wang NM, Tsai CH, Li H (2001) Treatment of spinal muscular atrophy by sodium butyrate. Proc Natl Acad Sci U S A 98:9808–9813
Corrionero A, Minana B, Valcarcel J (2011) Reduced fidelity of branch point recognition and alternative splicing induced by the anti-tumor drug spliceostatin A. Genes Dev 25:445–459
Cuajungco MP, Leyne M, Mull J, Gill SP, Lu W, Zagzag D, Axelrod FB, Maayan C, Gusella JF, Slaugenhaupt SA (2003) Tissue-specific reduction in splicing efficiency of IKBKAP due to the major mutation associated with familial dysautonomia. Am J Hum Genet 72:749–758
Darman RB, Seiler M, Agrawal AA, Lim KH, Peng S, Aird D, Bailey SL, Bhavsar EB, Chan B, Colla S, Corson L, Feala J, Fekkes P, Ichikawa K, Keaney GF, Lee L, Kumar P, Kunii K, MacKenzie C, Matijevic M, Mizui Y, Myint K, Park ES, Puyang X, Selvaraj A, Thomas MP, Tsai J, Wang JY, Warmuth M, Yang H, Zhu P, Garcia-Manero G, Furman RR, Yu L, Smith PG, Buonamici S (2015) Cancer-associated SF3B1 hotspot mutations induce cryptic 3′ splice site selection through use of a different branch point. Cell Rep 13:1033–1045
De Conti L, Baralle M, Buratti E (2013) Exon and intron definition in pre-mRNA splicing. Wiley Interdiscip Rev RNA 4:49–60
DeBoever C, Ghia EM, Shepard PJ, Rassenti L, Barrett CL, Jepsen K, Jamieson CH, Carson D, Kipps TJ, Frazer KA (2015) Transcriptome sequencing reveals potential mechanism of cryptic 3′ splice site selection in SF3B1-mutated cancers. PLoS Comput Biol 11:e1004105
Dvinge H, Kim E, Abdel-Wahab O, Bradley RK (2016) RNA splicing factors as oncoproteins and tumour suppressors. Nat Rev Cancer 16:413–430
Ellis PD, Smith CW, Kemp P (2004) Regulated tissue-specific alternative splicing of enhanced green fluorescent protein transgenes conferred by alpha-tropomyosin regulatory elements in transgenic mice. J Biol Chem 279:36660–36669
Fairclough RJ, Wood MJ, Davies KE (2013) Therapy for Duchenne muscular dystrophy: renewed optimism from genetic approaches. Nat Rev Genet 14:373–378
Fan L, Lagisetti C, Edwards CC, Webb TR, Potter PM (2011) Sudemycins, novel small molecule analogues of FR901464, induce alternative gene splicing. ACS Chem Biol 6:582–589
Folco EG, Coil KE, Reed R (2011) The anti-tumor drug E7107 reveals an essential role for SF3b in remodeling U2 snRNP to expose the branch point-binding region. Genes Dev 25:440–444

Fu XD, Ares M Jr (2014) Context-dependent control of alternative splicing by RNA-binding proteins. Nat Rev Genet 15:689–701
Geuens T, Bouhy D, Timmerman V (2016) The hnRNP family: insights into their role in health and disease. Hum Genet 135:851–867
Hasegawa M, Miura T, Kuzuya K, Inoue A, Won Ki S, Horinouchi S, Yoshida T, Kunoh T, Koseki K, Mino K, Sasaki R, Yoshida M, Mizukami T (2011) Identification of SAP155 as the target of GEX1A (Herboxidiene), an antitumor natural product. ACS Chem Biol 6:229–233
Hastings ML, Berniac J, Liu YH, Abato P, Jodelka FM, Barthel L, Kumar S, Dudley C, Nelson M, Larson K, Edmonds J, Bowser T, Draper M, Higgins P, Krainer AR (2009) Tetracyclines that promote SMN2 exon 7 splicing as therapeutics for spinal muscular atrophy. Sci Transll Med 1:5ra12
Hoffman EP, Brown RH Jr, Kunkel LM (1987) Dystrophin: the protein product of the Duchenne muscular dystrophy locus. Cell 51:919–928
Howard JM, Sanford JR (2015) The RNAissance family: SR proteins as multifaceted regulators of gene expression. Wiley Interdiscip Rev RNA 6:93–110
Ibrahim EC, Hims MM, Shomron N, Burge CB, Slaugenhaupt SA, Reed R (2007) Weak definition of IKBKAP exon 20 leads to aberrant splicing in familial dysautonomia. Hum Mutat 28:41–53
Kaida D, Motoyoshi H, Tashiro E, Nojima T, Hagiwara M, Ishigami K, Watanabe H, Kitahara T, Yoshida T, Nakajima H, Tani T, Horinouchi S, Yoshida M (2007) Spliceostatin A targets SF3b and inhibits both splicing and nuclear retention of pre-mRNA. Nat Chem Biol 3:576–583
Kashima T, Manley JL (2003) A negative element in SMN2 exon 7 inhibits splicing in spinal muscular atrophy. Nat Genet 34:460–463
Keren H, Donyo M, Zeevi D, Maayan C, Pupko T, Ast G (2010) Phosphatidylserine increases IKBKAP levels in familial dysautonomia cells. PLoS One 5:e15884
Kotake Y, Sagane K, Owa T, Mimori-Kiyosue Y, Shimizu H, Uesugi M, Ishihama Y, Iwata M, Mizui Y (2007) Splicing factor SF3b as a target of the antitumor natural product pladienolide. Nat Chem Biol 3:570–575
Krawczak M, Reiss J, Cooper DN (1992) The mutational spectrum of single base-pair substitutions in mRNA splice junctions of human genes: causes and consequences. Hum Genet 90:41–54
Kuroyanagi H, Kobayashi T, Mitani S, Hagiwara M (2006) Transgenic alternative-splicing reporters reveal tissue-specific expression profiles and regulation mechanisms in vivo. Nat Methods 3:909–915
Kuroyanagi H, Ohno G, Sakane H, Maruoka H, Hagiwara M (2010) Visualization and genetic analysis of alternative splicing regulation in vivo using fluorescence reporters in transgenic Caenorhabditis elegans. Nat Protoc 5:1495–1517
Lee Y, Rio DC (2015) Mechanisms and regulation of alternative pre-mRNA splicing. Annu Rev Biochem 84:291–323
Lee SC, Dvinge H, Kim E, Cho H, Micol JB, Chung YR, Durham BH, Yoshimi A, Kim YJ, Thomas M, Lobry C, Chen CW, Pastore A, Taylor J, Wang X, Krivtsov A, Armstrong SA, Palacino J, Buonamici S, Smith PG, Bradley RK, Abdel-Wahab O (2016) Modulation of splicing catalysis for therapeutic targeting of leukemia with mutations in genes encoding spliceosomal proteins. Nat Med 22:672–678
Lefebvre S, Burglen L, Reboullet S, Clermont O, Burlet P, Viollet L, Benichou B, Cruaud C, Millasseau P, Zeviani M, Paslier DL, Frézal J, Cohen D, Weissenbach J, Munnich A, Melki J (1995) Identification and characterization of a spinal muscular atrophy-determining gene. Cell 80:155–165
Lunn MR, Wang CH (2008) Spinal muscular atrophy. Lancet 371:2120–2133
MacMorris MA, Zorio DA, Blumenthal T (1999) An exon that prevents transport of a mature mRNA. Proc Natl Acad Sci U S A 96:3813–3818
Manley JL, Krainer AR (2010) A rational nomenclature for serine/arginine-rich protein splicing factors (SR proteins). Genes Dev 24:1073–1074
Muraki M, Ohkawara B, Hosoya T, Onogi H, Koizumi J, Koizumi T, Sumi K, Yomoda J, Murray MV, Kimura H, Furuichi K, Shibuya H, Krainer AR, Suzuki M, Hagiwara M (2004) Manipulation of alternative splicing by a newly developed inhibitor of Clks. J Biol Chem 279:24246–24254

Nilsen TW, Graveley BR (2010) Expansion of the eukaryotic proteome by alternative splicing. Nature 463:457–463

Nishida A, Kataoka N, Takeshima Y, Yagi M, Awano H, Ota M, Itoh K, Hagiwara M, Matsuo M (2011) Chemical treatment enhances skipping of a mutated exon in the dystrophin gene. Nat Commun 2:308

Palacino J, Swalley SE, Song C, Cheung AK, Shu L, Zhang X, Van Hoosear M, Shin Y, Chin DN, Keller CG, Beibel M, Renaud NA, Smith TM, Salcius M, Shi X, Hild M, Servais R, Jain M, Deng L, Bullock C, McLellan M, Schuierer S, Murphy L, Blommers MJ, Blaustein C, Berenshteyn F, Lacoste A, Thomas JR, Roma G, Michaud GA, Tseng BS, Porter JA, Myer VE, Tallarico JA, Hamann LG, Curtis D, Fishman MC, Dietrich WF, Dales NA, Sivasankaran R (2015) SMN2 splice modulators enhance U1-pre-mRNA association and rescue SMA mice. Nat Chem Biol 11:511–517

Pan Q, Shai O, Lee LJ, Frey BJ, Blencowe BJ (2008) Deep surveying of alternative splicing complexity in the human transcriptome by high-throughput sequencing. Nat Genet 40:1413–1415

Rigo F, Chun SJ, Norris DA, Hung G, Lee S, Matson J, Fey RA, Gaus H, Hua Y, Grundy JS, Krainer AR, Henry SP, Bennett CF (2014) Pharmacology of a central nervous system delivered 2′-O-methoxyethyl-modified survival of motor neuron splicing oligonucleotide in mice and nonhuman primates. J Pharmacol Exp Ther 350:46–55

Sealy L, Chalkley R (1978) The effect of sodium butyrate on histone modification. Cell 14:115–121

Seo J, Howell MD, Singh NN, Singh RN (2013) Spinal muscular atrophy: an update on therapeutic progress. Biochim Biophys Acta 1832:2180–2190

Sheives P, Lynch KW (2002) Identification of cells deficient in signaling-induced alternative splicing by use of somatic cell genetics. RNA 8:1473–1481

Shirai CL, White BS, Tripathi M, Tapia R, Ley JN, Ndonwi M, Kim S, Shao J, Carver A, Saez B, Fulton RS, Fronick C, O'Laughlin M, Lagisetti C, Webb TR, Graubert TA, Walter MJ (2017) Mutant U2AF1-expressing cells are sensitive to pharmacological modulation of the spliceosome. Nat Commun 8:14060

Singh NK, Singh NN, Androphy EJ, Singh RN (2006) Splicing of a critical exon of human survival motor neuron is regulated by a unique silencer element located in the last intron. Mol Cell Biol 26:1333–1346

Slaugenhaupt SA, Blumenfeld A, Gill SP, Leyne M, Mull J, Cuajungco MP, Liebert CB, Chadwick B, Idelson M, Reznik L, Robbins C, Makalowska I, Brownstein M, Krappmann D, Scheidereit C, Maayan C, Axelrod FB, Gusella JF (2001) Tissue-specific expression of a splicing mutation in the IKBKAP gene causes familial dysautonomia. Am J Hum Genet 68:598–605

Slaugenhaupt SA, Mull J, Leyne M, Cuajungco MP, Gill SP, Hims MM, Quintero F, Axelrod FB, Gusella JF (2004) Rescue of a human mRNA splicing defect by the plant cytokinin kinetin. Hum Mol Genet 13:429–436

Stein CA (2016) Eteplirsen approved for Duchenne muscular dystrophy: the FDA faces a difficult choice. Mol Ther 24:1884–1885

Takeuchi A, Hosokawa M, Nojima T, Hagiwara M (2010) Splicing reporter mice revealed the evolutionally conserved switching mechanism of tissue-specific alternative exon selection. PLoS One 5:e10946

Tran VK, Takeshima Y, Zhang Z, Habara Y, Haginoya K, Nishiyama A, Yagi M, Matsuo M (2007) A nonsense mutation-created intraexonic splice site is active in the lymphocytes, but not in the skeletal muscle of a DMD patient. Hum Genet 120:737–742

Wahl MC, Will CL, Luhrmann R (2009) The spliceosome: design principles of a dynamic RNP machine. Cell 136:701–718

Wang Z, Rolish ME, Yeo G, Tung V, Mawson M, Burge CB (2004) Systematic identification and analysis of exonic splicing silencers. Cell 119:831–845

Wang ET, Sandberg R, Luo S, Khrebtukova I, Zhang L, Mayr C, Kingsmore SF, Schroth GP, Burge CB (2008) Alternative isoform regulation in human tissue transcriptomes. Nature 456:470–476

Will CL, Urlaub H, Achsel T, Gentzel M, Wilm M, Luhrmann R (2002) Characterization of novel SF3b and 17S U2 snRNP proteins, including a human Prp5p homologue and an SF3b DEAD-box protein. EMBO J 21:4978–4988

Witten JT, Ule J (2011) Understanding splicing regulation through RNA splicing maps. Trends Genet 27:89–97

Yoshida K, Sanada M, Shiraishi Y, Nowak D, Nagata Y, Yamamoto R, Sato Y, Sato-Otsubo A, Kon A, Nagasaki M, Chalkidis G, Suzuki Y, Shiosaka M, Kawahata R, Yamaguchi T, Otsu M, Obara N, Sakata-Yanagimoto M, Ishiyama K, Mori H, Nolte F, Hofmann WK, Miyawaki S, Sugano S, Haferlach C, Koeffler HP, Shih LY, Haferlach T, Chiba S, Nakauchi H, Miyano S, Ogawa S (2011) Frequent pathway mutations of splicing machinery in myelodysplasia. Nature 478:64–69

Yoshida M, Kataoka N, Miyauchi K, Ohe K, Iida K, Yoshida S, Nojima T, Okuno Y, Onogi H, Usui T, Takeuchi A, Hosoya T, Suzuki T, Hagiwara M (2015) Rectifier of aberrant mRNA splicing recovers tRNA modification in familial dysautonomia. Proc Natl Acad Sci U S A 112:2764–2769

Zanetta C, Nizzardo M, Simone C, Monguzzi E, Bresolin N, Comi GP, Corti S (2014) Molecular therapeutic strategies for spinal muscular atrophies: current and future clinical trials. Clin Ther 36:128–140

Zhu J, Mayeda A, Krainer AR (2001) Exon identity established through differential antagonism between exonic splicing silencer-bound hnRNP A1 and enhancer-bound SR proteins. Mol Cell 8:1351–1361

Zhu L, Zhang Y, Zhang W, Yang S, Chen JQ, Tian D (2009) Patterns of exon-intron architecture variation of genes in eukaryotic genomes. BMC Genomics 10:47

Chapter 9
The Production of Recombinant Proteins from Mammalian Cells Using RNA Element

Intisar Fouad Ali Mursi and Seiji Masuda

Abstract Producing recombinant proteins in a large scale for pharmaceutical use is a challenging process as these proteins must be posttranscriptionally modified. Mammalian cells have proven to be good candidates for this process to take place efficiently. In order to optimize gene expression of the required proteins in mammalian cells, good vectors must be used such as the viral vectors. Vectors must be chosen cautiously according to the type of the mammalian cell line being utilized. Importantly, strong promoters must be selected to ensure large amounts of the gene(s) of interest.

The export of the messenger ribonucleic acid (mRNA) is a complex process in which many proteins are involved. A strategy to enhance recombinant protein production is to use the mRNA export pathway efficiently. In the mRNA export pathway, key proteins include the NXF1-NXT1 heterodimer. Here we introduce the use of constitutive transport element in the expression system. Constitutive transport element directly recruits mRNA export proteins NXF1-NXT1, and these events facilitate the mRNA export containing constitutive transport element. The simultaneous overexpression of mRNA export factors in addition to the use of RNA element recruiting mRNA export proteins is a potential strategy to obtain satisfactory amounts of the required proteins.

Keywords Recombinant protein production · RNA export · RNA element · RNA-binding protein

---

I. F. A. Mursi · S. Masuda (✉)
Division of Integrated Life Sciences, Graduate School of Biostudies, Kyoto University, Kyoto, Japan
e-mail: masuda@kais.kyoto-u.ac.jp

S. Masuda, S. Izawa (eds.), *Applied RNA Bioscience*,
https://doi.org/10.1007/978-981-10-8372-3_9

9.1 Introduction

The production of recombinant proteins to meet the ever-increasing demand of the biopharmaceutical industry is becoming a multibillion-dollar enterprise. The sales of giant pharmaceutical companies are increasing every year. Several methods and strategies have been developed to ensure the production of heterologous proteins with high quality. Several model systems have been devised including both cell-based and cell-free systems. Cell-free systems, in spite of the major and incremental improvements done during the past few decades, still face major challenges (Rosenblum and Cooperman 2014). Nonetheless, they were used to produce several purified proteins, especially for structural studies (Mikami et al. 2006). As for cell-based systems, various bacterial and eukaryotic models for protein production have been developed. In addition to the classical strategies to produce pure recombinant proteins, some strategies have been developed using viral systems for gene therapy. These strategies are becoming more popular than previous. The new concept of using RNA element-mediated strategies to enhance mRNA export also shed a light for recombinant protein production. Here, recently developed strategy as well as classical methods will be introduced.

9.2 Different Cell Systems Are Used for the Production of Recombinant Proteins

9.2.1 Yeast Cell Systems

Among the eukaryotic cell-based systems, yeast and insect cells became more popular alternatives to the bacterial systems. Recombinant expression systems in yeast cells were more common than their more advanced counterparts, the mammalian systems. *Saccharomyces cerevisiae* were still an important cell factory particularly advantageous for biopharmaceutical production because they are generally recognized as safe (GRAS) (Kim et al. 2015). They were used to produce proteins with high molecular weight especially hepatitis B vaccine, serum albumin, and insulin. The global market of hepatitis B vaccine is expected to grow from US $3 billion in 2014 to US $3.5 billion by 2021, that of human serum albumin will reach US $828.2 million by 2020, and that of insulin is estimated to increase from USD 12B in 2011 to more than USD 32B by 2018. An additional advantage of yeast is that in almost all cases, it can accomplish proper posttranslational modifications of the proteins of interest, including disulfide bond formation and glycosylation. One of the major drawbacks of using yeast, however, is that it adds high-mannose-type N-glycosylation to the proteins of interest, therefore rendering them less efficient for therapeutic use (Nielsen 2013).

9.2.2 Insect Cell Systems

Lepidopteran insect cell lines have gained much attention since the 1970s when the baculovirus expression vector was used to produce large amounts of recombinant proteins. Baculoviruses have an excellent cloning capacity and have a good record of biosafety (since they do not infect humans) with perceivably high levels of produced proteins. A similar problem here is that glycosylation in insect cells differs in many attributes from mammalian cells. A major problem when using the baculovirus expression vector system (BEVS) is that it causes cell lysis because the expression is initiated by a strong promotor (p10) and this has tragic effects compromising the quality of the posttranslated proteins (Hu 2005). This problem has been mitigated by substituting the p10 promoter with early baculovirus promoters (e.g., IE1) which results in more efficient production of glycosylated proteins but unfortunately leads to suboptimal expression levels. It's worth mentioning that BEVS are also used to transfect a wide range of mammalian cell lines with varying transfection efficiencies.

9.2.3 Mammalian Cell Systems

While each of the abovementioned cell lines has its own liabilities and advantages, mammalian cell-based systems have gained much interest and therefore undergone major dramatic improvements to satisfy the demands of the biopharmaceutical industry during the past three decades (Almo and Love 2014; Hu and Aunins 1997; Wurm 2004). Popular mammalian cell lines that have been used to produce proteins with great efficiency in maintaining their proper folding include, but are not limited to, Chinese hamster ovary (CHO) cells, HeLa cells, NS0 cells, HEK293T cells, as well as others. CHO cells gained popularity and have been used in research for a long time where an enhanced selectivity system using DHFR was developed. There are several steps which must be done to achieve an industrial scale of protein production; the major steps are shown in Fig. 9.1.

This chapter aims to provide a comprehensive review of the current use of mammalian cell lines, selected vectors used in transfecting the genes of interest, and the strategies used by researchers to improve the yield of expressed proteins. Special interest will be directed toward the latest advances aiming to improve the export mechanisms of mRNA coding for desired proteins.

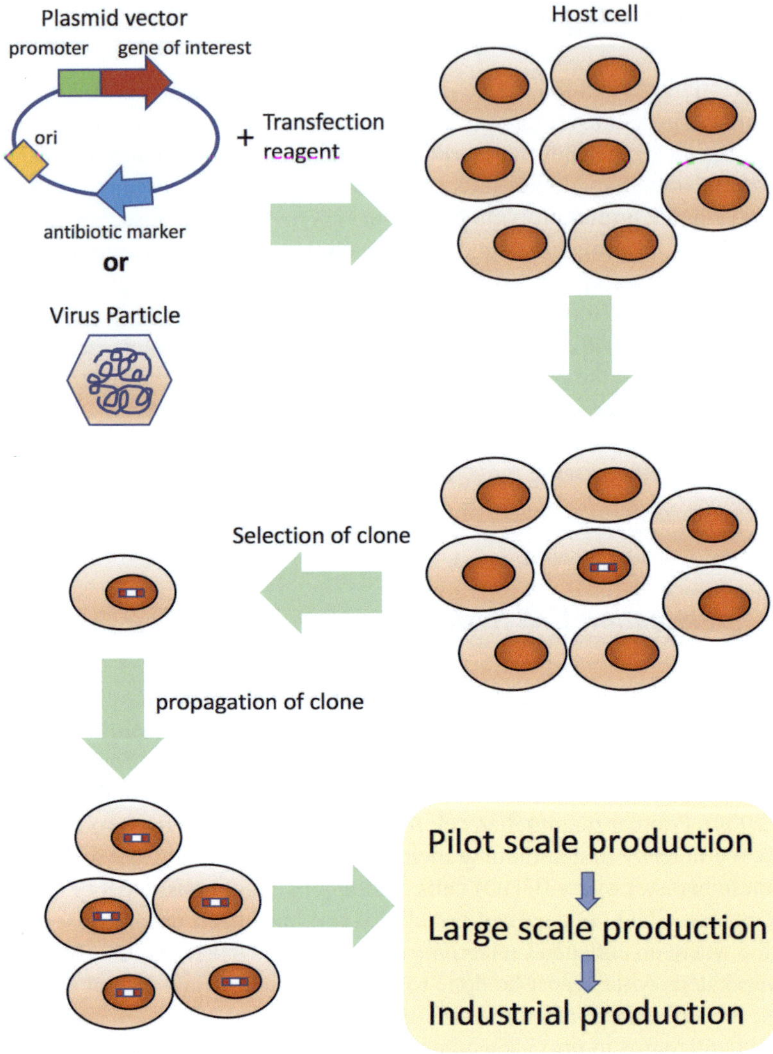

Fig. 9.1 The generation and the development of mammalian cell line culture contain several steps. First, the expression vector which contains the gene of interest is introduced into the cells. Then, the cells are cultured in appropriate cell culture media containing antibiotics. Choose the best clone producing recombinant proteins. Bioreactors which contain optimized cell culture media are used for larger-scale production

9.3 Basics of Recombinant Protein Production

In the recombinant protein production, expression vectors are used. Expression vectors require a promotor, a gene of interest, polyadenylation signals, and multicloning sites to insert the gene of interest. Among them, the promotor is most important because it determines how much transcription occurs. There are two categories. One is the constitutively active promoter. The other is the inducible promoter. The constitutively active promoters are predominantly used for the recombinant protein production in every case including the industrial scale. Alternatively, the inducible promoters are widely used in the laboratory scale.

9.3.1 Constitutively Active Promoters

Gene expression in mammalian cells requires, in addition to carefully chosen vectors, strong promoters. In this section, we will be focusing on the importance of the promoter region exclusively. There are several effective promotors available to design plasmids with both constitutive and inducible systems. The use of constitutive promoters is routinely used to drive the ectopic expression of genes in various in vivo and in vitro contexts. These promoters are introduced into the cells by using liposomes or by being integrated into the cells by virus particles. Promoters consist of a combination of response elements and an RNA polymerase II (RNAP) binding site, but in the case of using lentiviral vectors, viral long terminal repeats can be utilized as RNA promoters (McCarty et al. 2003). Promoters in general can be up to 100–1000 base pairs long. The levels of gene expression using different promoters may vary significantly depending on the type of the cells used and the cell culture conditions (Li et al. 2011). Some promoters are constitutively active, while others require conditional induction by several methods by heat shock or exogenous biotic factors. In spite of their success, the use of promoters which need induction may lead to collateral damage in the form of toxic or pleiotropic side effects on the cells. Below are a few examples of the most widely used constitutively active promoters and a brief account of the uses of each of them and what makes each of them distinctively suitable for certain application while not for others.

9.3.2 Types of Constitutively Active Promoters

9.3.2.1 CMV Promoter

This is most widely and frequently used promoter derived from cytomegalovirus (CMV) which contains immediate early enhancer and promoter for the recombinant protein production in mammalian cells. There are plenty of variations adapted for the recombinant protein production. This promoter was proven to be very efficient in enhancing gene expression both in transient and stable transfections.

9.3.2.2 EF-1a Promoter

The human elongation factor-1 alpha (EF-1 alpha) is also frequently used for the recombinant protein production in mammalian cells. It contains a constitutive promoter of human origin that can be used to drive ectopic gene expression. EF-1 alpha promoter is often useful in conditions where other promoters (such as CMV) have diminished activity or have been silenced (as in embryonic stem cells).

9.3.2.3 Beta Actin Promoter

This is a constitutively active ubiquitous promoter. Attention was drawn to this promoter after some studies in both undifferentiated and mature myoblasts. This promoter contains a recognition sequence of *5′cis* element which forms a DNA-protein complex that is essential for full promoter activity. The binding activity is widely distributed in a variety of cells and tissues and could be functional in the transcriptional activation of numerous promoters containing the factor recognition sequence (Quitschkes et al. 1989). A newly designed fusion promoter of both CMV and beta actin proved to direct higher expression levels of human factor X in mice liver (Xu et al. 2001).

9.3.2.4 UBC Promoter

As the name implies, this is a constitutive mammalian promoter obtained from the human ubiquitin C gene (Schorpp et al. 1996). It is widely known that this promoter results in low levels of expression of target test genes including GFP in different mammalian cell lines such as human embryonic stem cells (hESCs) compared to other promoters (Qin et al. 2010). This promoter is useful to avoid overexpression of transfected genes.

9.3.3 Inducible Promoters

These are promotors which require biotic factors to be added to the culture medium to trigger and/or regulate their transcriptional activity. Several systems are available and the most widely used in this category are the Tet systems.

The early prototypes in inducible systems were achieved in bacterial systems. The *lac* system along with the isopropyl β-D-1-thiogalactopyranoside (IPTG) inducer was tested at the beginning, but it resulted in low levels of induction (Labow et al. 1990). These were replaced by the tetracycline-regulated promoters. The tetracycline-induced systems consist of a tetracycline response element (TRE). A TRE is 7 repeats of a 19-nucleotide tetracycline operator (tetO) sequence. It is recognized by the tetracycline repressor (tetR). This system works by a certain mechanism: if a stable analogue of tetracycline, like doxycycline, is available, the tetR will interact with the tetracycline, but not to the TRE. This results in transcription (Das et al. 2016).

9.4 The Integration of Ectopic Genes into Genome

In the stable expression of ectopic genes, it is of extreme importance that they should be incorporated into the host genomes. To ensure genome integration, several methods were developed. The most frequently used method is the plasmid-based transfection. The plasmid is introduced into the nucleus and integrated into genome after transfection. In the transfection of genes into cells, viral-mediated methods have become more popular recently.

9.4.1 Viral Vectors

Viral vectors are tools which are commonly used to deliver genetic materials into cells and have a long history of applications in basic research, generation of vaccines, and gene therapy (where they have been extensively used). When compared to other methods of gene delivery, viral vectors ensure high percentage of propagation and facilitate stable expression of genes because some viruses integrate their genome within that of the host cell. From an evolutionary point of view, viruses, throughout many steps of evolution, have evolved certain molecular mechanisms to transport their genomes inside the cells they transfect (Deyle and Russell 2009; Schagen et al. 2000). Molecular biologists have exploited this mechanism to transduce genes of interest into cells, and their pioneering trail was done by Paul Berg in the breakthrough paper of when he used an altered simian virus 40 (SV40) which contains *λ phage* gene and the galactose operon in *E. coli* to infect monkey kidney cells (Jackson et al. 1972).

The selection of the proper viral vector is tailored to their specific application, but, in general, they share common features listed below.

1. Cell-type specificity: most viral vectors are engineered to be compatible with various mammalian cell types including transgenic ES cells and different cell types in the nervous system.
2. Identification: viral vectors can contain drug selection “marker” genes which aid in the isolation process of the cells which took up the viral vectors. Examples include blasticidin, hygromycin, zeocin, and others.
3. Stability: some viruses are unstable genetically and are capable of rearranging their genomes rapidly, and this is considered to be very disadvantageous to the reproducibility and predictability of the research data.

Safety and low toxicity: several concerns arise including induced oncogenesis through retrovirus integration and induced immunogenicity. A recent advantage is that current generations of lentiviruses are self-inactivating and consequently compatible with high biosafety levels (Zhou et al. 2010).

4. Viral vectors have gone a long way since the prototypic vectors of retroviruses were initially used and they have evolved to suit a wide array of applications in the areas of vaccine production, gene therapy, and basic research. This part of the book chapter will introduce the different types of viral vectors which are used in the different arenas mentioned above.

9.4.2 *Types of Viral Vectors*

9.4.2.1 Retroviruses

Retroviruses are the cornerstones of gene therapy. They contain a reverse transcriptase to ensure the integration of their genome into the host cell in a stable fashion such as the Moloney murine leukemia virus (MLV) (Varela-Echavarría et al. 1993). The viral vector contains viral and cellular gene promoters (such as CMV, to enhance the expression of genes of interest in the cells) (Kurian et al. 2000). Among some of their potential disadvantages is their inability to infect cells that are not actively dividing; hence, they cannot infect neuronal cells, although some success was achieved with infecting astrocytes using HIV-1-based lentivirus expression vector.

9.4.2.2 Lentiviruses

Lentiviruses are a subclass of retroviruses. On the contrary to other retroviruses, they are competent in integrating themselves into the genomes of nondividing cells, which made them favorable in the viral gene therapy for various neurological disorders (Fassler et al. 2013; Parr-Brownlie et al. 2015; Jakobsson and Lundberg 2006; Li et al. 2012). One of the drawbacks of lentiviruses is the unpredictability of the exact site of their integration into the host genome, and this might cause a problem because the provirus (which contains the viral DNA) can lead to the activation of several oncogenes resulting in tumorigeneses.

9.4.2.3 Adenoviruses

On the contrary to retroviruses and lentiviruses, the DNA of adenoviruses is not replicated during cell division because it is not integrated into the host cell genome, i.e., they are replication-defective and are retained in cells as episomes. This feature limited their use to vaccination and gene therapy. On the other hand, they have advantages such as being capable of infecting different types of both dividing and nondividing cells and their capability of being grown into stable stocks of high titer. One of the drawbacks of adenoviruses is their strong immunogenicity which has been studied extensively in mice models (Hendrickx et al. 2014). This pinpoints that analogous health hazards also occur when using these vectors for gene therapy and vaccination. In almost all of the individuals who received gene therapy using adenoviral vectors, high levels of both $CD4^+$ T and $CD8^+$ T cells were detected (Onion et al. 2007).

9.4.2.4 Adeno-Associated Viruses

Several characteristics make this group of single-stranded DNA viruses appealing to be good vector candidates for stable gene expression (Coura and Nardi 2008). These have low immunogenicity and ability to replicate independently as episomal particles. On the other hand, they can only incorporate up to 5 kb of genetic material which is considered to be small compared to the other viral vectors (Daya and Berns 2008).

9.4.2.5 Self-Complementary Adeno-Associated Viruses (scAAV)

Several researchers have designed the scAAV. It has similar characteristics to its adeno-associated virus (McCarty et al. 2001, 2003). One advantage of this virus vector over the adeno-associated virus is the existence of its genome in the form of double-stranded DNA; therefore it skips the process of synthesizing the second strand. Several studies evaluated the efficiency of using scAAV to improve gene expression in several cell lines, as it was recently proven to be a better alternative the lentiviral vectors for improving gene expression in corneal endothelial cells (Gruenert et al. 2016).

9.5 Messenger Ribonucleic Acid Processing and Export

9.5.1 Overview of the Process

Up to this point, approaches to enhance the recombinant gene expression in mammalian cells were introduced: utilizing different powerful promoters (each one of them has its own advantages and setbacks) and making the introduction of the genes of interest to the cells improved using effective viral vectors. Another important approach to better the expression of ectopic genes is to utilize messenger ribonucleic acid (mRNA) elements. The overexpression of these elements can improve gene expression more powerful. At the beginning of this part of the chapter, we will, initially, shed some light on the process of mRNA processing and export in general and then introduce those elements of interest which can potentially be good targets to enhance the gene expression.

mRNA export is a critical process in gene expression which involves a huge set of proteins to insure the ribonucleoparticles of interest are ready for export through the greatly sophisticated NPC. But, before the export through the nuclear pore complex, export-competent mRNAs should be manufactured. There are several steps to make the pre-mRNA export competent, and these steps are widely known as "the posttranscriptional modifications." Although each of these steps is multifaceted and differs, they are absolutely required before export.

The first step is known as "capping." This process involves the addition of a modified guanine nucleotide to the 5′ of the pre-mRNA immediately after the commencement of transcription. This cap is very important as its sole existence is considered as a critical sign to protect the pre-mRNA from degradation by the ribosomes as well as the initiation of translation by the ribosomes in the cytoplasm. The addition of the 5′ cap happens in a co-transcriptional fashion. Splicing is another important step for editing the pre-mRNA transcript. Splicing occurs co-transcriptionally. In this process, the introns (regions in the pre-mRNA which don't code for proteins) are eliminated, and exons (regions in the pre-mRNA which code for proteins) are joined together. The spliceosome is responsible for splicing. It is a huge RNA-protein complex and consists of five small nuclear ribonucleoproteins (snRNPs): U1, U2, U4, U5, and U6. Splicing happens three dimensionally in space and time as these small subunits take parts to cut the introns out of the ribonucleic transcript; this way of splicing is known as lariat splicing. The last major step to modify the mRNA posttranscriptionally is the polyadenylation. Polyadenylation as the name implies means the addition of a sequence of poly adenosine nucleotides, alternatively known as poly (A) tails at the 3′ end of the mRNA. This tail is important for several reasons: it protects the mRNA from degradation by the exosome and aids in transcription termination, export through the nuclear pore complex, and the initiation of translation in the cytoplasm. There are endonuclease complexes which cut the mRNA free at known sites, and the poly (A) tail, which is in average 250 nucleotides, is attached mainly through the action of the poly (A) polymerase enzyme. The different steps in the mRNA metabolism and export are shown in Fig. 9.2.

The abovementioned description is a brief account for the main steps of mRNA processing. Nonetheless, several reviews have addressed the steps and the various mechanisms involved in them in details (Köhler and Hurt 2007; Wickramasinghe and Laskey 2015). It is also worth mentioning that the numerous, diverse events of modifying pre-mRNA are tightly linked. An increasing body of knowledge during the past decade endorsed the importance of coupling the splicing process with export and that is because several proteins which function in splicing interact strongly with the export machinery (Reed 2003).

When the mRNA has undergone all these modifications successfully, and hasn't been degraded by endo- or exonucleases, it can be assumed to be export-competent. The mRNA export requires a family of RNP conserved proteins from yeast to mammals which bind and modify the mRNA cargo. Because the export process is

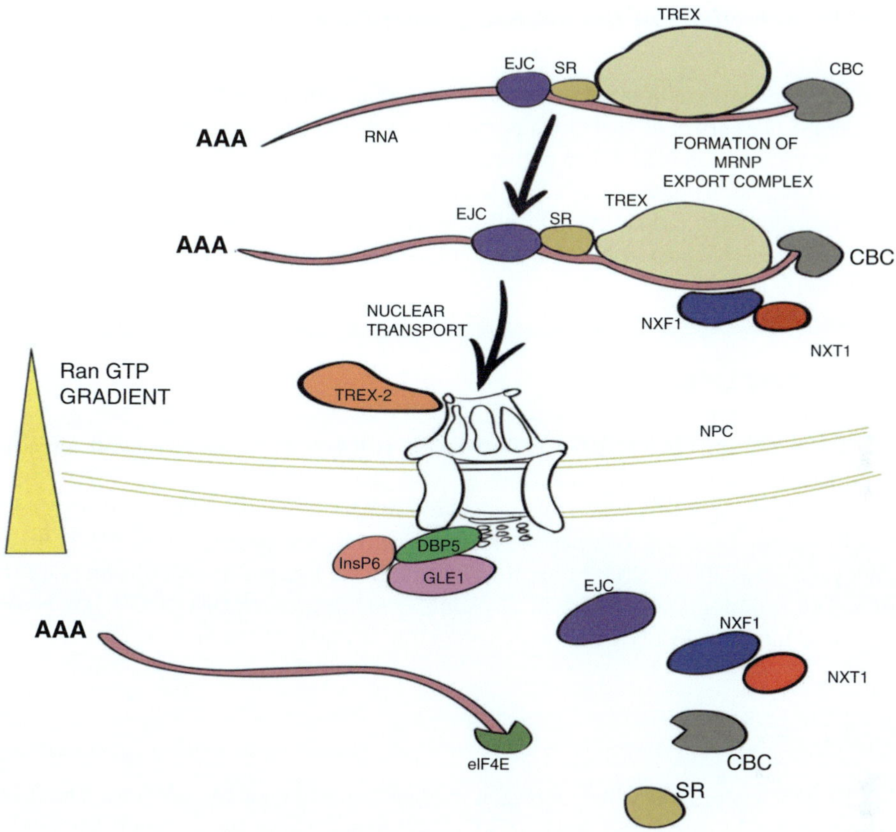

Fig. 9.2 The mRNA export pathway in cell. There are different posttranscriptional modifications the messenger ribonucleoprotein particle (mRNP) undergoes. Capping is achieved by the attachment of the cap binding complex (CBC) while the exon junction complex (EJC) just after the splicing. The mRNP is also modeled via the recruitment of the TREX and TREX-2 complexes, as well as other export-facilitating proteins. NXF1 and its cofactor, p15, are recruited to the mature mRNP

complicated, several complexes are involved in it. The transcription export complexes (known as TREX and TREX-2 complexes) play an important role in the export of mRNA. They are composed of several proteins (e.g., THOC1, THOC2, UAP56) as well as many others (Strässer et al. 2002; Masuda et al. 2005; Reed and Cheng 2005; Katahira 2012). Any defects or genetic modifications in any of the proteins which compose these complexes result in the accumulation of mRNA in the nucleus. This indicates the importance of a competent mRNA export process and that it is considered to be an important part of the cell homeostasis.

9.5.2 Selectivity of the mRNA Export Pathway

mRNA transcripts require a specialized export machinery which was previously covered (Okamura et al. 2015). At present, two export pathways are identified. The primary pathway is NXF1/Tap pathway. It is required for the bulk mRNA export. The second export pathway is CRM1/XPO1 pathway. The specific set of mRNAs are exported using CRM1/XPO1 pathway. Below is a brief account for the major effectors in mRNA export in eukaryotic cells together with recent research advances in the field.

9.5.2.1 NXF1/TAP

It is known as the nuclear RNA export factor 1. It is a major and essential mRNA export factor which acts by attaching to fully processed mRNAs (Fig. 9.3). Upon the interaction of NXF1 with fully processed mRNAs, the export of the transcripts bearing them becomes competent. NXF1 forms a heterodimer with NXT1/p15 (Wiegand et al. 2002). NXF1 functions in bulk mRNA export together with NXT1/p15. The orthologue of NXF1 in *Drosophila melanogaster* DM NXF1 has been shown to interact closely with the cytoskeleton elements (Mamon et al. 2017).

9.5.2.2 NXT1/p15

NXT1/p15 is a protein which acts as a nuclear export factor for mRNAs. The discovery of NXT1 came after that of NXF1/TAP. Studies have shown that NXF1/TAP is activated by coupling with this NXT1. This protein has been shown to affect the export of many tissue-specific mRNAs in various species (Caporilli et al. 2013).

9.5.2.3 REF/ALY

It is heat-stable adapter protein for mRNA export. It is a member of the TREX complex. It functions in several areas of RNA metabolism including transcription, stability, and export. While its role in mRNA export is extensively investigated, recent evidence shows that it affects the fate of nascent mRNA upstream to its export.

9.5.2.4 SR Proteins

These are abundant, revolutionary conserved phosphoproteins which regulate several steps in mRNA metabolism (Jeong 2017). Upon hyperphosphorylation, they are recruited to the splicing machinery; ex post facto, they become hypophosphorylated allowing the binding of NXT1/NXF1 (Kędzierska and Piekiełko-Witkowska 2017; Müller-Mcnicoll et al. 2016). SR proteins shuttle between the nucleus and cytoplasm.

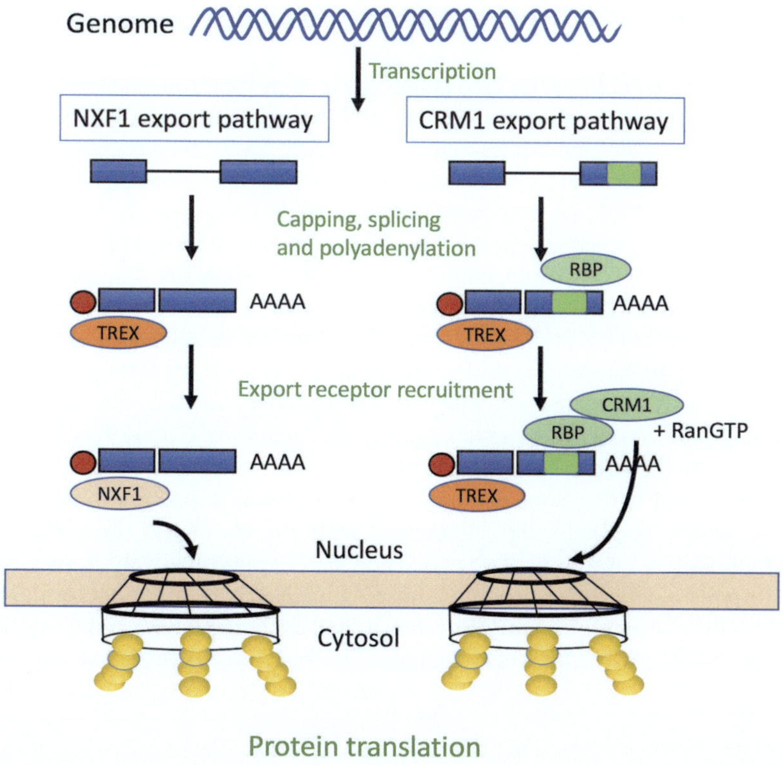

Fig. 9.3 The mechanism of mRNA export. This is a simplified diagram showing the main proteins which are considered to be targets for mRNA export. Introns are removed by the splicing process. The heterodimer NXF1-NXT1 binds to fully processed mRNA. NXF1-NXT1 has a role in bulk mRNA export. There is another mechanism by which mRNA can be exported in eukaryotic cells. This pathway requires unique RNA element and its binding protein(s). Binding of RNA binding protein(s) triggers the recruitment of CRM1, a protein exporter, to the unique RNA element. Such mRNA is exported by the CRM1

9.5.2.5 The Exon Junction Complex

The exon junction complex (EJC) is a complex of proteins that is installed on mRNA (~20–24 nucleotides upstream of exon-exon junctions) during splicing. It is composed of core proteins such as EIF4A3 (DDX48), RBM8A (Y14), and MAGOH and several associated proteins. This complex remains bound to mRNA during export. The exon junction complex is critical to many processes in mRNA metabolism including the distribution of the newly spliced mRNA, mRNA surveillance, and translation as well as export. The formation of EJC is a signal that splicing has proceeded correctly.

9.5.2.6 CRM1/XPO1

It is a member of the importin beta superfamily of nuclear transport receptors and an export adapter of specific sets of mRNAs. It was initially identified in *Saccharomyces cerevisiae* (Hutten and Kehlenbach 2007). CRM1/XPO1 can bind substantially to various leucine-rich nuclear export signals (NES) on target proteins (Köhler and Hurt 2007; Koyama and Matsuura 2012; Fung et al. 2017). Therefore, it is primarily recognized as a protein exporter. However, it also supports the export of a specific set of RNAs which contains unique RNA elements. The association of RNA binding protein(s) with NES signals onto specific RNA element triggers the recruitment of CRM1/XPO1 to RNA via RNA binding protein(s). It exports RNA molecule via RNA binding protein(s). While it assists in the export of various transcripts in strictly regulated manner, CRM1-dependent export pathway becomes overly active in various metastatic cancers, and therefore, CRM1 has become an enticing target for several anticancer therapies (Turner et al. 2012; Turner et al. 2014; Higby et al. 2017; Shao et al. 2017). CRM1 forms a dimer with an extensive interface which heightens the interaction with the Rev-RRM in a Ran GTP-dependent manner (Booth et al. 2014). The CRM1 export pathway is very crucial and is responsible for the transport of subset of mRNA. It is important for the transport of a subset of important mRNAs such as cell cycle regulators and transcription factors, albeit NXF1 mediates the transport of the bulk mRNA (Delaleau and Borden 2015).

9.6 Enhancing Protein Production Using Virus *cis* RNA Element

9.6.1 Packaging of Full-Length RNA Genome in Viruses

Several of the recent advances in the RNA export machinery was initiated by pioneering works on viruses (especially RNA viruses) which manage to replicate their genome through a proviral DNA intermediate that is initially synthesized by reverse transcriptase and integrated in the host genome. This proviral intermediate is an initial genome-length version of RNA that is deferentially spliced to yield full, partial, and unspliced transcripts which serve for various biological functions (Sandri-Goldin 2004). This mechanism is in complete divergence with that adopted in cellular systems where several thousands of transcripts are synthesized coding for a colossal number of proteins (Kornblihtt et al. 2013; Roy et al. 2013; Lee and Rio 2015; Raj and Blencowe 2015). The ability of viruses to export those "differentially" spliced RNA got several research groups to study splicing in a much concrete approach. During the early stages of the research on mRNA splicing and export, the conventional belief in the field was that introns should be removed from pre-mRNA; otherwise, the newly synthesized transcripts would be retained in the nucleus and quickly degraded. In virus-infected cells, these are unspliced genomic RNAs. To avoid the degradation of full-length genomic viral RNAs with intron from the cellular defense system, viruses have evolved various mechanisms to ensure the export of these intron containing transcripts which include cis-acting RNA elements and RNA export regulatory proteins (Pasquinelli et al. 1997). One of such elements is the constitutive transport element (CTE). Type D retroviruses such as Mason-Pfizer monkey virus contain a conserved *cis* RNA element called CTE that is necessary for virus replication (Bray et al. 1994; Tabernero et al. 1996; Zolotukhin et al. 2001). CTE recruits NXF1/Tap directly its RNA sequence. CTE-containing RNA(s) can be exported to cytosol with or without introns. Several studies followed this research pattern to prove that the viral genomic RNA (gRNA) is transcribed by polymerase II, differentially spliced and polyadenylated, and exported to the cytoplasm afterward to act as viral messenger RNAs encoding for both structural proteins and genetic substrates.

9.6.2 *Strategies to Enhance Recombinant Protein Production in Mammalian Cells*

Several reviews discussed the various ways in which the expression of recombinant proteins can be enhanced in mammalian cells (Khan 2013; Li et al. 2013; Hacker and Balasubramanian 2016). In the last part of the chapter, we will focus on the expression of recombinant proteins using virus RNA element.

The most commonly used strategy to enhance recombinant protein production is the use of strong promoters (i.e., designing effective plasmids) as described in the former part. Here, we describe the use of cis RNA element and its binding partner or support proteins which will support better mRNA export that contains cis RNA element in addition to the effect of strong promoter (Fig. 9.4). The addition of CTE element to the 3′ UTR portion of target genes (luciferase, galactosidase, and erythropoietin) efficiently produced more recombinant proteins than the expression vector without CTE in all human cell lines tested (Aihara et al. 2011). This is the compelling finding because erythropoietin used in this article is used as pharmaceutical agent. The forced expression of NXF1-NFT1 to this expression system can be expected to promote target mRNA export and thereafter recombinant protein production. The forced expression of NXF1-NFT1 associating proteins like ARY/REF or SR proteins also promotes target mRNA export and thereafter recombinant protein production. This is explained by the reason/notion that more expression of these associating factors supports more NXF1-NFT1 recruitment onto target mRNA and results in enhanced protein production. Combing all of the three abovementioned approaches will result in enhanced expression of recombinant proteins (Fig. 9.4). These indicate that mRNAs containing CTE rapidly arrived at cytoplasm after its transcription and act as a template for protein translation. This achievement makes us to examine in CHO cells. As previously mentioned in the beginning of this chap-

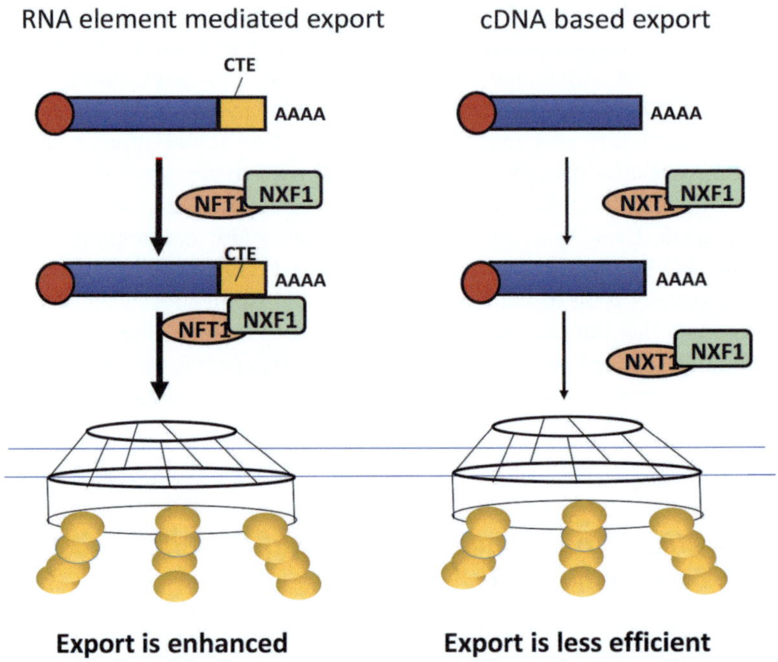

Fig. 9.4 Methods for enhancing recombinant proteins in mammalian cells. The genes coding for the proteins of interest are introduced to the cells in the form of a complementary DNA (cDNA) using a suitable vector. The cDNA is a synthesized form of a mRNA which codes for a specific protein of interest. In order to enhance the export of these carefully constructed genetic materials, the addition of CTE will have a positive effect. Additional expression of the mRNA exporter, NXF1-NXT1 heterodimer, further supports the efficient export of CTE-containing mRNA. Therefore, the incorporation of the CTE into the expression vector synergistically with the overexpression of NXF1-NXT1 results in improved export and consequently higher amount of the desired recombinant protein in human cell lines

ter, production of the recombinant proteins using the Chinese hamster ovary (CHO) cells becomes the major stream in mammalian cell lines in an industrial scale; this strategy should adopt CHO cell lines. Unfortunately, CTE element did not work well in CHO cells probably because of the deficiency of the interaction of hamster NXF1 to virus CTE element. In the next step, the production system using RNA element needs to be adopted in CHO cells with the combined strategies. It will achieve satisfactory levels of the desired protein of interest even in CHO cells.

9.7 Conclusion

Gene expression to produce recombinant proteins is still undergoing major changes and is on the brink of becoming an autonomous production system. Several strategies can be adopted, either solely or in combination to enhance the expression of genes. Either way, the condition of the mammalian cell line in use must be closely monitored to avoid any possible interference with any of the steps involved in the gene expression.

Acknowledgments We would like to express our thanks to Lwando Moshani from Kyoto Seika University, Japan, for creating some of the drawings for this book chapter. This work was partially supported by JSPS KAKENHI Grant Number 17K19232 to SM.

References

Aihara Y, Fujiwara N, Yamazaki T, Kambe T, Nagao M, Hirose Y, Masuda S (2011) Enhancing recombinant protein production in human cell lines with a constitutive transport element and mRNA export proteins. J Biotechnol 153:86–91

Almo SC, Love JD (2014) Better and faster: improvements and optimization for mammalian recombinant protein production. Curr Opin Struct Biol 26:39–43

Booth DS, Cheng Y, Frankel AD (2014) The export receptor Crm1 forms a dimer to promote nuclear export of HIV RNA. elife 3:e04121

Bray M, Prasad S, Dubay JW, Hunter E, Jeang KT, Rekosh D, Hammarskjold ML (1994) A small element from the Mason-Pfizer monkey virus genome makes human immunodeficiency virus type 1 expression and replication rev-independent. Proc Natl Acad Sci U S A 91:1256–1260

Caporilli S, Yu Y, Jiang J, White-Cooper H (2013) The RNA export factor, Nxt1, is required for tissue specific transcriptional regulation. PLoS Genet 9:e1003526

Coura S, Nardi NB (2008) A role for adeno-associated viral vectors in gene therapy. Genet Mol Biol 31:1–11

Das AT, Zhou X, Metz SW, Vink MA, Berkhout B (2016) Selecting the optimal Tet-On system for doxycycline-inducible gene expression in transiently transfected and stably transduced mammalian cells. Biotechnol J 11:71–79

Daya S, Berns KI (2008) Gene therapy using adeno-associated virus vectors. Clin Microbiol Rev 21:583–593

Delaleau M, Borden KLB (2015) Multiple export mechanisms for mRNAs. Cell 4:452–473

Deyle DR, Russell DW (2009) Adeno-associated virus vector integration. Curr Opin Mol Ther 11:442–447

Fassler M, Weissberg I, Levy N, Diaz-Griffero F, Monsonego A, Friedman A (2013) Preferential lentiviral targeting of astrocytes in the central nervous system. PLoS One 8:e76092

Fung HYJ, Fu S, Chook YM (2017) Nuclear export receptor CRM1 recognizes diverse conformations in nuclear export signals. elife 6:e23961

Gruenert AK, Czugala M, Mueller C, Schmeer M, Schleef M (2016) Self-complementary adeno-associated virus vectors improve transduction efficiency of corneal endothelial cells. PLoS One 11:e0152589

Hacker DL, Balasubramanian S (2016) Recombinant protein production from stable mammalian cell lines and pools. Curr Opin Struct Biol 38:129–136

Hendrickx R, Stichling N, Koelen J (2014) Innate immunity to adenovirus. Hum Gene Ther 25:265–284

Higby KJ, Bischak MM, Campbell CA, Anderson RG, Broskin SA, Foltz LE, Koper JA, Nickle AC, Resendes KK (2017) 5-Flurouracil disrupts nuclear export and nuclear pore permeability in a calcium dependent manner. Apoptosis 22:393–405
Hu Y (2005) Baculovirus as a highly efficient expression vector in insect and mammalian cells. Acta Pharmacol Sin 26:405–416
Hu WS, Aunins JG (1997) Large-scale mammalian cell culture. Curr Opin Biotechnol 8:148–153
Hutten S, Kehlenbach RH (2007) CRM1-mediated nuclear export: to the pore and beyond. Trends Cell Biol 17:193–201
Jackson DA, Symonst RH, Berg P (1972) Biochemical method for inserting new genetic information into DNA of Simian Virus 40: circular SV40 DNA molecules containing lambda phage genes and the galactose operon of *Escherichia coli*. Proc Natl Acad Sci U S A 69:2904–2909
Jakobsson J, Lundberg C (2006) Lentiviral vectors for use in the central nervous system. Mol Ther 13:484–493
Jeong S (2017) SR proteins : binders, regulators, and connectors of RNA. Mol Cells 40:1–9
Katahira J (2012) mRNA export and the TREX complex. Biochim Biophys Acta 819:507–513
Kędzierska H, Piekiełko-Witkowska A (2017) Splicing factors of SR and hnRNP families as regulators of apoptosis in cancer. Cancer Lett 396:53–65
Khan KH (2013) Gene expression in mammalian cells and its applications. Adv Pharm Bull 3:257–263
Kim H, Yoo SJ, Kang HA (2015) Yeast synthetic biology for the production of recombinant therapeutic proteins. FEMS Yeast Res 15:1–16
Köhler A, Hurt E (2007) Exporting RNA from the nucleus to the cytoplasm. Nat Rev Mol Cell Biol 8:761–773
Kornblihtt AR, Schor IE, Alló M, Dujardin G, Petrillo E, Muñoz MJ (2013) Alternative splicing : a pivotal step between eukaryotic transcription and translation. Nat Rev Mol Cell Biol 14:153–166
Koyama M, Matsuura Y (2012) Mechanistic insights from the recent structures of the CRM1 nuclear export complex and its disassembly intermediate. Biophysics 8:145–150
Kurian KM, Watson CJ, Wyllie AH, Currie SA (2000) Retroviral vectors. Mol Pathol 53:173–176
Labow MA, Baim SB, Shenk T, Levine AJ (1990) Conversion of the lac repressor into an allosterically regulated transcriptional activator for mammalian cells. Mol Cell Biol 10:3343–3356
Lee Y, Rio DC (2015) Mechanisms and regulation of alternative pre-mRNA splicing. Annu Rev Biochem 84:291–323
Li M, Husic N, Lin Y, Christensen H, Malik I, McIver S, LaPash Daniels CM, Harris DA, Kotzbauer PT, Goldberg MP, Snider BJ (2011) Optimal promoter usage for lentiviral vector-mediated transduction of cultured central nervous system cells. J Neurosci Methods 189:56–64
Li M, Husic N, Lin Y, Snider BJ (2012) Production of lentiviral vectors for transducing cells from the central nervous system. J Vis Exp 24:E4031
Li Z, Michael IP, Zhou D, Nagy A, Rini JM (2013) Simple piggyBac transposon-based mammalian cell expression system for inducible protein production. Proc Natl Acad Sci U S A 110:5004–5009
Mamon LA, Ginanova VR, Kliver SF, Yakimova AO, Atsapkina AA, Golubkova EV (2017) RNA-binding proteins of the NXF (nuclear export factor) family and their connection with the cytoskeleton. Cytoskeleton (Hoboken) 74:161–169
Masuda S, Das R, Cheng H, Hurt E, Dorman N, Reed R (2005) Recruitment of the human TREX complex to mRNA during splicing. Genes Dev 19:1512–1517
McCarty DM, Monahan PE, Samulski RJ (2001) Self-complementary recombinant adeno-associated virus (scAAV) vectors promote efficient transduction independently of DNA synthesis. Gene Ther 8:1248–1254
McCarty DM, Fu H, Monahan PE, Toulson CE, Naik P, Samulski RJ (2003) Adeno-associated virus terminal repeat (TR) mutant generates self-complementary vectors to overcome the rate-limiting step to transduction in vivo. Gene Ther 10:2112–2118

Mikami S, Masutani M, Sonenberg N, Yokoyama S, Imataka H (2006) An efficient mammalian cell-free translation system supplemented with translation factors. Protein Expr Purif 46:348–357

Müller-Mcnicoll M, Botti V, Domingues AMDJ, Brandl H, Schwich OD, Steiner MC, Curk T, Poser I, Zarnack K, Neugebauer KM (2016) SR proteins are NXF1 adaptors that link alternative RNA processing to mRNA export. Genes Dev 30:553–566

Nielsen J (2013) Production of biopharmaceutical proteins by yeast: advances through metabolic engineering. Bioengineered 4:207–211

Okamura M, Inose H, Masuda S (2015) RNA export through the NPC in eukaryotes. Genes 6:124–149

Onion D, Crompton LJ, Milligan DW, Moss PA, Lee SP, Mautner V (2007) The CD4+ T-cell response to adenovirus is focused against conserved residues within the hexon protein. J General Virol 88:2417–2425

Parr-Brownlie LC, Bosch-Bouju C, Schoderboeck L, Sizemore RJ, Abraham WC, Hughes SM (2015) Lentiviral vectors as tools to understand central nervous system biology in mammalian model organisms. Front Mol Neurosci 8:14

Pasquinelli AE, Ernst RK, Lund E, Grimm C, Zapp ML, Rekosh D, Hammarskjöld M, Dahlberg JE (1997) The constitutive transport element (CTE) of Mason-Pfizer monkey virus (MPMV) accesses a cellular mRNA export pathway. EMBO J 16:7500–7510

Qin JY, Zhang L, Clift KL, Hulur I, Xiang AP, Ren BZ, Lahn BT (2010) Systematic comparison of constitutive promoters and the doxycycline-inducible promoter. PLoS One 5:e10611

Quitschkes WW, Lin Z, Deponti-Zilli L, Paterson BM, Tris M (1989) The beta actin promoter. J Biol Chem 264:9539–9546

Raj B, Blencowe BJ (2015) Alternative splicing in the mammalian nervous system : recent insights into mechanisms and functional roles. Neuron 87:14–27

Reed R (2003) Coupling transcription, splicing and mRNA export. Curr Opin Cell Biol 15:326–331

Reed R, Cheng H (2005) TREX, SR proteins and export of mRNA. Curr Opin Cell Biol 17:269–273

Rosenblum G, Cooperman BS (2014) Engine out of the chassis: cell-free protein synthesis and its uses. FEBS Lett 588:261–268

Roy B, Haupt LM, Griffiths LR (2013) Alternative splicing (AS) of genes as an approach for generating protein complexity. Curr Genomics 14:182–194

Sandri-goldin RM (2004) Viral regulation of mRNA export. J Virol 78:4389–4396

Schagen FH, Rademaker HJ, Fallaux FJ, Hoeben RC (2000) Insertion vectors for gene therapy. Gene Ther 7:271–272

Schorpp M, Jäger R, Schellander K, Schenkel J, Wagner EF, Weiher H, Angel P (1996) The human ubiquitin C promoter directs high ubiquitous expression of transgenes in mice. Nucleic Acids Res 24:1787–1788

Shao WY, Yang YL, Yan H, Huang Q, Liu KJ, Zhang S (2017) Phenethyl isothiocyanate suppresses the metastasis of ovarian cancer associated with the inhibition of CRM1-mediated nuclear export and mTOR-STAT3 pathway. Cancer Biol Ther 18:26–35

Strässer K, Masuda S, Mason P, Pfannstiel J, Oppizzi M, Rodriguez-Navarro S, Rondón AG, Aguilera A, Struhl K, Reed R, Hurt E (2002) TREX is a conserved complex coupling transcription with messenger RNA export. Nature 417:304–308

Tabernero C, Zolotukhin AS, Valentin A, Pavlakis GN, Felber BK (1996) The posttranscriptional control element of the simian retrovirus type 1 forms an extensive RNA secondary structure necessary for its function. J Virol 70:5998–6011

Turner JG, Dawson J, Sullivan DM (2012) Nuclear export of proteins and drug resistance in cancer. Biochem Pharmacol 83:1021–1032

Turner JG, Dawson J, Cubitt CL, Baz R, Sullivan DM (2014) Inhibition of CRM1-dependent nuclear export sensitizes malignant cells to cytotoxic and targeted agents. Semin Cancer Biol 27:62–73

Varela-Echavarría A, Prorock CM, Ron Y, Dougherty JP (1993) High rate of genetic rearrangement during replication of a Moloney murine leukemia virus-based vector. J Virol 67:6357–6364
Wickramasinghe VO, Laskey RA (2015) Control of mammalian gene expression by selective mRNA export. Nat Rev Mol Cell Biol 16:431–442
Wiegand HL, Coburn GA, Zeng Y, Kang Y, Bogerd HP, Cullen BR (2002) Formation of Tap/NXT1 heterodimers activates Tap-dependent nuclear mRNA export by enhancing recruitment to nuclear pore complexes. Mol Cell Biol 22:245–256
Wurm FM (2004) Production of recombinant protein therapeutics in cultivated mammalian cells. Nat Biotechnol 22:1393–1398
Xu L, Daly T, Gao C, Flotte TR, Song S, Byrne BJ, Sands MS, Parker Ponder K (2001) CMV-beta-actin promoter directs higher expression from an adeno-associated viral vector in the liver than the cytomegalovirus or elongation factor 1 alpha promoter and results in therapeutic levels of human factor X in mice. Human Gene Ther 12:563–573
Zhou S, Mody D, DeRavin SS, Hauer J, Lu T, Ma Z, Abina SH, Gray JT, Greene MR, Cavazzana-Calvo M, Malech HL, Sorrentino BP (2010) A self-inactivating lentiviral vector for SCID-X1 gene therapy that does not activate LMO2 expression in human T cells. Blood 116:900–908
Zolotukhin S, Michalowski D, Smulevitch S, Felber BK (2001) Retroviral constitutive transport element evolved from cellular TAP(NXF1)-binding sequences. J Virol 75:5567–5575

Chapter 10
Recent Progress Toward RNA Manipulation with Engineered Pentatricopeptide Repeat Proteins

Takayoshi Imai, Yusuke Yagi, and Takahiro Nakamura

Abstract Pentatricopeptide repeat (PPR) proteins are RNA-binding proteins that are widely distributed in plants. They contain 2 to 30 repeating units of ~35-amino acid PPR motifs. They are known to play important roles in RNA processing, RNA editing, and translational regulation. Recent studies on the RNA recognition mode of PPR proteins revealed that one PPR motif interacts with one nucleotide. In addition, it was revealed that amino acids at three specific positions in a single motif serve to specify its binding base. Thus, mutation of these amino acids can cause a modification of the binding specificity of PPR motifs. Indeed, the engineered PPR motifs fused with various effector domains are shown to bind to and manipulate RNAs in a controlled manner. In this review, we summarize the recent progress in structural studies on PPR motifs. We focus on their RNA recognition mode and discuss the potentials of PPR as novel, versatile tools for RNA manipulation.

Keywords RNA-binding protein · Translation activation · Pentatricopeptide repeat (PPR)

10.1 The Available RNA-Binding Proteins for RNA Manipulation

RNAs associate with RNA-binding proteins in cells to form ribonucleoprotein complexes. These complexes govern the structure and functions of RNAs and play important roles in the biogenesis, stability, function, and localization of the RNAs (Burd and Dreyfuss 1994; Ross 1995; Caponigro and Parker 1996; Wilusz et al. 2001).

T. Imai (✉) · Y. Yagi · T. Nakamura
EditForce Inc, Fukuoka, Japan

Faculty of Agriculture, Kyushu University, Fukuoka, Japan
e-mail: taka-ima@editforce.jp

S. Masuda, S. Izawa (eds.), *Applied RNA Bioscience*,
https://doi.org/10.1007/978-981-10-8372-3_10

RNA-binding proteins therefore can function as tools to alter the properties of RNAs and may be useful for genetic studies and related industrial applications.

Since the discovery of the heterogeneous nuclear ribonucleoproteins (hnRNPs) and other pre-mRNA/mRNA-binding proteins, a number of different varieties of RNA-binding proteins have been characterized (Glisovic et al. 2008). Recent studies have not only revealed how RNA-binding proteins select their targets in a cellular pool of RNAs with various structures and functions but also provided opportunities to design RNA-binding proteins fused with various effecter domains that can bind to and manipulate RNAs in a controlled manner. So far, the application of RNA-binding proteins has been based on the unique properties of known RNA-binding domains (RBDs). For example, the RBD of the spliceosomal protein, U1A, has been used to facilitate crystallization and structure determination of RNAs and RNA-protein complexes. Co-crystallization with U1A helped resolve the structure of the tetracycline aptamer (Ferré-D'Amaré 2010, 2016). MS2 coat protein and λN protein have been used as scaffolds to tether RNA sequences for characterizing protein function or tracking mRNA localization (Pankert et al. 2017; Baron-Benhamou et al. 2004). In addition, the trans-activation (Tat) peptide from the bovine immunodeficiency virus and the coat protein of the bacteriophage PP7 have been used in a similar way to visualize and investigate mRNA localization (Frankel et al. 1989; Lim and Peabody 2002). The iron-responsive element (IRE) RNA-binding protein IRP-1 was fused with the C-terminal region of the human translation initiation factor, eIF4GI, to directly drive translation of the reporter gene bearing the IRE RNA sequence (Gray and Hentze 1994; Gregorio et al. 1999). In such applications, the RNA specificity of the proteins is not readily changeable. Thus, these RNA manipulation tools require the preexistence or insertion of binding sequences into the target RNAs, limiting the use of these RNA-binding proteins (Frankel et al. 1989; Gray and Hentze 1994; Lim and Peabody 2002; Baron-Benhamou et al. 2004; Buxbaum et al. 2015; Pankert et al. 2017). RNA-binding motifs, with their modular architecture and programmable binding specificity, could facilitate the establishment of more efficient, versatile RNA manipulation tools (Glisovic et al. 2008; Chen and Varani 2013; Baron-Benhamou et al. 2004). A group of such motifs is the Pumilio-Fem3-binding factors (PUFs). PUFs can recognize RNAs on a one-motif-to-one-nucleotide basis (Wang et al. 2013). The RNA recognition mode of PUFs is supported by their crystal structure complexed with their target RNAs (Wilinski et al. 2015). When their structure was further studied, it was revealed that the PUF module adopts a helical repeat motif, and amino acids at particular positions govern the base recognition. Based on the RNA recognition mode of PUF proteins, many researchers have constructed artificial PUF proteins with various recognition sequences (Abil et al. 2014). These custom PUFs can be functional when conjugated with various effector modules. For example, an artificial sequence-specific ribonuclease has been successfully constructed by fusing a PUF with a PilT N-terminus (PIN) domain, which is a nonspecific endonuclease domain (Choudhury et al. 2012). Native PUF proteins are usually composed of only eight modules (Wang et al. 2013) and so were the artificial ones, until PUFs with 11 repeats were reported (Zhang et al. 2016). The crystal structures of these 11-motif

PUFs revealed that the individual repeats were similar to those of the canonical PUF repeats, with three α-helices; however, they adopted an overall L-shaped structure, in contrast to the crescent-shaped canonical PUF proteins (Edwards et al. 2001; Wang et al. 2001). Furthermore, the 11-motif PUFs could bind single- or double-stranded nucleic acids without any apparent sequence specificity (Zhang et al. 2016). Therefore, it is difficult to construct PUF proteins that can bind to longer target sequences, especially more than 9 nt.

10.2 Pentatricopeptide Repeat Motifs

A promising RNA-binding motif without any of the limitations mentioned in the previous section is the pentatricopeptide repeat (PPR) protein. PPR proteins are eukaryote-specific modular RNA-binding proteins that are widely distributed in plants (Kotera et al. 2005; Fujii et al. 2010; Nakamura et al. 2012). PPR motifs are composed of approximately 35 amino acids. In nature, they exist as arrays of 2 to 30 repeats per protein (Howard et al. 2012; Ringel et al. 2011). PPR motifs are categorized into three groups, based on the number of amino acids: S-type, with 31 residues; L-type, with 36 residues; and P-type with 35 residues. Similarly, PPR proteins are also classified into two categories: P-type PPR proteins, which are mainly composed of P-type PPR motifs, and PLS-type PPR proteins, which comprise P-, L-, and S-type PPR motifs (Kotera et al. 2005; Fujii et al. 2010; Nakamura et al. 2012; Manna 2015).

PPR proteins are involved in different processes of RNA metabolism, including translational regulation, RNA splicing, and RNA cleavage. For example, many P-type PPR proteins regulate translation by binding to the noncoding sequence of an RNA, such as introns and 5′ or 3′ UTRs. A small class of P-type PPR proteins, encoded by the restorer-of-fertility (Rf) genes, are associated with the genetic expression of mitochondrial proteins (Fujii et al. 2010; Dahan and Mireau 2013). Kazama et al. reported that the PPR protein, Rf1, inhibits the translation of the cytotoxic peptide, ORF79, which causes cytoplasmic male sterility (CMS), by degrading RNA transcripts of *atp6-orf79* (Kazama et al. 2008). Evidence from another study also suggests that Rf1 directly binds to the *atp6-orf79* RNA transcripts (Uyttewaal et al. 2008), thereby regulating CMS. PPR proteins have also been implicated in RNA splicing through RNA remodeling (Colas des Francs-Small et al. 2014), as demonstrated by the translational activation of *atp*H by PPR10 in maize chloroplasts (Prinkry et al. 2011). PPR10 binds to one half of a palindromic sequence that extends over the *atp*H initiation codon and prevents the formation of a stem-loop structure that would otherwise impede translation initiation. Further, a small class of naturally occurring PPR proteins contains characteristic C-terminal domains with various enzymatic functions. Processing of the 5′ end tRNA precursors in plants requires the proteinaceous RNase P (PRORP) family of proteins, a small subfamily of PPR proteins that contains a Nedd4/YacP-like nucleases (NYN) domain (Anantharaman and Aravind 2006; Gobert et al. 2010). Gobert et al. found that *Arabidopsis thaliana* expresses PRORP1, PRORP2, and PRORP3; PRORP1 is localized in the mitochondria and chloroplasts, while PRORP2 and PRORP3 are found in the nucleus (Gobert et al. 2010; Gutmann et al. 2012). Cross-linking and mutation studies showed that G19 and C56 of tRNAs are involved in the interaction with PRORP1, indicating that their PPR motifs provide the tRNA recognition domain (Gobert et al. 2013; Imai et al. 2014). Another PPR family, the PPR-SMR family, comprises site-specific endonucleases. The PPR-SMR family contains a small MutS-related (SMR) domain at the C-terminal of typical PPR motifs (Liu et al. 2013). The SMR domain is a single-stranded DNA endonuclease, and at least one SMR protein is known to possess RNA endonuclease activity in nature (Bhandari et al. 2011). Recently, Zhou et al. have reported that the suppressor of thylakoid formation1 (SOT1), containing PPR and SMR domains, has endonuclease activity and contributes to the endonucleolytic maturation of the 23S-4.5S rRNA precursor (Zhou et al. 2017). These findings have provided opportunities to construct site-specific RNA endonucleases that can, for example, be used as an antiviral defense against many RNA viruses, which are disastrous to agriculture. Although the properties of PPR motifs had been characterized, the molecular basis of RNA-binding by PPR was unknown, until the discovery of its RNA recognition codes.

10.3 RNA Recognition Mode of PPR Motifs

Bioinformatic and biochemical approaches have elucidated many putative RNA recognition codes (Barkan et al. 2012; Yagi et al. 2013a, b, 2014). These codes consist of three amino acid residues at positions 1, 4, and ii (indicating the second last residue) in each PPR motif; each motif recognizes an RNA base (Fig. 10.1a) (Barkan et al. 2012; Yagi et al. 2013a, b, 2014; Manna 2015). The amino acid residues at positions 4 and ii play crucial roles in the recognition of RNA bases (Barkan et al. 2012; Yagi et al. 2013a, b, 2014), while the amino acid residue at position 1 fine-tunes the base recognition (Yagi et al. 2013a, b, 2014; Shen et al. 2016). The well-characterized RNA recognition codes include (Val, Asn, and Asp) at positions (1, 4, and ii) for U, (Val, Thr, and Asp) for G, (Val, Asn, and Ser) for C, and (Phe, Thr, and Asn) for A (Fig. 10.1b).

The crystal structures of designed PPR (dPPR) proteins complexed with their specific single-stranded RNAs were recently resolved (Fig. 10.1c) (Shen et al. 2016). These structures support the RNA recognition codes of PPR and reveal the molecular base for specific modular recognition of bases (Shen et al. 2016). The polar residue at position 4 is the major participant in the base recognition via hydrogen bonding. The residues at this position, frequently Ser and Thr, interact with purine bases (A and G); Asp at this position is correlated with a preference for pyrimidine bases (U and C) (Fig. 10.2) (Shen et al. 2016). The amide group of the Asn4 side chain donates a hydrogen bond to the O2 atom of the corresponding pyrimidine, whereas the N3 atom of purine accepts a hydrogen bond form the hydroxyl group of Ser/Thr, leading to purine/pyrimidine selectivity by the amino

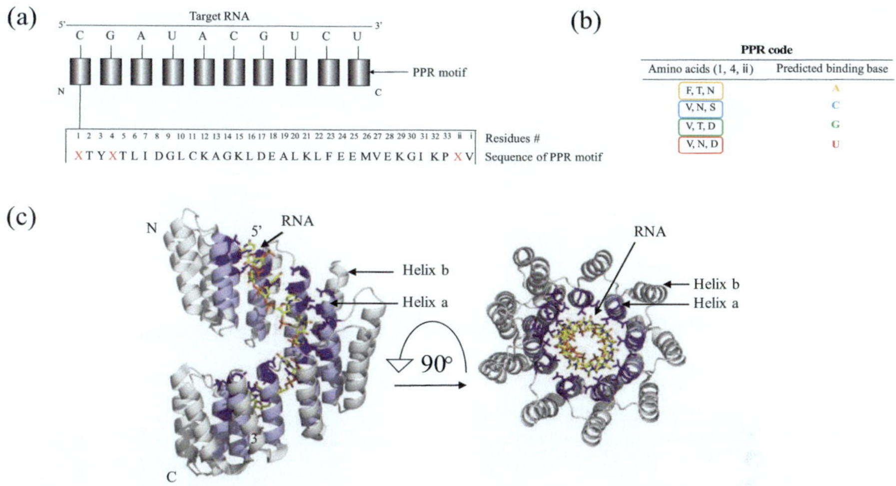

Fig. 10.1 Pentatricopeptide repeat (PPR) motif. (**a**) Schematic representation of PPR-RNA interaction. A single PPR motif recognizes a specific RNA base in a modular manner. An example of the amino acids that configures a PPR motif is shown below. The amino acids in RNA recognition code (referred as "PPR code") are indicated by red "X." (**b**) Example of PPR codes for RNA recognition. (**c**) Overall structure of the designed PPR complexed with its target RNA (Protein Data Bank: 5I9F) (Adapted from Shen et al. 2016). Helix a that exists in the cavity of the superhelix is colored by blue, and helix b that appears in the outline the external layer of the superhelix is colored by gray. The amino acids positioned at the PPR codes in each motif are highlighted by purple stick representation. All structures were prepared by PyMOL

acid residue at position 4. The polar residue at position ii is also important for RNA base recognition through either direct or water-mediated hydrogen bonds (Shen et al. 2016). When residues positioned at ii bind with the purine bases, the N1 atom of adenine acts as a hydrogen bond receptor, while the N1 and N2 atoms of guanine act as the hydrogen bond donors, resulting in the amino acids Asn and Asp at position ii serving as the differentiators to the purine bases (Fig. 10.2). For pyrimidine bases, the selectivity is mainly determined by water-bridged polarity. The N3 atom of uracil acts as a hydrogen bond donor, while the N3 atom of cytosine acts as the hydrogen bond acceptor (Fig. 10.2).

10.4 PPR Motifs as RNA Manipulation Tools

Since a large number of RNA-binding proteins play crucial roles in controlling biological reactions, it should be possible to control a vast spectrum of RNA metabolism using engineered RNA-binding proteins (Burd and Dreyfuss 1994; Ross 1995; Caponigro and Parker 1996; Wilusz et al. 2001). PPR proteins bind nucleic acids with a ratio of 1:1 between protein motifs and RNA bases; in addition, a definite RNA recognition code determines a couple of key sites in each motif. Such a

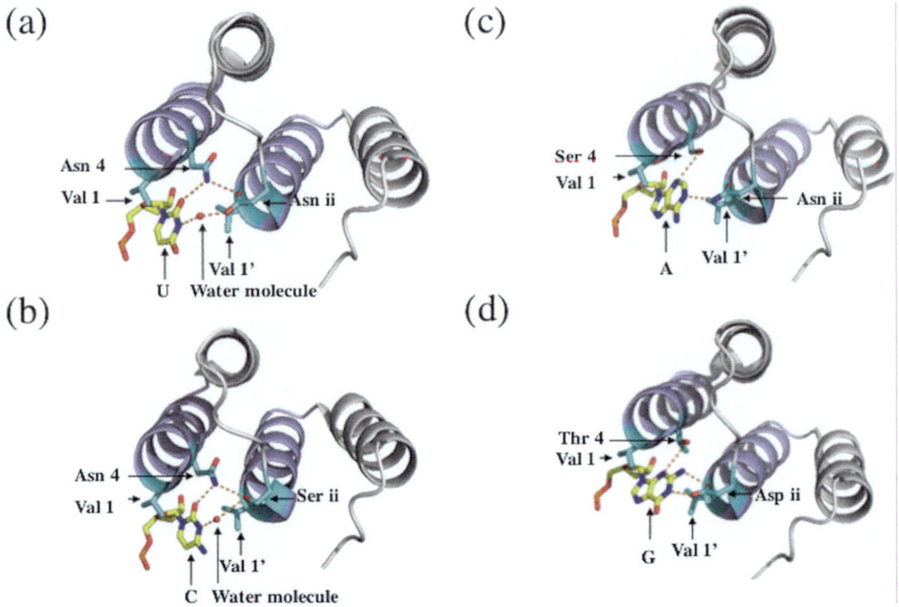

Fig. 10.2 RNA recognition mode of PPR proteins. The specific recognition pattern of the base U, C, A, and G are shown in (**a**), (**b**), (**c**), and (**d**), respectively. The side chains of the amino acids at the positions1, 4, and ii in each PPR motif are shown by cyan stick representation. The RNA base interacted by each PPR is labeled and shown with yellow stick representation. The hydrogen bonding is represented by orange dotted line and the water molecules by red spheres. The crystal structures deposited in the Protein Data Bank with the accession code (5I9F, 5I9G, 5I9D, and 5I9H) were used as references

binding mode allows rational design of artificial PPR proteins with controlled sequence specificity.

It is now clear that engineered PPR proteins with desired functions are feasible. According to Shen et al., the important aspect of PPR engineering is that it is possible to construct functional PPR motifs with an amino acid sequence not found in nature; further, the sequence specificity of PPR proteins can be altered by mutating the amino acid residues that comprise the PPR code (Shen et al. 2016). Regarding the functions of PPR tools, a protein composed of PPR motifs found in nature and the human translation initiation factor, eIF4G, has been constructed, which specifically binds to the target RNA sequence predicted by the established PPR code and regulates translation (Kobayashi et al. 2016). Additionally, various PPR tools for up- and downregulation, splicing, and RNA editing are being constructed (Fig. 10.3). These novel tools will eventually enable the editing of transcriptomes for bio-industrial and therapeutic purposes.

Although the future looks bright for PPR-based technology for transcriptome editing, many challenges remain to be resolved before routine application of these tools is possible. Further investigations into RNA recognition by PPR through various approaches are required in order to engineer PPR proteins as versatile and

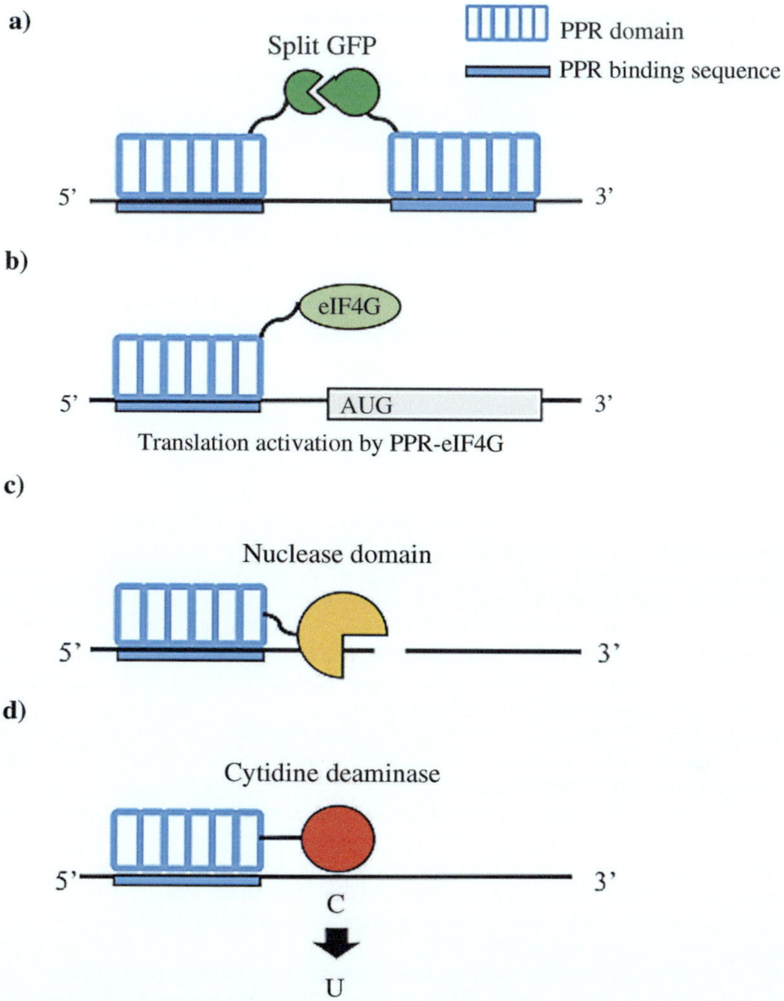

Fig. 10.3 Engineered PPR proteins. (**a**) Fluorescent probe for in vivo RNA detection. Combination of split GFP and PPR proteins enables visualizing of RNA in living cells. (**b**) Translation regulator. Fusion of eIF4G and a PPR domain can activate translation of the targeted mRNA. (**c**) Artificial site-specific RNA endonuclease. Combination of PPR proteins and nonspecific RNA endonucleases, such as PIN domain, makes a novel class of enzymes that sequence-specifically recognizes and cleaves RNA. (**d**) Site-specific RNA editing tools. The designed PPR proteins conjugated with cytidine deaminase domain catalyze the hydrolytic reaction of cytosine to uracil at the specific positions

powerful RNA manipulation tools. Improving specificity and minimizing off-target effects are the most important issues. RNA-binding tools are ideally specific to their target sequences, but they can also bind to similar sequences present elsewhere. These off-target effects can seriously limit the use of RNA-binding tools. The sequence specificity of the modular RNA-binding tools is linked to the base selectivity of each motif and to the length of the recognition motifs. Barkan et al. reported that PPR motifs have higher selectivity to purine bases than to the pyrimidine bases, and so, they might tolerate mismatches at C/U sites (Barkan et al. 2012). To overcome the limitations due to such mismatches, the length of the target sequence should be optimized to increase specificity and decrease off-target effects. Although the mean length of PPR motifs is approximately 12, the modularity of the PPR allows the variability of motif length and target sequence length. The contribution of each PPR motif-nucleotide interaction to the total affinity and specificity of the whole protein to the target RNA is, however, yet to be completely elucidated. Systematic quantitative evaluation of PPR-RNA interactions, based on single motifs and nucleotides will be an important area of future research.

References

Abil Z, Denard CA, Zhao H (2014) Modular assembly of designer PUF proteins for specific post-transcriptional regulation of endogenous RNA. J Biol Eng 8:7

Anantharaman V, Aravind L (2006) The NYN domains, novel predicted RNases with a PIN domain-linker fold. RNA Biol 3:18–27

Barkan A, Rojas M, Fujii S, Yap A, Chong YS, Bond CS, Small I (2012) A combinatorial amino acids code for RNA recognition by pentatricopeptide repeat proteins. PLoS Genet 8:e1002910

Baron-Benhamou J, Gehring NH, Kulozik AE, Hetze MW (2004) Using the λN peptide to tether proteins to RNAs. Methods Mol Biol 257:135–153

Bhandari D, Guha K, Bhaduri N, Saha P (2011) Ubiquitination of mRNA cycling sequence binding protein from *Leishmania donovani* (LdCSBP) modulates the RNA endonuclease activity of its Smr domain. FEBS Lett 585:809–813

Burd CG, Dreyfuss G (1994) Conserved structures and diversity of functions of RNA-binding proteins. Science 265:615–621

Buxbaum AR, Haimovich G, Singer RH (2015) In the right place at the right time: visualizing and understanding mRNA localization. Nat Rev Mol Cell Biol 16:95–109

Caponigro G, Parker R (1996) mRNA turnover in yeast promoted by the MATalpha1 instability element. Nucleic Acids Res 24:4304–4312

Chen Y, Varani G (2013) Engineering RNA-binding proteins for biology. FEBS J 280:3734–3754

Choudhury R, Tsai YS, Dominguez D, Wang Y, Wand Z (2012) Engineering RNA endonucleases with customized sequence specificities. Nat Commun 3:1147

Colas des Francs-Small C, Falcon de Longevialle A, Li Y, Lowe E, Tanz SK, Smith C, Bevan MW, Small I (2014) The pentatricopeptide repeat proteins TANG2 and ORGANELLE TRANSCRIPT PROCESSING439 are involved in the splicing of the multipartite *nad5* transcript encoding a subunit of mitochondrial complex 1. Plant Physiol 165:1409–1416

Dahan J, Mireau H (2013) The Rf and Rf-linker PPR in higher plants, a fast-evolving subclass of PPR genes. RNA Biol 19:1469–1476

Edwards TA, Pyle SE, Whartsom RP, Aggarwal AK (2001) Structure of Pumilio reveals similarity between RNA and peptide binding motifs. Cell 105:281–289

Ferré-D'Amaré AR (2010) Use of the spliceosomal protein U1A to facilitate crystallization and structure determination of complex RNAs. Methods 52:159–167

Ferré-D'Amaré AR (2016) Use of the U1A protein to facilitate crystallization and structure determination of large RNAs. Methods Mol Biol 1320:67–76

Frankel AD, Biancalana S, Hudson D (1989) Activity of synthetic peptide from the Tat protein of human immunodeficiency virus type 1. Proc Natl Acad Sci U S A 86:7397–7401

Fujii S, Bond CS, Small ID (2010) Selection patterns on restorer-linker genes reveal a conflict between nuclear and mitochondrial genomes throughout angiosperm evolution. Proc Natl Acad Sci U S A 108:1723–1728

Glisovic T, Bachorik JL, Yong J, Dreyfuss G (2008) RNA-binding proteins and post-transcriptional gene regulation. FEBS Lett 582:1977–1986

Gobert A, Gutmann B, Taschner A, Gößringer M, Holzmann J, Hartmann R, Rossmanith W, Giege P (2010) A single *Arabidopsis* organelle protein has RNase P activity. Nat Struct Mol Biol 17:740–744

Gobert A, Pinker F, Fuchsbauser NO, Gutmann B, Boutin R, Robin P, Sauter C, Giege P (2013) Structural insights into protein-only RNase P complexed with tRNA. Nat Commun 4:1353

Gray NK, Hentze MW (1994) Iron regulatory protein prevents binding of the 43S translation pre-initiation complex ferritin and eALAS mRNAs. EMBO J 13:3882–3891

Gregorio ED, Preiss T, Henize MV (1999) Translation drive by an eIF4G core domain in vivo. EMBO J 18:4865–4874

Gutmann B, Gobert A, Giege P (2012) PRORP proteins support RNase P activity in both organelles and the nucleus in *Arabidopsis*. Genes Dev 26:1022–1027

Howard MJ, Lim WH, Fierke CA, Koutmos M (2012) Mitochondrial ribonuclease P structure provides insight into the evolution of catalytic strategies for precursor-tRNA 5′ processing. Proc Natl Acad Sci U S A 109:16149–16154

Imai T, Nakmura T, Maeda T, Nakayama K, Gao X, Nakashima T, Kakuta Y, Kimura M (2014) Pentatricopeptide repeat motifs in the processing enzyme PRORP1 in Arabidopsis thaliana play a crucial role in recognition of nuclease bases at TΨC loop in precursor tRNAs. Biochem Biophys Res Commun 450:1541–1546

Kazama T, Nakamura T, Watanabe M, Sugita M, Toriyama K (2008) Suppression mechanism of mitochondrial ORF79 accumulation by Rf1 protein in BT-type cytoplasmic male sterile rice. Plant J 55:619–628

Kobayashi T, Yagi Y, Nakamura T (2016) Development of genome engineering tools from plant-specific PPR proteins using animal cultured cells. In: Murata M (ed) Chromosome and genomic engineering in plants, vol 1469. Springer, Heidelberg, pp 147–155

Kotera E, Tasaka M, Shinkanai T (2005) A pentatricopeptide repeat protein is essential for RNA editing in chloroplasts. Nature 433:326–330

Lim F, Peabody DS (2002) RNA recognition site of PP 7 coat protein. Nucleic Acids Res 30:4138–4144

Liu S, Melonek J, Boykin LM, Small I, Howel KA (2013) PPR-SMRs: ancient proteins with enigmatic functions. RNA Biol 10:1501–1511

Manna S (2015) An overview of pentatricopeptide repeat proteins and their applications. Biochimie 113:93–99

Nakamura T, Yagi Y, Kobayashi K (2012) Mechanistic insight into pentatricopeptide repeat proteins as sequence-specific RNA binding proteins for organellar RNAs in plants. Plant Cell Physiol 53:1171–1179

Pankert T, Jegou T, Caudro-Herger M, Rippe K (2017) Tethering RNA to chromatin for fluorescence microscopy based analysis of nuclear organization. Methods S1046–2023:30471–30476

Prinkry J, Rojas M, Schuster G, Barkan A (2011) Mechanism of RNA stabilization and translational activation by a pentatricopeptide repeat protein. Proc Natl Acad Sci U S A 108:415–420

Ringel R, Sologub M, Morozov YI, Litonin D, Cramer P, Temiakov D (2011) Structure of human mitochondrial RNA polymerase. Nature 478:269–273

Ross J (1995) mRNA stability in mammalian cells. Microbiol Rev 59:423–450

Shen C, Zhang D, Guan Z, Liu Y, Yang Z, Yang Y, Wang X, Wang Q, Zhang Q, Fan S, Zou T, Yin P (2016) Structural basis for specific single-stranded RNA recognition by designer pentatricopeptide repeat proteins. Nat Commun 7:11285

Uyttewaal M, Arnal N, Quadrabo M, Martin-Canadell A, Vrielynck N, Hiard S, Gherbi H, Bendahmane A, Budar F (2008) Characterization of *Raphanus sativus* pentatricopeptide repeat proteins encoded by the fertility restorer locus for Ogura cytoplasmic male sterility. Plant Cell 20:3331–3345

Wang X, McLachlan J, Zamore PD, Hall TM (2001) Modular recognition of RNA by a human pumilio-homology domain. Cell 110:501–512

Wang Y, Wang Z, Tanaka Hall TM (2013) Engineered proteins with Pumilio/fem-3 mRNA binding factor scaffold to manipulate RNA metabolism. FEBS J 280:3755–3767

Wilinski D, Qiu C, Lapointe CP, Nevil M, Champbell ZT, Tanaka Hall TM, Wicken M (2015) RNA regulatory networks diversified through curvature of the PUF protein scaffold. Nat Commun 6:8213

Wilusz CJ, Gao M, Jones CL, Wilusz J, Peltz SW (2001) Poly(A)-binding proteins regulate mRNA degradation and decapping in yeast cytoplasmic extracts. RNA 7:11416–11424

Yagi Y, Hayashi S, Kobayashi K, Hirayama T, Nakamura T (2013a) Elucidation of the RNA recognition code for pentatricopeptide repeat proteins involved in organelle RNA editing in plants. PLoS One 8:e57286

Yagi Y, Tachikawa M, Noguchi H, Satoh S, Obokata J, Nakamura T (2013b) Pentatricopeptide repeat proteins involved in plant organellar RNA editing. RNA Biol 10:1236–1242

Yagi Y, Nakamura T, Small I (2014) The potential for manipulation RNA with pentatricopeptide repeat proteins. Plant J 78:772–782

Zhang J, McCann KL, Qiu C, Gonzalez LE, Baserga SJ, Tanaka Hall TM (2016) Nop9 is a PUF-like protein that prevents premature cleavage to correctly process pre-18S rRNA. Nat Commun 7:13085

Zhou W, Lu Q, Li Q, Wang L, Ding S, Zhang A, Wen X, Zhang L, Lu C (2017) PPR-SMR protein SOT1 has RNA endonuclease activity. Proc Natl Acad Sci U S A 114:1554–1563

Chapter 11
Long Noncoding RNAs and Their Applications: Focus on Architectural RNA (arcRNA), a Class of lncRNA

Tomohiro Yamazaki

Abstract Transcriptome analyses have revealed large numbers of non-protein coding transcripts called noncoding RNAs (ncRNAs), which are produced from most genomic regions in mammalian cells. These ncRNAs include many thousands of long noncoding RNAs (lncRNAs) more than 200 nucleotides in length. Although our knowledge of these lncRNAs remains limited, recent studies have revealed their diverse roles under physiological and pathological conditions, as well as their mechanisms of action in a variety of cellular processes including epigenetic regulation, transcriptional regulation, posttranscriptional processing, and intracellular organization. In addition, multiple studies show that aberrant expression of lncRNAs is associated with various diseases, including cancer and neurodegenerative disorders, suggesting that lncRNAs represent promising target molecules for biomedical applications. Here, I review lncRNAs and several related applications and in particular an emerging class of lncRNAs termed architectural RNA (arcRNA). I describe and discuss arcRNAs in mammals, focusing on their biogenesis, mechanisms of action, and potential applications. In addition, I highlight our newly established methods for discovering arcRNA candidates. Finally, I emphasize the importance of identifying the RNA elements embedded in lncRNAs that dictate their functions; these elements provide opportunities for future applications in biotechnology, biomarkers, and therapeutics.

Keywords Long noncoding RNA (lncRNA) · Architectural RNA (arcRNA) · Nuclear body · Semi-extractable RNA (seRNA) · NEAT1 · Satellite RNA · SINEUP · Prion-like domain · Low-complexity domain · Intrinsically disordered domain

T. Yamazaki (✉)
Institute for Genetic Medicine, Hokkaido University, Sapporo, Hokkaido, Japan
e-mail: tomohiro_yamazaki@igm.hokudai.ac.jp

S. Masuda, S. Izawa (eds.), *Applied RNA Bioscience*,
https://doi.org/10.1007/978-981-10-8372-3_11

11.1 Introduction

Only 2% of the human genome consists of protein-coding genes, and the rest of the genome was once considered "junk." However, over the past decades, advances in sequencing technologies and analytical methods have led to the discovery of tens of thousands of RNAs that are pervasively transcribed from mammalian genomes but are apparently not translated into proteins (Bertone et al. 2004; Carninci et al. 2005; Kapranov et al. 2007). The subset of these RNAs greater than 200 nucleotides in length is collectively termed long noncoding RNAs (lncRNAs), many of which have been shown to have functions just as rRNA, tRNA, and snRNA do (Cech and Steitz 2014). Moreover, lncRNAs have been recognized as key regulatory molecules in gene expression programs (Guttman and Rinn 2012; Geisler and Coller 2013; Quinn and Chang 2016), and recent studies have identified lncRNAs that play critical roles in cellular function, development, and disease (Ponting et al. 2009; Rinn and Chang 2012; Schmitt and Chang 2016). Different types of lncRNAs, such as circRNA and sno-lncRNA, are generated by distinct processing mechanisms (Chen 2016b; Quinn and Chang 2016).

lncRNAs are expressed in a more tissue- and development-specific manner than mRNAs are, and this characteristic makes them suitable as biomarkers for diagnostic and therapeutic applications (Ulitsky and Bartel 2013). Although the vast majority of lncRNAs remain uncharacterized, a recent CRISPRi screen of 16,401 lncRNA loci (using the CRiNCL guide RNA library, available to academic users via Addgene) in seven human cell lines identified 499 lncRNAs that affect cell viability (Liu et al. 2017). Interestingly, most of them (89%) affected cell growth in only one of the seven cell lines, suggesting that lncRNA functions are cell type-specific and demonstrating that this method could be used as a platform to study these functions. These observations support the utility of lncRNAs for therapeutic applications.

One important feature of lncRNAs is that they carry positional information within the nucleus (Batista and Chang 2013; Engreitz et al. 2016b). By contrast, protein-coding RNAs lose such information after transcription because they are exported to the cytoplasm before translation. Consistent with this, lncRNAs regulate expression of nearby genes *in cis* (Engreitz et al. 2016a). In addition, lncRNAs have been proposed to act as spatial amplifiers that control gene expression and three-dimensional genome architecture (Engreitz et al. 2016b). lncRNAs function in several biological processes, including epigenetics, histone modification, locus-specific gene regulation, enhancers, chromatin remodeling, transcriptional regulation, and posttranscriptional regulation (Quinn and Chang 2016). Mechanistically, lncRNAs can act as guides, scaffolds, architectures, decoys, or enhancers (Guttman and Rinn 2012; Hirose et al. 2014a). For example, several lncRNAs act as chromatin regulators by recruiting and integrating chromatin regulatory proteins at specific chromatin sites (Rinn and Chang 2012).

In some cases, the specific RNA sequences themselves are not necessary; instead, the process of transcription itself at a specific locus is an important regulatory cue for the expression of nearby genes (Engreitz et al. 2016a; Paralkar et al. 2016).

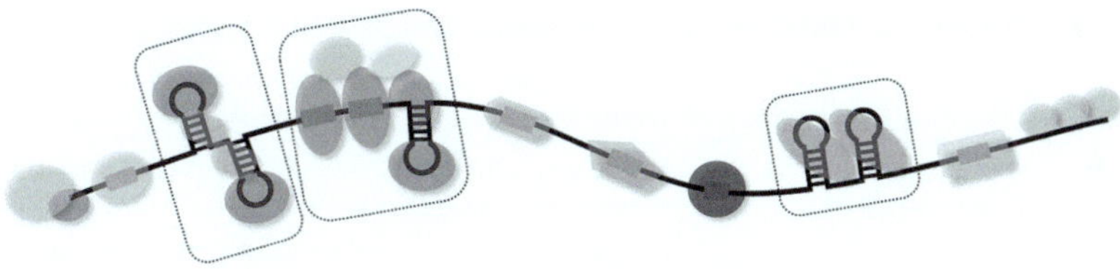

Fig. 11.1 A concept for modular RNA domains and RNA motifs in lncRNAs. lncRNAs are surrounded by numerous RBPs, some of which recognize specific RNA sequence or structural motifs. Some RBP binding sites and combinations of RBPs on the specific binding sites (e.g., dashed boxes) form functional modular RNA domains that play specific cellular functions

Notably, some RNAs annotated as lncRNAs are in fact translated to produce small polypeptides that are biologically active (Anderson et al. 2015; Nelson et al. 2016; Matsumoto et al. 2017). In skeletal and heart muscle cells, small peptides conserved among species are produced from lncRNAs and modulate the activities of membrane-bound proteins.

The functional characterization of lncRNAs is ongoing, and one of the next important challenges is to classify lncRNAs according to their functional RNA elements found in lncRNAs that are RNA sequences or structures for interactions with specific RNA-binding proteins (RBPs) (Hirose et al. 2014a) (Fig. 11.1). Understanding the RNA elements hidden in lncRNAs would open the door to new applications of lncRNAs, e.g., interference with specific functions of lncRNAs and engineering of artificial lncRNAs. Similar approaches using modular domains of RNAs and proteins have already been used to engineer ribozymes, RNPs, and proteins. Thus, identification and characterization of the modular domains of lncRNAs would expand the toolbox for a wide variety of experimental applications in molecular biology, biotechnology, and therapeutics.

11.2 Examples of lncRNAs with Applications

Here, I summarize several examples of lncRNAs, along with potential applications, focusing on their modular RNA functional domains, biogenesis, mechanisms of actions, and in vivo functions (Table 11.1).

The lncRNA called SINEUP (RNA containing *SINE* elements that *UP*regulate translation) is an antisense lncRNA that promotes translation of its target mRNAs (Takahashi and Carninci 2014; Zucchelli et al. 2015a). This class of lncRNAs was originally identified in mouse as an antisense transcript of the *Uchl1* gene, AS *Uchl1* (Carrieri et al. 2012). AS *Uchl1* spans a 73 nt region overlapping the 5′ untranslated region (UTR) and translational start site of the *Uchl1* mRNA and promotes translation of UCHL1 protein without affecting mRNA stability. Outside the overlapping region, AS *Uchl1* consists of inverted SINEB2 (short interspersed nuclear element B2) elements. Promotion of translation depends on the overlapping region, called

Table 11.1 Examples of lncRNAs with applications

lncRNA	Mechanism of action	Intervention methods	Outcomes and potential applications
SINEUP	Translational upregulation	Introduce RNAs or expression vectors to cells	Induced production of target proteins
			Therapeutic use (e.g., for diseases caused by haploinsufficiency of the causative genes)
			Protein manufacturing
			Research use
NATs	Repress overlapped sense mRNA partner (mainly)	Inhibition (e.g., antagoNATs)	Upregulation of sense mRNAs
			Therapeutic use
circRNA	Molecular sponge for proteins and RNAs	Introduce RNAs or expression vectors to cells	Sequestration of proteins and RNAs
		Inhibition of circRNAs	Therapeutic use
HOTAIR, UPAT	Targeted protein degradation	Introduce RNAs or expression vectors to cells	Degradation of target proteins
½-sbsRNA, TINCR	Targeted mRNA degradation	Introduce RNAs or expression vectors to cells	Degradation of target mRNAs
eRNA	Transcriptional enhancement	Inhibition of eRNAs	Transcriptional inhibition on the specific genomic loci and in specific cell lineages
			Therapeutic use

the binding domain (BD), and an inverted SINEB2 element called the effector domain (ED) (Zucchelli et al. 2015b). By changing the sequence of the BD into the antisense sequence of another mRNA, synthetic SINEUPs can be designed that target a mRNA of interest and function in trans (Zucchelli et al. 2015a). These synthetic SINEUPs can promote translation in multiple species and cell types, including human, monkey, hamster, and mouse cells in vitro and mice in vivo (Zucchelli et al. 2015a; Indrieri et al. 2016; Zucchelli et al. 2016). In general, this approach yields protein upregulation from 1.5- to 3-fold. miniSINEUPs that exclusively contain the BD and ED (~250 nt in length) are also active, and these shorter sequences are easier to deliver by viral vectors, or as naked synthetic SINEUPs, for therapeutic use (Zucchelli et al. 2015a, b). Among the available methods for increasing protein production, SINEUPs have two advantages: (1) they increase protein production without introducing stable genomic changes and (2) induction is typically moderate and within the physiological range (~2-fold). These features make SINEUPs appropriate for use in research, protein manufacturing, and therapeutics (Zucchelli et al. 2015a, b). As an example of a therapeutic approach, some inborn diseases are caused by haploinsufficiency of the causative gene, and in such cases, synthetic

SINEUPs could be used to upregulate the sole remaining wild-type allele. As SINEUP, another strategy has been developing to upregulate the gene expression based on the finding of natural antisense transcripts (NATs) (Wahlestedt 2013). NATs are transcribed in an antisense direction, in proximity to or overlapping with their sense mRNA partners, and usually repress the expression of the partner mRNAs. Therefore, inhibition of NATs can derepress their partner mRNAs (Katayama et al. 2005). Many overlapping NATs have been identified in human (Engström et al. 2006). Oligonucleotides targeting NATs, called antagoNATs, hold promise for therapeutic applications aimed at upregulating the expression of target mRNAs (Wahlestedt 2013).

Circular RNA (circRNA), another class of lncRNAs, is produced by back-splicing from pre-mRNAs (Chen 2016a; Salzman 2016). Because they cannot be targeted by exoribonucleases, circRNAs are generally quite stable. Thousands of circRNAs have been identified in several species, including human, mouse, and *C. elegans*. In functional terms, these RNAs act as molecular sponges for miRNAs and RBPs. For example, CDR1as sequesters miR-7, and circMBL sequesters RBP MBNL (Hansen et al. 2013; Memczak et al. 2013; Ashwal-Fluss et al. 2014). The sequences and structures of circRNAs are critical for this sequestration. Accordingly, understanding the rules underlying these interactions would lead to novel applications of circRNA for specific sequestration of RNAs and RBPs of interest. Over the course of efforts to understand the biogenesis of circRNA, systems have been developed for expression of circRNAs in cells, thus expanding their potential usage (Liang and Wilusz 2014). In addition to the noncoding features of circRNAs, they have recently been shown to be translated into proteins. For example, Circ-ZNF609 is a circRNA that can be translated into a protein with a function in myogenesis (Legnini et al. 2017). Furthermore, circRNAs are abundant in specific conditions and cell types, suggesting that they might be suitable as biomarkers or therapeutic targets (Chen 2016a). Moreover, fusion-circRNAs generated from oncogenic translocation contribute to cancer cell survival and oncogenic potential in vivo (Guarnerio et al. 2016).

lncRNAs such as HOTAIR and UPAT regulate the degradation of specific proteins. HOTAIR plays roles in target gene repression and cancer-induced ubiquitin-mediated proteolysis by providing scaffolding for E3 ubiquitin ligases such as Dzip3 and Mex3b and ubiquitination substrates such as Ataxin-1 and Snurportin-1 (Yoon et al. 2013). These E3 ligases and substrates associate with specific RNA domains of HOTAIR, suggesting its potential application in targeted degradation. On the other hand, UPAT lncRNA prevents proteolysis of epigenetic factor UHRF1 by blocking its association with the E3 ubiquitin ligase β-TrCP (Taniue et al. 2016). The specific RNA domains on UPAT responsible for its binding to partner proteins have not yet been identified.

1/2-sbsRNA (half STAU1-binding site) promotes the degradation of target mRNAs via STAU1-mediated mRNA decay (SMD) by providing binding sites for STAU1, an RBP that binds double-stranded RNAs (Gong and Maquato 2011). 1/2-sbsRNA forms imperfect base pairs with target mRNAs. In contrast to 1/2-sbsRNA, TINCR lncRNA directly associates with STAU1 and interacts with its target mRNAs via the 25 nt TINCR box motif, which is enriched in its target sequences (Kretz et al. 2013).

Enhancer RNAs (eRNAs), which are transcribed from enhancers, are important determinants for cell lineages (Kaikkonen et al. 2013). Consequently, inhibition of eRNAs influences specific genes in specific cell lineages, leading to the idea of "enhancer therapy." Many lncRNAs play crucial roles in epigenetic regulation. XIST lncRNA is one of the most extensively studied lncRNAs, and its RNA domains and interacting proteins have been identified (Chu et al. 2015; McHugh et al. 2015; Chen et al. 2016). In addition, lncRNAs play major roles in cancer biology. Multiple lncRNAs, including PVT1, CCAT2, PCAT-1, SAMMSON, MALAT1, and NEAT1, are associated with cancer progression and repression (Schmitt and Chang 2016). For example, NORAD sequesters PUMILIO2 proteins in the cytoplasm and plays a critical role in genome stability (Lee et al. 2016). Therefore, elucidation of the in vivo cancer-related functions of lncRNAs could lead to therapeutic and diagnostic applications.

11.3 Architectural RNAs (arcRNAs) and Their Potential Applications

In this section, I will focus on architectural RNAs (arcRNAs), a class of lncRNAs that serve as architectural components of nuclear bodies, i.e., cellular bodies within the nucleus. Eukaryotic cells compartmentalize cellular materials to organize and promote essential cellular functions. Cells possess two types of compartments: cellular organelles, which are surrounded by lipid bilayers, and membraneless organelles, also known as cellular bodies (Courchaine et al. 2016; Banani et al. 2017). The latter compartments, which are typically composed of specific sets of proteins and RNAs, are fundamental cellular compartments required for specific biochemical reactions, RNP assembly, storage of proteins and RNAs, and sequestration of proteins and nucleic acids. Numerous cellular bodies have been identified to date, including the nucleolus, perinucleolar compartment, nuclear speckles, paraspeckle, Cajal body, gems, PML body, histone locus body, Sam68 nuclear body, stress granules, and P-bodies. Typically, cellular bodies exchange their constituents dynamically. Recent work showed that phase separation between distinct material states (i.e., liquid, hydrogel, and solid) is a key mechanism underlying the formation of these compartments (Wu 2013; Alberti and Hyman 2016; Banani et al. 2017). Proteins containing prion-like domains (PLDs) or low-complexity domains (LCDs), which are unstructured and prone to aggregate, play essential roles in the formation of phase-separated cellular bodies (Aguzzi and Altmeyer 2016; Uversky 2016).

Although many cellular bodies are proteinaceous, some nuclear bodies have RNAs as their architectural cores (Chujo et al. 2016). This concept was originally proposed based on the identification of NEAT1 (nuclear paraspeckle assembly transcript 1), a nuclear-retained lncRNA that is an essential component of the paraspeckle, a nuclear body (Clemson et al. 2009; Sasaki et al. 2009; Sunwoo et al. 2009). In addition to NEAT1, other lncRNAs play similar architectural roles in the construction of nuclear bodies in various species, suggesting that this is a general function of lncRNAs. For example, heat shock RNA (hsr) omega is the lncRNA for the omega speckle in *Drosophila melanogaster*, and meiRNA is the lncRNA for the Mei2 dot in *S. pombe* (Chujo et al. 2016). Accordingly, we refer to such RNAs as architectural RNAs (arcRNAs) (Chujo et al. 2016). An arcRNA can be defined as a lncRNA that localizes in a specific nuclear body and is essential for its integrity. In this section, I focus on the arcRNAs in mammals and describe the known arcRNAs, focusing on their biogenesis, biological functions, relationship to diseases, and potential applications (Table 11.2).

11.3.1 NEAT1 lncRNA

Several groups identified NEAT1 lncRNA as an essential architectural component of the paraspeckle, which was originally identified as a distinct nuclear body localized adjacent to nuclear speckles (Clemson et al. 2009; Sasaki et al. 2009; Sunwoo et al. 2009). The paraspeckle is a massive (~360 nm diameter), highly ordered RNP structure comprising more than 60 kinds of proteins, several of which are required for paraspeckle biogenesis (Souquere et al. 2010; Naganuma et al. 2012; Fong et al. 2013; Yamazaki and Hirose 2015). The DBHS (*Drosophila melanogaster* behavior human splicing) family of proteins, SFPQ, NONO, and PSPC1, have coiled-coil structures that form homo- or heterodimers; two of these, SFPQ and NONO, are essential for expression of NEAT1 and formation of paraspeckles (Sasaki et al. 2009; Naganuma et al. 2012; Passon et al. 2012; Lee et al. 2015). In addition, several paraspeckle proteins, including FUS, DAZAP1, HNRNPH3, and the SWI/SNF complex components BRG1 and BRM, are essential for paraspeckle integrity (Naganuma et al. 2012; Kawaguchi et al. 2015). Many paraspeckle proteins are RBPs with a PLD or LCD, some of which are essential for paraspeckle formation (Naganuma et al. 2012; Yamazaki and Hirose 2015). In addition, the PLDs of FUS and RBM14 are essential for paraspeckle integrity (Hennig et al. 2015).

The *NEAT1* gene is located on chromosome 11q13 in human and chromosome 19qA in mouse. In both species, the gene encoding another abundant nuclear lncRNA, MALAT1, is adjacent to *NEAT1*. The NEAT1 lncRNA has two isoforms, NEAT1_1 (~3.7 kb) and NEAT1_2 (~22.7 kb), which are produced from the same transcription start site under the control of the same promoter and then subjected to alternative 3′-end processing (Naganuma et al. 2012). The NEAT1_1 lncRNA has a poly(A) tail, whereas NEAT1_2 has a unique triple-helix structure at its 3′ end that stabilizes cognate RNAs (Wilusz et al. 2012). Similar cis-acting RNA structures, some of which are called ENE (element for nuclear expression), are found in MALAT1 lncRNA, genomic RNAs of diverse viruses including Kaposi's sarcoma-associated herpesvirus, and ~200 transposable element RNAs in plants and fungi

Table 11.2 Mammalian arcRNAs

arcRNA	Nuclear body	Inducing stress or condition	Protein components	Biological functions	Relationship to diseases	Potential applications
NEAT1	Paraspeckle	Proteasome inhibition	>60 proteins including PSPC1, SFPQ, NONO, RBM14, HNRNPK, FUS, DAZAP1, HNRNPH3, HNRNPH1, HNRNPF, HNRNPA1, HNRNPR, HNRNPUL1,TDP-43, BRG1, BRM, BAF155, CPSF7, CPSF6, NUDT21, EWSR1,TAF15, SS18L1	Sequestration of specific proteins (mainly RBPs) and RNAs for gene regulation	Cancer	Disease biomarkers
		Viral infection		Suppression of apoptosis in response to certain stresses	Virus infection	Therapeutic target
		p53 activation		Induction of antiviral genes upon virus infection	Neurodegenerative diseases (e.g., ALS, FTD)	
		Hypoxia		Establishment of pregnancy		
				Corpus luteum formation		
				Mammary gland formation		
				Lactation		

rIGS	A-body (also known as nucleolar detention center)	Acidosis	Various proteins including VHL, DNMT1, POLD1, POLA1, MDM2, HAT1, UAP56, Cdk1, Ku70, HDAC2, HSP27, HSP70, HSP90	Induction of physiological amyloidogenesis to store large quantities of proteins	Cancer	Disease biomarkers
		Hypoxia		Ribosome biogenesis arrest		Therapeutic target
		Heat shock		Activation of hypoxic response Entrance to a dormant state in response to stressors		
		Serum starvation				
		DNA damage				
		Aspirin treatment				
		Transcriptional stress				
Satellite III	Nuclear stress body	Heat shock	HSF1, HSF2, SAFB, SRSF1, SRSF7, SRSF9, TDP-43, BRG1, Sam68, TonEBP, NFAT5	Sequestration of specific RBPs and transcription factors	Cancer	Disease biomarkers
		Heavy metal		Stress response and recovery?		Therapeutic target
		Various stresses				
HSAT II	CAST body	Aberrantly expressed in many tumors	MeCP2	Sequestration of MeCP2 for epigenetic regulations	Cancer	Diagnostic use for cancer
						Therapeutic target

(Conrad and Steitz 2005; Brown et al. 2012; Tycowski et al. 2012; Wilusz et al. 2012; Tycowski et al. 2016). In addition to its role in the stabilization of RNAs, the triple-helix structure increases their translation rates (Wilusz et al. 2012). An important role in 3′-end processing of NEAT1_2 is played by a tRNA-like structure, located just after the triple-helix structure, that is required for 3′-end cleavage of the NEAT1_2 transcript (Wilusz et al. 2008). This tRNA-like structure is processed by RNase P, which is involved in tRNA maturation.

Importantly, the long isoform NEAT1_2 is essential for paraspeckle formation, whereas the short isoform NEAT1_1 is dispensable (Naganuma et al. 2012). RNA polymerase II inhibition rapidly disrupts the paraspeckle, suggesting that paraspeckles form co-transcriptionally and are highly dynamic in nature (Fox et al. 2002). This idea is supported by the direct observation of de novo formation of the paraspeckle during transcription of the *NEAT1* locus (Mao et al. 2011). An elegant electron microscopic study demonstrated that NEAT1 is spatially organized within paraspeckles (Souquere et al. 2010). Specifically, the 5′ and 3′ ends of NEAT1_2 are located on the periphery of the paraspeckle, whereas the middle portion is located in the interior. These data indicate that NEAT1 is folded and arranged within the paraspeckle, suggesting that the paraspeckle has a highly ordered structure that may contribute to the formation and functions of this nuclear body. A recent super-resolution microscopic study showed that paraspeckles are typically spherical and that specific proteins are localized to specific domains within a paraspeckle, implying a core–shell spheroidal structure (West et al. 2016).

Several studies have described the molecular functions of NEAT1. For example, NEAT1 regulates several specific types of RNAs, including IR*Alu* (inverted repeated *Alu* elements)-containing RNAs, of which 333 are present in human. mRNAs that contain IR*Alu* in their 3′ UTR are thought to be retained in paraspeckles (Chen and Carmichael 2009). In mouse, CTN mRNAs are retained in a manner dependent upon the paraspeckle component NONO/p54nrb and are exported in response to certain stimuli (Prasanth et al. 2005). Although the biological importance of this phenomenon is unknown, AG-rich RNAs are enriched in paraspeckles at their surface (West et al. 2016). In addition to regulating RNAs, the paraspeckle sequesters proteins and thus controls the free availability of these proteins in the nucleoplasm. Paraspeckle proteins such as SFPQ, which functions as a transcription activator or repressor dependent upon context, are sequestered in paraspeckles, thereby controlling expression of their target genes (Hirose et al. 2014b; Imamura et al. 2014). Together, paraspeckles function in gene regulation as molecular sponges for both RNAs and proteins. A study using the CHART (capture hybridization analysis of RNA targets) method showed that NEAT1 binds actively transcribed genes in specific chromosome loci, suggesting possible roles in direct regulation of these genes (West et al. 2014).

NEAT1 is induced by several stress-related, developmental, and pathological conditions. Proteasome inhibition by compounds such as MG132 and bortezomib induces NEAT1 expression, and NEAT1 knockout (KO) mouse embryonic fibroblasts are sensitive to MG132 treatment (Hirose et al. 2014b). NEAT1_2 is expressed in many cell lines, but not in ES cells. In mice, however, extensive investigation of

NEAT1 expression by in situ hybridization revealed that NEAT1_2 is only expressed in a subset of cell types in tissues, whereas NEAT1_1 is expressed in the majority of cell types (Nakagawa et al. 2011, 2014). Consistent with this observation, paraspeckles are absent from most cells. Among the tissues that do have paraspeckles, NEAT1_2 is highly expressed in the corpus luteum. Consistent with this expression pattern, defects in pregnancy are observed in NEAT1 KO mice (Nakagawa et al. 2014). Also, paraspeckles are assembled in luminal epithelial cells in the mammary gland during development; accordingly, NEAT1 KO mice also exhibit defects in mammary gland development and lactation (Standaert et al. 2014).

In addition to the in vivo functions of NEAT1 under normal conditions, multiple studies have revealed its role in diseases. In particular, several reports have demonstrated the critical importance of NEAT1 in cancer. NEAT1 lncRNA is dysregulated (mainly upregulated) in many types of cancer. Moreover, *NEAT1* is a prominent target gene of tumor suppressor p53 (Blume et al. 2015; Adriaens et al. 2016). NEAT1 expression is induced, and paraspeckles are assembled, by pharmacologically activating p53; alternatively, p53 can also be activated by oncogene-induced replication stress (Adriaens et al. 2016). In this context, NEAT1 prevents DNA damage resulting from replication stress (Adriaens et al. 2016). Paraspeckles form in tumor tissues, whereas normal tissues adjacent to cancer lack the paraspeckles. Strikingly, NEAT1 KO mice exhibit impaired skin tumorigenesis, indicating that NEAT1 promotes tumorigenesis (Adriaens et al. 2016). Furthermore, depletion of NEAT1 sensitizes cancer cells to chemotherapy by modulating DNA damage responses such as the ATR–CHK1 signaling pathway, suggesting a synthetic lethal interaction between NEAT1 and chemotherapeutic agents (Adriaens et al. 2016). Together, these observations strongly suggest that the NEAT1 lncRNA is a prominent target for increasing the genotoxicity of cancer chemotherapeutics.

In addition to being regulated by p53, paraspeckle formation is also induced in response to tumor hypoxia via transcriptional upregulation of NEAT1_2 by HIF2-alpha, thereby promoting cancer cell survival (Choudhry et al. 2014, 2015). Consistent with this, NEAT1 is the most upregulated lncRNA in prostate cancer. NEAT1 expression is regulated by estrogen receptor alpha and is associated with prostate cancer progression (Chakravarty et al. 2014). NEAT1 has been proposed to drive oncogenic growth by promoting epigenetic changes in the promoters of its target genes (Chakravarty et al. 2014). In addition, the short isoform NEAT1_1, but not NEAT1_2, promotes gene expression by inducing active chromatin states in prostate cancer, indicating that the two isoforms have distinct functions in this context. In addition, expression of NEAT1_2, but not NEAT1_1, predicts the response of ovarian cancer to platinum-based chemotherapy, providing further evidence that the isoforms play different roles (Adriaens et al. 2016). Very recently, NEAT1_1 was detected outside of the paraspeckles, where it forms numerous nucleoplasmic "microspeckles," which were proposed to have paraspeckle-independent functions (Li et al. 2017). Accordingly, the precise dissection of NEAT1_1 and NEAT1_2 functions is essential for the development of applications of NEAT1. In addition to the dysregulation of its expression, NEAT1 is also highly mutated in several cancers, including liver cancer, although it remains unclear how these mutations affect NEAT1 functions (Fujimoto et al. 2016).

NEAT1 expression is also induced by viruses, including Japanese encephalitis virus, influenza virus, herpes simplex virus, measles, rabies virus, and hantavirus, as well as in HIV-infected cells (Saha et al. 2006; Zhang et al. 2013; Imamura et al. 2014; Ma et al. 2017). In the case of influenza virus, NEAT1 is induced via the Toll-like receptor 3 (TLR3) pathway, resulting in elongation of paraspeckles (Imamura et al. 2014). The resultant elongated paraspeckles regulate immune-responsive genes, including interleukin-8 (IL-8), by sequestering paraspeckle proteins including SFPQ away from their target gene promoters, as in the case of paraspeckles induced by proteasome inhibition. Taken together, these observations highlight the broad importance of NEAT1 in antiviral responses.

NEAT1 is also implicated in several neurodegenerative diseases. For example, NEAT1 is upregulated in the early stage of amyotrophic lateral sclerosis (ALS), but is not expressed in the same cells under normal conditions (Nishimoto et al. 2013). In addition, NEAT1 is significantly upregulated in the brains of frontotemporal dementia (FTD) patients (Tollervey et al. 2011; Tsuiji et al. 2013). NEAT1 is upregulated in Huntington's disease, and NEAT1_1 expression prevents neuronal death in cell culture, suggesting that NEAT1 plays a protective role in Huntington's (Sunwoo et al. 2017). By contrast, NEAT1 is acutely downregulated in response to neuronal activity (Barry et al. 2017). Many other paraspeckle proteins are linked to neurodegenerative diseases, including TDP-43, FUS, SS18L1, HNRNPA1, SFPQ, EWSR1, TAF15, and HNRNPH1 (Yamazaki and Hirose 2015; Taylor et al. 2016).

As described above, NEAT1 has pleiotropic functions in diseases and functions in a context-dependent manner. Precise dissection of the functions of NEAT1_1 and NEAT1_2 and their roles in physiological conditions is important for the development of applications. Notably in this regard, phosphorothioate-modified antisense oligonucleotides (ASOs) can induce NEAT1-free paraspeckle-like foci in the nucleus (Shen et al. 2014). Similar approaches might enable intervention in the formation of various nuclear bodies, including paraspeckles, in vivo.

11.3.2 rIGS lncRNAs

Stresses induce dramatic changes in silent genomic regions. The nucleolar intergenic spacer (IGS), where heterochromatin normally forms and which is therefore transcriptionally silent, produces ribosomal IGS (rIGS) lncRNAs in response to several stresses, including acidosis, heat shock, hypoxia, serum starvation, aspirin treatment, and DNA damage (Audas et al. 2012a, b, 2016; Jacob et al. 2012, 2013). Knockdown of IGS lncRNAs disrupts the recruitment of target proteins, implying that IGS lncRNAs are arcRNAs (Audas et al. 2012a; Jacob et al. 2013). Under stress, rIGS lncRNAs form large subnucleolar structures called amyloid bodies (A-bodies, also called nucleolar detention centers [DCs]) (Audas et al. 2016). Unlike other nuclear bodies, A-bodies sequester and immobilize key cellular proteins in the nucleolus, as demonstrated by the immobilization of protein components revealed by FRAP (fluorescent recovery after photo-bleaching), proteinase K insensitivity, and staining with amyloid dyes such as 8-anilino-1-naphthalenesulfonate,

Congo red, and Amylo-Glo (Audas et al. 2012a, 2016, Jacob et al. 2013). Proteomic analysis revealed that A-bodies are characterized by physiological amyloids mediated by rIGS RNAs and the amyloid-converting motif (ACM), an arginine/histidine-rich sequence that is enriched in the A-body proteome (Audas et al. 2016). Proteomic analysis also showed that A-bodies have heterogeneous protein compositions and that their components include heat shock proteins such as HSP27, HSP70, and HSP90. The activity of heat shock proteins is required for disassembly of A-bodies, suggesting that the amyloid-like state of its protein components is reversible (Audas et al. 2016). Accordingly, it has been proposed that A-bodies serve to store large quantities of proteins in a dormant state.

There are several IGS lncRNAs, including $rIGS_{28}RNA$, $rIGS_{16}RNA$, $rIGS_{22}RNA$, and $rIGS_{20}RNA$, which are transcribed from regions ~28, 16, 22, and 20 kilobases (kb), respectively, downstream of the rRNA gene loci under specific stress conditions (Audas et al. 2012a). A-bodies constructed by these rIGS lncRNAs are functionally and compositionally distinct. For instance, the stress-responsive transcription factor HIF1A is degraded by the von Hippel–Lindau (VHL) ubiquitin E3 ligase under normal conditions. However, under oxidative stress conditions, $rIGS_{28}RNA$ is induced and sequesters VHL, along with other cellular proteins, in the nucleolus. Knockdown of $rIGS_{28}RNA$ prevents localization of VHL to the nucleolus under oxidative stress (Audas et al. 2012a). In addition to VHL, many nuclear proteins, including DNA methyltransferase 1 (DNMT1) and DNA polymerase delta 1 (POLD1), are also sequestered (Audas et al. 2012a). Under heat shock conditions, HSP70 is sequestered in the nucleolus in a $rIGS_{16}RNA$- and $rIGS_{22}RNA$-dependent manner (Audas et al. 2012a). Moreover, either heat shock or acidosis immobilizes RNA polymerase I subunits and rRNA processing proteins in A-bodies, thereby halting ribosome biogenesis. In addition, MDM2 is sequestered in the nucleolus by $rIGS_{20}RNA$ under transcriptional stress (Audas et al. 2012a; Jacob et al. 2013). These acidic and hypoxic conditions are thought to be prevalent in cancer microenvironments. Consistent with this, in human cancer tissues, $IGS_{28}RNA$ is induced, and several A-body components have been detected (Audas et al. 2016). In addition, $rIGS_{28}RNA$-depleted MCF7 and PC3 cells form larger tumor masses in nude mouse xenograft assay (Audas et al. 2016).

The ability of rIGS lncRNAs to convert proteins into physiological amyloids by inducing a phase transition to a solid state is highly intriguing. Further dissection of specific RNA elements in rIGS lncRNAs may lead to the development of a method to induce such phase transitions in cellular contexts.

11.3.3 Satellite III lncRNA

Satellite III (Sat III) lncRNAs are primate-specific and essential components of the nuclear stress body (nSB) (Cotto et al. 1997; Chiodi et al. 2000; Denegri et al. 2002; Biamonti 2004; Jolly and Lakhotia 2006; Biamonti and Vourc'h 2010). nSBs, which have a diameter of 0.3–3 micrometers, form in response to heat shock and several chemical stress conditions and are present in humans and monkeys, but not in rodents. Their formation is initiated by the transcription of Sat III lncRNAs, which are not expressed in non-stressed conditions, from pericentromeric heterochromatic regions (primarily 9q12 in human) (Chiodi et al. 2000; Denegri et al. 2002; Jolly and Lakhotia 2006; Biamonti and Vourc'h 2010). As with nSBs, Sat III lncRNAs are strongly induced under various stress conditions, including heat shock (Valgardsdottir et al. 2008; Biamonti and Vourc'h 2010). They are thought to be polyadenylated RNAs containing repetitive sequences with multiple tandem GGAAU or GGAGU repeats connected by linker sequences (Choo et al. 1990; Biamonti 2004; Biamonti and Vourc'h 2010). The protein components of nSBs include several transcription factors, including HSF1, HSF2, TonEBP, TDP-43, and Sam68, as well as several splicing factors, including SAFB, SRSF1, SRSF7, and SRSF9 (Denegri et al. 2002; Biamonti 2004; Biamonti and Vourc'h 2010). Formation of nSBs is initiated through a direct interaction between HSF1 and Sat III lncRNA (Biamonti and Vourc'h 2010). Like paraspeckles, the nSB is a dynamic structure, as determined by FRAP measuring HSF1 dynamics (Audas et al. 2016). nSBs are thought to function in stress response and recovery by globally influencing gene expression via sequestration of transcription and splicing factors (Biamonti 2004; Biamonti and Vourc'h 2010). Our group reported that the SWI/SNF complex is required for the formation of nSBs, as well as the paraspeckles, suggesting its general importance in formation of nuclear bodies containing arcRNAs (Kawaguchi and Hirose 2015). When Sat III lncRNAs are knocked down, SRSF1 and SAFB cannot localize to nSBs, although HSF1 still localizes in nSBs, suggesting that Sat III can act as a scaffold for RBPs (Valgardsdottir et al. 2008). HSF1 is essential for the integrity of nSBs (Goenka et al. 2016). Consistent with this, RRM2 of SRSF1 is required for its targeting to nSBs, suggesting that splicing factors are recruited to Sat III lncRNA via direct RNA binding (Chiodi et al. 2004). Furthermore, artificial tethering experiments showed that Sat III lncRNA can initiate de novo formation of nSBs, indicating that it plays an architectural role (Shevtsov and Dundr 2011). In addition to stress conditions, SAT III is upregulated in A431 epidermoid carcinoma cells, as well as in several senescent cells, implying that it plays a role in cancer and senescence (Enukashvily et al. 2007).

11.3.4 HSATII RNA

In addition to the Sat III lncRNA, other satellite DNAs are transcribed into noncoding RNAs (ncRNAs) under specific conditions. High-copy satellite II (HSATII) DNA is normally methylated and silenced, but DNA methylation of HSATII is frequently lost in cancer (Ting et al. 2011). Hence, HSATII RNA is aberrantly expressed in many tumors. Small blocks of HSATII are found in the pericentromeres of 11 human chromosomes, including chromosomes 2, 5, 7, 10, 13, 14, 15, 17, 21, 22, and Y (Hall et al. 2017). The transcribed HSATII RNA forms distinct large nuclear RNA foci, termed cancer-associated satellite transcript (CAST) bodies (Hall et al. 2017). These CAST bodies act as molecular sponges to sequester master epigenetic regulatory proteins, including MeCP2 (methyl-CpG-binding protein 2) and are thought to influence the epigenome in cancer cells. In that context, HSATII DNA and RNA form nuclear foci called cancer-associated Polycomb (CAP) bodies from the 1q12 mega-satellite, which also act as molecular sponges (Hall et al. 2017). CAP bodies sequester Polycomb group complex component PRC1. Sequestrations by CSAT and CAP bodies cause epigenetic instability, which is recognized as a hallmark of cancer (Hall et al. 2017). HSATII RNA is highly expressed in cancers, including pancreatic cancer, and is also expressed in preneoplastic pancreatic lesions, suggesting that it could serve as a biomarker for pancreatic cancer (Ting et al. 2011). HSATII RNA is also found in circulating blood of cancer patients; accordingly, a sensitive detection system has been developed to identify this biomarker (Kishikawa et al. 2016). In general, cancer-specific nuclear bodies constructed by HSATII RNA and DNA may serve as biomarkers of epigenetic instability, with diagnostic utility in several cancers.

11.3.5 Other arcRNAs

Our group has screened for nuclear structures that are diminished by RNase treatment. This approach has identified known and novel nuclear bodies, including the Sam68 nuclear body and a novel structure called the DBC1 body (Mannen et al. 2016). These findings suggest that these nuclear bodies, which are present in a subset of cancer cells, are constructed by unidentified arcRNAs. Interestingly, the Sam68 and DBC1 nuclear bodies are joined by the adaptor protein HNRNPL, warranting investigation of the underlying mechanism for this feature (Mannen et al. 2016). The number of validated arcRNAs is still limited, but we recently reported a method for identifying arcRNA candidates on a genome-wide scale (Chujo et al. 2017). This technique takes advantage of a characteristic feature of NEAT1, namely, that it is difficult to extract by conventional RNA extraction methods using acid–guanidine–phenol–chloroform (AGPC) reagents (e.g., Thermo Scientific, TRIzol). Shearing with a needle or heating improves the extraction efficiency of NEAT1. We termed this feature “semi-extractability” and RNAs with this property as “semi-extractable RNAs (seRNAs)” (Chujo et al. 2017). In addition to NEAT1, another arcRNA, rIGS$_{16}$RNA, was semi-extractable, suggesting that this feature might help identify novel arcRNAs. A comparison of the expression levels of RNAs extracted by the conventional and improved methods revealed seRNAs throughout the genome, including 50 seRNAs in HeLa cells, as determined by RNA-seq (Chujo et al. 2017) (Fig. 11.2). Some of the seRNAs formed distinct nuclear foci that were distant from the transcription sites, indicating the formation of nuclear bodies (Chujo et al. 2017). The list of seRNAs contains several putative arcRNAs, including the LINE-1 and SPA, supporting the utility of this method for exploring arcRNAs at a genome-wide scale.

In addition to endogenous nuclear bodies, repeat RNAs form nuclear foci and play critical roles in pathogenesis in microsatellite expansion diseases such as myotonic dystrophy, a genetic disorder with multisystemic symptoms (Wojciechowska and Krzyzosiak 2011; Belzil et al. 2013; Nelson et al. 2013; Ramaswami et al. 2013;

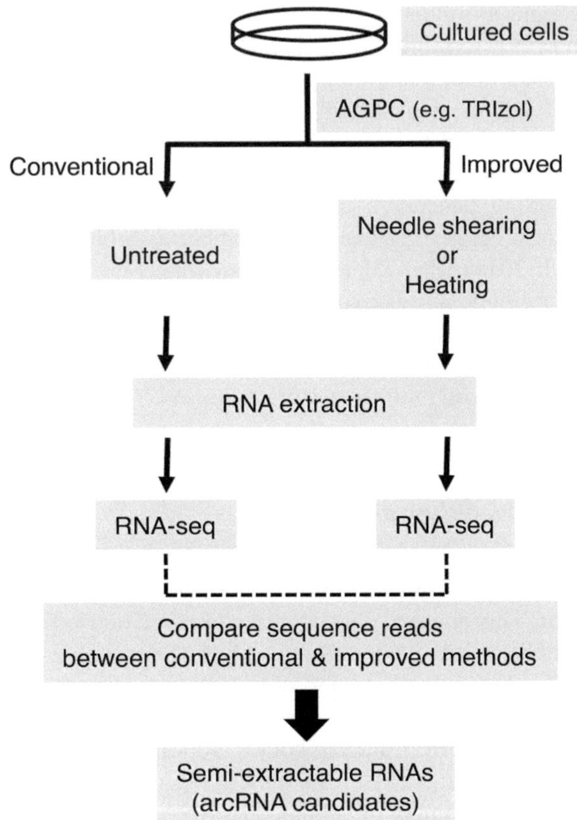

Fig. 11.2 An experimental procedure to search semi-extractable RNAs. An example of workflows to search semi-extractable RNAs from cultured cells is shown. The cells are suspended in AGPC reagents such as TRIzol and divided into two groups for conventional and improved extraction methods. In the conventional method, the samples are directly subjected to RNA extraction. In the improved method, prior to RNA extraction, the samples are subjected to the needle shearing or heating. Both of the extracted RNAs are subjected to RNA-seq analyses. After the sequencing, the sequence reads in both samples are compared. Semi-extractable RNAs will be enriched in the RNAs extracted by the improved method

Mohan et al. 2014). Typically, such RNA granules sequester proteins. For example, expanded CTG or CCTG repeats are present in genomic DNA of patients with myotonic dystrophy type 1 or type 2, respectively (Wojciechowska and Krzyzosiak 2011). These repeats are transcribed into RNAs and form RNA foci that sequester RBP and MBNL; in mice, KOs of the corresponding genes cause myotonic dystrophy-like phenotypes (Kanadia et al. 2003). These data indicate that disease symptoms originate from toxic RNAs. Thus, understanding the mechanisms that dictate the biogenesis of these structures, which could be related to endogenous nuclear bodies, will help to develop therapeutic methods for these disorders.

11.3.6 Commonalities and Potential Applications of arcRNAs

In this section, I summarize the commonalities of arcRNAs and arcRNA-constructed nuclear bodies and discuss the reasons why RNA is used as a scaffold or platform for nuclear bodies. In addition, I address potential applications of arcRNAs.

First, most arcRNAs serve as the molecular sponges for proteins and, in some cases, RNAs (Biamonti and Vourc'h 2010; Hirose et al. 2014b; Imamura et al. 2014; Audas et al. 2016; Hall et al. 2017). It is possible that biochemical reactions and RNP assembly can also take place in these compartments, as observed for proteinaceous nuclear bodies. The components of nuclear bodies are usually dynamic (Fox et al. 2002). In the case of rIGS lncRNAs, the component proteins are completely detained within the bodies (Audas et al. 2012a, 2016). These data suggest that the arcRNAs induce phase transitions among material states (i.e., liquid, hydrogel, and solid) and sequester different factors into distinct states.

Second, many protein components of arcRNA-constructed nuclear bodies contain proteins harboring PLDs, LCDs, or intrinsically disordered domains, suggesting that aggregation-prone sequences play a role in assembly of these bodies (Yamazaki and Hirose 2015). In fact, PLD-containing paraspeckle proteins are essential for the integrity of paraspeckles (Hennig et al. 2015). Whereas most proteins localizing in arcRNA-constructed nuclear bodies are dispensable for the integrity of the bodies, a few of them play an essential role in maintaining nuclear bodies, suggesting that they represent essential core factors.

Third, transcription of arcRNAs is essential for nucleation, leading to assembly of nuclear bodies. Shutdown of transcription quickly eliminates these bodies, indicating that arcRNA-regulated nuclear bodies are dynamically controlled and that sequestration by arcRNAs is reversible (Fox et al. 2002). Consistent with this, the arcRNAs characterized so far are all nuclear ncRNAs induced under stress and disease conditions (Chujo et al. 2016). The rIGS, Sat III, and HSATII lncRNAs are transcriptionally silenced under normal conditions but are dramatically induced in response to stress or disease (Biamonti and Vourc'h 2010; Audas et al. 2012a, 2016; Hall et al. 2017). Several arcRNAs are aberrantly expressed in disease such as cancer and neurodegenerative disorders (Yamazaki and Hirose 2015; Hall et al. 2017). Therefore, it is conceivable that the primary physiological roles of arcRNAs are related to stress and disease, raising the possibility that identification of arcRNA candidates as seRNAs under various conditions might reveal important regulatory lncRNAs (Chujo et al. 2017).

Fourth, lncRNAs have an advantage as the architectural core of nuclear bodies: they do not require a frame for protein translation. Consequently, their sequences are presumably only constrained by the requirement to form RNA–protein interaction, and arcRNAs have more flexibility to add and change their sequences than proteins, in which changes might cause aggregation. This advantage allows lncRNAs to increase the diversity of their binding partners and the combinations of proteins that are incorporated into nuclear bodies, likely via interactions with unique sequences or structures irrelevant to their protein–protein interaction potential (Chujo et al. 2016). This feature enables the formation of a wide range of nuclear bodies, allowing more rapid adaptation to demands for circumstantial change. Because there are more than 1000 RBPs in human cells, the RBPs that bind to arcRNAs can in turn sequester proteins that interact with RBPs (Baltz et al. 2012; Castello et al. 2012, 2016). The nature of this scaffolding should be suitable for integrating diverse regulatory proteins into specific sites.

Fifth, various repetitive sequences in NEAT1 lncRNA and other arcRNAs derived from satellite regions contain, or consist primarily of, repetitive sequences (Ulitsky and Bartel 2013; Chujo et al. 2016). Half of our genome comprises repeat sequences, including SINEs, LINEs, pseudogenes, endogenous viruses, and other repeats (Steitz 2015). Although these repetitive sequences are usually ignored in genomic and transcriptome analyses, arcRNAs derived from these repetitive sequences could be important regulators of the genome.

Overall, many recent studies have demonstrated that nuclear bodies are phase-separated into liquid, hydrogel, or solid states in cells (Courchaine et al. 2016; Banani et al. 2017). RNA is a suitable molecule for nucleating nuclear bodies by inducing phase separation. Specifically, arcRNAs are thought to induce phase separation by increasing the local concentration of proteins containing PLDs or LCDs and/or via direct RNA–RNA interaction (Chujo et al. 2016). The arcRNAs validated to date induce phase separation into different material states, as mentioned above. These states are likely to be defined by the sequences of arcRNAs, which serve as scaffolds for specific RBPs. By understanding the principles underlying the regulation of nuclear bodies by RNA elements in arcRNAs, it might be possible to artificially control phase separation among several material states that have different molecular dynamics in cells. This could lead in turn to the development of biochemical micro-reactors, cellular compartments that can achieve specific biochemical reactions by recruiting specific sets of protein and nucleic acids. This strategy could enable intervention in cellular functions such as gene and epigenome regulation. In addition, as noted above, phosphorothioate-modified ASOs can induce NEAT1-free paraspeckle-like foci in the nucleus, and ASOs have recently been approved for therapeutic uses in diseases such as spinal muscular atrophy (SMA). Accordingly, ASO or related methods hold promise for control of phase separation and nuclear body formation in vivo (Shen et al. 2014).

arcRNAs are an emerging class of lncRNAs for potential therapeutic targets and other applications. RNA-seq taking advantage of semi-extractability could launch a new era in arcRNA research and lead to the identification of new biomarkers, therapeutic targets for diseases, and methods for intervening in cellular functions (Chujo et al. 2017).

11.4 Conclusion and Future Perspectives

Decoding RNA elements and their protein partners is analogous to understanding the domains or motifs of proteins (Hirose et al. 2014a). In the case of proteins, this understanding has led to the development of inhibitors or activators of specific functions of multifunctional proteins. Similarly, the understanding of lncRNA elements would help develop methods to repress or activate specific functions of lncRNAs, although currently we are only capable of modulating all functions of a given lncRNA. Locked nucleic acid (LNA) has been used to target domains of XIST lncRNA that is essential for X chromosome inactivation and can repress domain-specific functions by blocking the interaction between XIST lncRNA and specific proteins (Sarma et al. 2010). Furthermore, RNA can dynamically change its secondary structure. Therefore, specific inhibition or activation of lncRNA domains could be achieved by modulating the corresponding RNA structures. It will be important to focus on the RNA elements and to accumulate knowledge about these elements.

We now have access to CRISPR interference (CRISPRi), a new method for exploring the functions of lncRNA on a genome-wide scale, as described above (Liu et al. 2017). This is a very powerful tool for exploring functionally important disease-linked lncRNAs. Because lncRNAs are expressed in specific tissues and cells, they are suitable targets for therapeutics with minimal side effects. In addition, many new approaches have been developed for investigating the biological importance of lncRNAs. One of them, CRISPR-mediated KO of lncRNAs in model organisms, is very useful for investigating the roles of these RNAs under physiological conditions. Although CRISPR is powerful, some lncRNAs are generated from multiple genomic loci, and therefore it might be difficult to repress their functions. In this case, an alternative strategy would be direct knockdown of lncRNAs. Because many lncRNAs are localized in the nucleus, siRNA-based methods are usually not capable of targeting them efficiently. However, knockdown using ASOs represents a powerful alternative approach for exploring the functions of nuclear-retained lncRNAs (Ideue et al. 2009). In addition, several methods have been designed to investigate the RNA elements in lncRNAs by elucidating the interactions between lncRNAs and proteins; these include RNA-centric technologies such as ChIRP (chromatin isolation by RNA purification), CHART, and RAP (RNA antisense purification) and protein-centric technologies such as CLIP (cross-linking immunoprecipitation) (Kashi et al. 2016). In addition, technologies for probing RNA secondary structure would also be useful (Lu and Chang 2016). Together, these new strategies will help to expand our understanding of RNA elements in lncRNAs.

References

Adriaens C, Standaert L, Barra J, Latil M, Verfaillie A, Kalev P, Boeckx B, Wijnhoven PW, Radaelli E, Vermi W, Leucci E, Lapouge G, Beck B, van den Oord J, Nakagawa S, Hirose T, Sablina AA, Lambrechts D, Aerts S, Blanpain C, Marine JC (2016) p53 induces formation of NEAT1 lncRNA-containing paraspeckles that modulate replication stress response and chemosensitivity. Nat Med 22:861–868

Aguzzi A, Altmeyer M (2016) Phase separation: linking cellular compartmentalization to disease. Trends Cell Biol 26:547–558

Alberti S, Hyman AA (2016) Are aberrant phase transitions a driver of cellular aging? BioEssays 38:959–968

Anderson DM, Anderson KM, Chang CL, Makarewich CA, Nelson BR, McAnally JR, Kasaragod P, Shelton JM, Liou J, Bassel-Duby R, Olson EN (2015) A micropeptide encoded by a putative long noncoding RNA regulates muscle performance. Cell 160:595–606

Ashwal-Fluss R, Meyer M, Pamudurti NR, Ivanov A, Bartok O, Hanan M, Evantal N, Memczak S, Rajewsky N, Kadener S (2014) circRNA biogenesis competes with pre-mRNA splicing. Mol Cell 56:55–66

Audas TE, Jacob MD, Lee S (2012a) Immobilization of proteins in the nucleolus by ribosomal intergenic spacer noncoding RNA. Mol Cell 45:147–517

Audas TE, Jacob MD, Lee S (2012b) The nucleolar detention pathway: a cellular strategy for regulating molecular networks. Cell Cycle 11:2059–2062

Audas TE, Audas DE, Jacob MD, Ho JJ, Khacho M, Wang M, Perera JK, Gardiner C, Bennett CA, Head T, Kryvenko ON, Jorda M, Daunert S, Malhotra A, Trinkle-Mulcahy L, Gonzalgo ML, Lee S (2016) Adaptation to stressors by systemic protein amyloidogenesis. Dev Cell 39:155–168

Baltz AG, Munschauer M, Schwanhäusser B, Vasile A, Murakawa Y, Schueler M, Youngs N, Penfold-Brown D, Drew K, Milek M, Wyler E, Bonneau R, Selbach M, Dieterich C, Landthaler M (2012) The mRNA-bound proteome and its global occupancy profile on protein-coding transcripts. Mol Cell 46:674–690

Banani SF, Lee HO, Hyman AA, Rosen MK (2017) Biomolecular condensates: organizers of cellular biochemistry. Nat Rev Mol Cell Biol 18:285–298

Barry G, Briggs JA, Hwang DW, Nayler SP, Fortuna PR, Jonkhout N, Dachet F, Maag JL, Mestdagh P, Singh EM, Avesson L, Kaczorowski DC, Ozturk E, Jones NC, Vetter I, Arriola-Martinez L, Hu J, Franco GR, Warn VM, Gong A, Dinger ME, Rigo F, Lipovich L, Morris MJ, O'Brien TJ, Lee DS, Loeb JA, Blackshaw S, Mattick JS, Wolvetang EJ (2017) The long non-coding RNA NEAT1 is responsive to neuronal activity and is associated with hyperexcitability states. Sci Rep 7:40127

Batista PJ, Chang HY (2013) Long noncoding RNAs: cellular address codes in development and disease. Cell 152:1298–1307

Belzil VV, Gendron TF, Petrucelli L (2013) RNA-mediated toxicity in neurodegenerative disease. Mol Cell Neurosci 56:406–419

Bertone P, Stolc V, Royce TE, Rozowsky JS, Urban AE, Zhu X, Rinn JL, Tongprasit W, Samanta M, Weissman S, Gerstein M, Snyder M (2004) Global identification of human transcribed sequences with genome tiling arrays. Science 306:2242–2246

Biamonti G (2004) Nuclear stress bodies: a heterochromatin affair? Nat Rev Mol Cell Biol 5:493–498

Biamonti G, Vourc'h C (2010) Nuclear stress bodies. Cold Spring Harb Perspect Biol 2:a000695

Blume CJ, Hotz-Wagenblatt A, Hüllein J, Sellner L, Jethwa A, Stolz T, Slabicki M, Lee K, Sharathchandra A, Benner A, Dietrich S, Oakes CC, Dreger P, te Raa D, Kater AP, Jauch A, Merkel O, Oren M, Hielscher T, Zenz T (2015) p53-dependent non-coding RNA networks in chronic lymphocytic leukemia. Leukemia 29:2015–2023

Brown JA, Valenstein ML, Yario TA, Tycowski KT, Steitz JA (2012) Formation of triple-helical structures by the 3′-end sequences of MALAT1 and MENβ noncoding RNAs. Proc Natl Acad Sci U S A 109:19202–19207

Carninci P, Kasukawa T, Katayama S, Gough J, Frith MC, Maeda N, Oyama R, Ravasi T, Lenhard B, Wells C, Kodzius R, Shimokawa K, Bajic VB, Brenner SE, Batalov S, Forrest AR, Zavolan M, Davis MJ, Wilming LG, Aidinis V, Allen JE, Ambesi-Impiombato A, Apweiler R, Aturaliya RN, Bailey TL, Bansal M, Baxter L, Beisel KW, Bersano T, Bono H, Chalk AM, Chiu KP, Choudhary V, Christoffels A, Clutterbuck DR, Crowe ML, Dalla E, Dalrymple BP, de Bono B, Della Gatta G, di Bernardo D, Down T, Engstrom P, Fagiolini M, Faulkner G, Fletcher CF, Fukushima T, Furuno M, Futaki S, Gariboldi M, Georgii-Hemming P, Gingeras TR, Gojobori T, Green RE, Gustincich S, Harbers M, Hayashi Y, Hensch TK, Hirokawa N, Hill D, Huminiecki L, Iacono M, Ikeo K, Iwama A, Ishikawa T, Jakt M, Kanapin A, Katoh M, Kawasawa Y, Kelso J, Kitamura H, Kitano H, Kollias G, Krishnan SP, Kruger A, Kummerfeld SK, Kurochkin IV, Lareau LF, Lazarevic D, Lipovich L, Liu J, Liuni S, McWilliam S, Madan Babu M, Madera M, Marchionni L, Matsuda H, Matsuzawa S, Miki H, Mignone F, Miyake S, Morris K, Mottagui-Tabar S, Mulder N, Nakano N, Nakauchi H, Ng P, Nilsson R, Nishiguchi S, Nishikawa S, Nori F, Ohara O, Okazaki Y, Orlando V, Pang KC, Pavan WJ, Pavesi G, Pesole G, Petrovsky N, Piazza S, Reed J, Reid JF, Ring BZ, Ringwald M, Rost B, Ruan Y, Salzberg SL, Sandelin A, Schneider C, Schönbach C, Sekiguchi K, Semple CA, Seno S, Sessa L,

Sheng Y, Shibata Y, Shimada H, Shimada K, Silva D, Sinclair B, Sperling S, Stupka E, Sugiura K, Sultana R, Takenaka Y, Taki K, Tammoja K, Tan SL, Tang S, Taylor MS, Tegner J, Teichmann SA, Ueda HR, van Nimwegen E, Verardo R, Wei CL, Yagi K, Yamanishi H, Zabarovsky E, Zhu S, Zimmer A, Hide W, Bult C, Grimmond SM, Teasdale RD, Liu ET, Brusic V, Quackenbush J, Wahlestedt C, Mattick JS, Hume DA, Kai C, Sasaki D, Tomaru Y, Fukuda S, Kanamori-Katayama M, Suzuki M, Aoki J, Arakawa T, Iida J, Imamura K, Itoh M, Kato T, Kawaji H, Kawagashira N, Kawashima T, Kojima M, Kondo S, Konno H, Nakano K, Ninomiya N, Nishio T, Okada M, Plessy C, Shibata K, Shiraki T, Suzuki S, Tagami M, Waki K, Watahiki A, Okamura-Oho Y, Suzuki H, Kawai J, Hayashizaki Y, FANTOM Consortium, RIKEN Genome Exploration Research Group, GenomeScience Group (Genome Network Project Core Group) (2005) The transcriptional landscape of the mammalian genome. Science 309:1559–1563

Carrieri C, Cimatti L, Biagioli M, Beugnet A, Zucchelli S, Fedele S, Pesce E, Ferrer I, Collavin L, Santoro C, Forrest AR, Carninci P, Biffo S, Stupka E, Gustincich S (2012) Long non-coding antisense RNA controls Uchl1 translation through an embedded SINEB2 repeat. Nature 491:454–457

Castello A, Fischer B, Eichelbaum K, Horos R, Beckmann BM, Strein C, Davey NE, Humphreys DT, Preiss T, Steinmetz LM, Krijgsveld J, Hentze MW (2012) Insights into RNA biology from an atlas of mammalian mRNA-binding proteins. Cell 149:1393–1406

Castello A, Fischer B, Frese CK, Horos R, Alleaume AM, Foehr S, Curk T, Krijgsveld J, Hentze MW (2016) Comprehensive identification of RNA-binding domains in human cells. Mol Cell 63:696–710

Cech TR, Steitz JA (2014) The noncoding RNA revolution-trashing old rules to forge new ones. Cell 157:77–94

Chakravarty D, Sboner A, Nair SS, Giannopoulou E, Li R, Hennig S, Mosquera JM, Pauwels J, Park K, Kossai M, MacDonald TY, Fontugne J, Erho N, Vergara IA, Ghadessi M, Davicioni E, Jenkins RB, Palanisamy N, Chen Z, Nakagawa S, Hirose T, Bander NH, Beltran H, Fox AH, Elemento O, Rubin MA (2014) The oestrogen receptor alpha-regulated lncRNA NEAT1 is a critical modulator of prostate cancer. Nat Commun 5:5383

Chen LL (2016a) The biogenesis and emerging roles of circular RNAs. Nat Rev Mol Cell Biol 17:205–211

Chen LL (2016b) Linking long noncoding RNA localization and function. Trends Biochem Sci 41:761–772

Chen LL, Carmichael GG (2009) Altered nuclear retention of mRNAs containing inverted repeats in human embryonic stem cells: functional role of a nuclear noncoding RNA. Mol Cell 35:467–478

Chen CK, Blanco M, Jackson C, Aznauryan E, Ollikainen N, Surka C, Chow A, Cerase A, McDonel P, Guttman M (2016) Xist recruits the X chromosome to the nuclear lamina to enable chromosome-wide silencing. Science 354:468–472

Chiodi I, Biggiogera M, Denegri M, Corioni M, Weighardt F, Cobianchi F, Riva S, Biamonti G (2000) Structure and dynamics of hnRNP-labelled nuclear bodies induced by stress treatments. J Cell Sci 113(Pt 22):4043–4053

Chiodi I, Corioni M, Giordano M, Valgardsdottir R, Ghigna C, Cobianchi F, Xu RM, Riva S, Biamonti G (2004) RNA recognition motif 2 directs the recruitment of SF2/ASF to nuclear stress bodies. Nucleic Acids Res 32:4127–4136

Choo KH, Earle E, Mcquikkan C (1990) A homologous subfamily of satellite III DNA on human chromosomes 14 and 22. Nucleic Acids Res 18:5641–5648

Choudhry H, Schödel J, Oikonomopoulos S, Camps C, Grampp S, Harris AL, Ratcliffe PJ, Ragoussis J, Mole DR (2014) Extensive regulation of the non-coding transcriptome by hypoxia: role of HIF in releasing paused RNApol2. EMBO Rep 15:70–76

Choudhry H, Albukhari A, Morotti M, Haider S, Moralli D, Smythies J, Schödel J, Green CM, Camps C, Buffa F, Ratcliffe P, Ragoussis J, Harris AL, Mole DR (2015) Tumor hypoxia induces nuclear paraspeckle formation through HIF-2alpha dependent transcriptional activation of NEAT1 leading to cancer cell survival. Oncogene 34:4546

Chu C, Zhang QC, da Rocha ST, Flynn RA, Bharadwaj M, Calabrese JM, Magnuson T, Heard E, Chang HY (2015) Systematic discovery of Xist RNA binding proteins. Cell 161:404–416
Chujo T, Yamazaki T, Hirose T (2016) Architectural RNAs (arcRNAs): a class of long noncoding RNAs that function as the scaffold of nuclear bodies. Biochim Biophys Acta 1859:139–146
Chujo T, Yamazaki T, Kawaguchi T, Kurosaka S, Takumi T, Nakagawa S, Hirose T (2017) Unusual semi-extractability as a hallmark of nuclear body-associated architectural noncoding RNAs. EMBO J 36:1447–1462
Clemson CM, Hutchinson JN, Sara SA, Ensminger AW, Fox AH, Chess A, Lawrence JB (2009) An architectural role for a nuclear noncoding RNA: NEAT1 RNA is essential for the structure of paraspeckles. Mol Cell 33:717–726
Conrad NK, Steitz JA (2005) A Kaposi's sarcoma virus RNA element that increases the nuclear abundance of intronless transcripts. EMBO J 24:1831–1841
Cotto J, Fox S, Morimoto R (1997) HSF1 granules: a novel stress-induced nuclear compartment of human cells. J Cell Sci 110(Pt 23):2925–2934
Courchaine EM, Lu A, Neugebauer KM (2016) Droplet organelles? EMBO J 35:1603–1612
Denegri M, Moralli D, Rocchi M, Biggiogera M, Raimondi E, Cobianchi F, De Carli L, Riva S, Biamonti G (2002) Human chromosomes 9, 12, and 15 contain the nucleation sites of stress-induced nuclear bodies. Mol Biol Cell 13:2069–2079
Engreitz JM, Haines JE, Perez EM, Munson G, Chen J, Kane M, McDonel PE, Guttman M, Lander ES (2016a) Local regulation of gene expression by lncRNA promoters, transcription and splicing. Nature 539:452–455
Engreitz JM, Ollikainen N, Guttman M (2016b) Long non-coding RNAs: spatial amplifiers that control nuclear structure and gene expression. Nat Rev Mol Cell Biol 17:756–770
Engström PG, Suzuki H, Ninomiya N, Akalin A, Sessa L, Lavorgna G, Brozzi A, Luzi L, Tan SL, Yang L, Kunarso G, Ng EL, Batalov S, Wahlestedt C, Kai C, Kawai J, Carninci P, Hayashizaki Y, Wells C, Bajic VB, Orlando V, Reid JF, Lenhard B, Lipovich L (2006) Complex Loci in human and mouse genomes. PLoS Genet 2:e47
Enukashvily NI, Donev R, Waisertreiger IS, Podgornaya OI (2007) Human chromosome 1 satellite 3 DNA is decondensed, demethylated and transcribed in senescent cells and in A431 epithelial carcinoma cells. Cytogenet Genome Res 118:42–54
Fong KW, Li Y, Wang W, Ma W, Li K, Qi RZ, Liu D, Songyang Z, Chen J (2013) Whole-genome screening identifies proteins localized to distinct nuclear bodies. J Cell Biol 203:149–164
Fox AH, Lam YW, Leung AK, Lyon CE, Andersen J, Mann M, Lamond AI (2002) Paraspeckles: a novel nuclear domain. Curr Biol 12:13–25
Fujimoto A, Furuta M, Totoki Y, Tsunoda T, Kato M, Shiraishi Y, Tanaka H, Taniguchi H, Kawakami Y, Ueno M, Gotoh K, Ariizumi S, Wardell CP, Hayami S, Nakamura T, Aikata H, Arihiro K, Boroevich KA, Abe T, Nakano K, Maejima K, Sasaki-Oku A, Ohsawa A, Shibuya T, Nakamura H, Hama N, Hosoda F, Arai Y, Ohashi S, Urushidate T, Nagae G, Yamamoto S, Ueda H, Tatsuno K, Ojima H, Hiraoka N, Okusaka T, Kubo M, Marubashi S, Yamada T, Hirano S, Yamamoto M, Ohdan H, Shimada K, Ishikawa O, Yamaue H, Chayama K, Miyano S, Aburatani H, Shibata T, Nakagawa H (2016) Whole-genome mutational landscape and characterization of noncoding and structural mutations in liver cancer. Nat Genet 48:500–509
Geisler S, Coller J (2013) RNA in unexpected places: long non-coding RNA functions in diverse cellular contexts. Nat Rev Mol Cell Biol 14:699–712
Goenka A, Sengupta S, Pandey R, Parihar R, Mohanta GC, Mukerji M, Ganesh S (2016) Human satellite-III non-coding RNAs modulate heat-shock-induced transcriptional repression. J Cell Sci 129:3541–3552
Gong C, Maquato LE (2011) lncRNAs transactivate STAU1-mediated mRNA decay by duplexing with 3′ UTRs via Alu elements. Nature 470:284–288
Guarnerio J, Bezzi M, Jeong JC, Paffenholz SV, Berry K, Naldini MM, Lo-Coco F, Tay Y, Beck AH, Pandolfi PP (2016) Oncogenic role of fusion-circRNAs derived from cancer-associated chromosomal translocations. Cell 165:289–302

Guttman M, Rinn JL (2012) Modular regulatory principles of large non-coding RNAs. Nature 482:339–346

Hall LL, Byron M, Carone DM, Whitfield TW, Pouliot GP, Fischer A, Jones P, Lawrence JB (2017) Demethylated HSATII DNA and HSATII RNA foci sequester PRC1 and MeCP2 into cancer-specific nuclear bodies. Cell Rep 18:2943–2956

Hansen TB, Jensen TI, Clausen BH, Bramsen JB, Finsen B, Damgaard CK, Kjems J (2013) Natural RNA circles function as efficient microRNA sponges. Nature 495:384–388

Hennig S, Kong G, Mannen T, Sadowska A, Kobelke S, Blythe A, Knott GJ, Iyer KS, Ho D, Newcombe EA, Hosoki K, Goshima N, Kawaguchi T, Hatters D, Trinkle-Mulcahy L, Hirose T, Bond CS, Fox AH (2015) Prion-like domains in RNA binding proteins are essential for building subnuclear paraspeckles. J Cell Biol 210:529–539

Hirose T, Mishima Y, Tomari Y (2014a) Elements and machinery of non-coding RNAs: toward their taxonomy. EMBO Rep 15:489–507

Hirose T, Virnicchi G, Tanigawa A, Naganuma T, Li R, Kimura H, Yokoi T, Nakagawa S, Bénard M, Fox AH, Pierron G (2014b) NEAT1 long noncoding RNA regulates transcription via protein sequestration within subnuclear bodies. Mol Biol Cell 25:169–183

Ideue T, Hino K, Kitao S, Yokoi T, Hirose T (2009) Efficient oligonucleotide-mediated degradation of nuclear noncoding RNAs in mammalian cultured cells. RNA 15:1578–1587

Imamura K, Imamachi N, Akizuki G, Kumakura M, Kawaguchi A, Nagata K, Kato A, Kawaguchi Y, Sato H, Yoneda M, Kai C, Yada T, Suzuki Y, Yamada T, Ozawa T, Kaneki K, Inoue T, Kobayashi M, Kodama T, Wada Y, Sekimizu K, Akimitsu N (2014) Long noncoding RNA NEAT1-dependent SFPQ relocation from promoter region to paraspeckle mediates IL8 expression upon immune stimuli. Mol Cell 53:393–406

Indrieri A, Grimaldi C, Zucchelli S, Tammaro R, Gustincich S, Franco B (2016) Synthetic long non-coding RNAs [SINEUPs] rescue defective gene expression in vivo. Sci Rep 6:27315

Jacob MD, Audas TE, Mullineux ST, Lee S (2012) Where no RNA polymerase has gone before: novel functional transcripts derived from the ribosomal intergenic spacer. Nucleus 3:315–319

Jacob MD, Audas TE, Uniacke J, Trinkle-Mulcahy L, Lee S (2013) Environmental cues induce a long noncoding RNA-dependent remodeling of the nucleolus. Mol Biol Cell 24:2943–2953

Jolly C, Lakhotia SC (2006) Human sat III and Drosophila hsr omega transcripts: a common paradigm for regulation of nuclear RNA processing in stressed cells. Nucleic Acids Res 34:5508–5514

Kaikkonen MU, Spann NJ, Heinz S, Romanoski CE, Allison KA, Stender JD, Chun HB, Tough DF, Prinjha RK, Benner C, Glass CK (2013) Remodeling of the enhancer landscape during macrophage activation is coupled to enhancer transcription. Mol Cell 51:310–325

Kanadia RN, Johnstone KA, Mankodi A, Lungu C, Thornton CA, Esson D, Timmers AM, Hauswirth WW, Swanson MS (2003) A muscleblind knockout model for myotonic dystrophy. Science 302:1978–1980

Kapranov P, Cheng J, Dike S, Nix DA, Duttagupta R, Willingham AT, Stadler PF, Hertel J, Hackermüller J, Hofacker IL, Bell I, Cheung E, Drenkow J, Dumais E, Patel S, Helt G, Ganesh M, Ghosh S, Piccolboni A, Sementchenko V, Tammana H, Gingeras TR (2007) RNA maps reveal new RNA classes and a possible function for pervasive transcription. Science 316:1484–1488

Kashi K, Henderson L, Bonetti A, Carninci P (2016) Discovery and functional analysis of lncRNAs: methodologies to investigate an uncharacterized transcriptome. Biochim Biophys Acta 1859:3–15

Katayama S, Tomaru Y, Kasukawa T, Waki K, Nakanishi M, Nakamura M, Nishida H, Yap CC, Suzuki M, Kawai J, Suzuki H, Carninci P, Hayashizaki Y, Wells C, Frith M, Ravasi T, Pang KC, Hallinan J, Mattick J, Hume DA, Lipovich L, Batalov S, Engström PG, Mizuno Y, Faghihi MA, Sandelin A, Chalk AM, Mottagui-Tabar S, Liang Z, Lenhard B, Wahlestedt C, RIKEN Genome Exploration Research Group, Genome Science Group (Genome Network Project Core Group), FANTOM Consortium (2005) Antisense transcription in the mammalian transcriptome. Science 309:1564–1566

Kawaguchi T, Hirose T (2015) Chromatin remodeling complexes in the assembly of long noncoding RNA-dependent nuclear bodies. Nucleus 6:462–467

Kawaguchi T, Tanigawa A, Naganuma T, Ohkawa Y, Souquere S, Pierron G, Hirose T (2015) SWI/SNF chromatin-remodeling complexes function in noncoding RNA-dependent assembly of nuclear bodies. Proc Natl Acad Sci U S A 112:4304–4309

Kishikawa T, Otsuka M, Yoshikawa T, Ohno M, Yamamoto K, Yamamoto N, Kotani A, Koike K (2016) Quantitation of circulating satellite RNAs in pancreatic cancer patients. JCI Insight 1:e86646

Kretz M, Siprashvili Z, Chu C, Webster DE, Zehnder A, Qu K, Lee CS, Flockhart RJ, Groff AF, Chow J, Johnston D, Kim GE, Spitale RC, Flynn RA, Zheng GX, Aiyer S, Raj A, Rinn JL, Chang HY, Khavari PA (2013) Control of somatic tissue differentiation by the long non-coding RNA TINCR. Nature 493:231–235

Lee M, Sadowska A, Bekere I, Ho D, Gully BS, Lu Y, Iyer KS, Trewhella J, Fox AH, Bond CS (2015) The structure of human SFPQ reveals a coiled-coil mediated polymer essential for functional aggregation in gene regulation. Nucleic Acids Res 43:3826–3840

Lee S, Kopp F, Chang TC, Sataluri A, Chen B, Sivakumar S, Yu H, Xie Y, Mendell JT (2016) Noncoding RNA NORAD regulates genomic stability by sequestering PUMILIO proteins. Cell 164:69–80

Legnini I, Di Timoteo G, Rossi F, Morlando M, Briganti F, Sthandier O, Fatica A, Santini T, Andronache A, Wade M, Laneve P, Rajewsky N, Bozzoni I (2017) Circ-ZNF609 is a circular RNA that can be translated and functions in myogenesis. Mol Cell 66:22–37

Li R, Harvey AR, Hodgetts SI, Fox AH (2017) Functional dissection of NEAT1 using genome editing reveals substantial localization of the NEAT1_1 isoform outside paraspeckles. RNA 23:872–881

Liang D, Wilusz JE (2014) Short intronic repeat sequences facilitate circular RNA production. Genes Dev 28:2233–2247

Liu SJ, Horlbeck MA, Cho SW, Birk HS, Malatesta M, He D, Attenello FJ, Villalta JE, Cho MY, Chen Y, Mandegar MA, Olvera MP, Gilbert LA, Conklin BR, Chang HY, Weissman JS, Lim DA (2017) CRISPRi-based genome-scale identification of functional long noncoding RNA loci in human cells. Science 355:aah7111

Lu Z, Chang HY (2016) Decoding the RNA structurome. Curr Opin Struct Biol 36:142–148

Ma H, Han P, Ye W, Chen H, Zheng X, Cheng L, Zhang L, Yu L, Wu X, Xu Z, Lei Y, Zhang F (2017) The long noncoding RNA NEAT1 exerts antihantaviral effects by acting as positive feedback for RIG-I signaling. J Virol 91:e02250–e02216

Mannen T, Yamashita S, Tomita K, Goshima N, Hirose T (2016) The Sam68 nuclear body is composed of two RNase-sensitive substructures joined by the adaptor HNRNPL. J Cell Biol 214:45–59

Mao YS, Sunwoo H, Zhang B, Spector DL (2011) Direct visualization of the co-transcriptional assembly of a nuclear body by noncoding RNAs. Nat Cell Biol 13:95–101

Matsumoto A, Pasut A, Matsumoto M, Yamashita R, Fung J, Monteleone E, Saghatelian A, Nakayama KI, Clohessy JG, Pandolfi PP (2017) mTORC1 and muscle regeneration are regulated by the LINC00961-encoded SPAR polypeptide. Nature 541:228–232

McHugh CA, Chen CK, Chow A, Surka CF, Tran C, McDonel P, Pandya-Jones A, Blanco M, Burghard C, Moradian A, Sweredoski MJ, Shishkin AA, Su J, Lander ES, Hess S, Plath K, Guttman M (2015) The Xist lncRNA interacts directly with SHARP to silence transcription through HDAC3. Nature 521:232–236

Memczak S, Jens M, Elefsinioti A, Torti F, Krueger J, Rybak A, Maier L, Mackowiak SD, Gregersen LH, Munschauer M, Loewer A, Ziebold U, Landthaler M, Kocks C, le Noble F, Rajewsky N (2013) Circular RNAs are a large class of animal RNAs with regulatory potency. Nature 495:333–338

Mohan A, Goodwin M, Swanson MS (2014) RNA-protein interactions in unstable microsatellite diseases. Brain Res 1584:3–14

Naganuma T, Nakagawa S, Tanigawa A, Sasaki YF, Goshima N, Hirose T (2012) Alternative 3′-end processing of long noncoding RNA initiates construction of nuclear paraspeckles. EMBO J 31:4020–4034

Nakagawa S, Naganuma T, Shioi G, Hirose T (2011) Paraspeckles are subpopulation-specific nuclear bodies that are not essential in mice. J Cell Biol 193:31–39

Nakagawa S, Shimada M, Yanaka K, Mito M, Arai T, Takahashi E, Fujita Y, Fujimori T, Standaert L, Marine JC, Hirose T (2014) The lncRNA Neat1 is required for corpus luteum formation and the establishment of pregnancy in a subpopulation of mice. Development 141:4618–4627

Nelson DL, Orr HT, Warren ST (2013) The unstable repeats--three evolving faces of neurological disease. Neuron 77:825–843

Nelson BR, Makarewich CA, Anderson DM, Winders BR, Troupes CD, Wu F, Reese AL, McAnally JR, Chen X, Kavalali ET, Cannon SC, Houser SR, Bassel-Duby R, Olson EN (2016) A peptide encoded by a transcript annotated as long noncoding RNA enhances SERCA activity in muscle. Science 351:271–275

Nishimoto Y, Nakagawa S, Hirose T, Okano HJ, Takao M, Shibata S, Suyama S, Kuwako K, Imai T, Murayama S, Suzuki N, Okano H (2013) The long non-coding RNA nuclear-enriched abundant transcript 1_2 induces paraspeckle formation in the motor neuron during the early phase of amyotrophic lateral sclerosis. Mol Brain 6:31

Paralkar VR, Taborda CC, Huang P, Yao Y, Kossenkov AV, Prasad R, Luan J, Davies JO, Hughes JR, Hardison RC, Blobel GA, Weiss MJ (2016) Unlinking an lncRNA from its associated cis element. Mol Cell 62:104–110

Passon DM, Lee M, Rackham O, Stanley WA, Sadowska A, Filipovska A, Fox AH, Bond CS (2012) Structure of the heterodimer of human NONO and paraspeckle protein component 1 and analysis of its role in subnuclear body formation. Proc Natl Acad Sci U S A 109:4846–4850

Ponting CP, Oliver PL, Reik W (2009) Evolution and functions of long noncoding RNAs. Cell 136:629–641

Prasanth KV, Prasanth SG, Xuan Z, Hearn S, Freier SM, Bennett CF, Zhang MQ, Spector DL (2005) Regulating gene expression through RNA nuclear retention. Cell 123:249–263

Quinn JJ, Chang HY (2016) Unique features of long non-coding RNA biogenesis and function. Nat Rev Genet 17:47–62

Ramaswami M, Taylor JP, Parker R (2013) Altered ribostasis: RNA-protein granules in degenerative disorders. Cell 154:727–736

Rinn JL, Chang HY (2012) Genome regulation by long noncoding RNAs. Annu Rev Biochem 81:145–166

Saha S, Sugumar P, Bhandari P, Rangarajan PN (2006) Identification of Japanese encephalitis virus-inducible genes in mouse brain and characterization of GARG39/IFIT2 as a microtubule-associated protein. J Gen Virol 87:3285–3289

Salzman J (2016) Circular RNA expression: its potential regulation and function. Trends Genet 32:309–316

Sarma K, Levasseur P, Aristarkhov A, Lee JT (2010) Locked nucleic acids (LNAs) reveal sequence requirements and kinetics of Xist RNA localization to the X chromosome. Proc Natl Acad Sci U S A 107:22196–22201

Sasaki YT, Ideue T, Sano M, Mituyama T, Hirose T (2009) MENepsilon/beta noncoding RNAs are essential for structural integrity of nuclear paraspeckles. Proc Natl Acad Sci U S A 106:2525–2530

Schmitt AM, Chang HY (2016) Long noncoding RNAs in cancer pathways. Cancer Cell 29:452–463

Shen W, Liang XH, Crooke ST (2014) Phosphorothioate oligonucleotides can displace NEAT1 RNA and form nuclear paraspeckle-like structures. Nucleic Acids Res 42:8648–8662

Shevtsov SP, Dundr M (2011) Nucleation of nuclear bodies by RNA. Nat Cell Biol 13:167–173

Souquere S, Beauclair G, Harper F, Fox A, Pierron G (2010) Highly ordered spatial organization of the structural long noncoding NEAT1 RNAs within paraspeckle nuclear bodies. Mol Biol Cell 21:4020–4027

Standaert L, Adriaens C, Radaelli E, Van Keymeulen A, Blanpain C, Hirose T, Nakagawa S, Marine JC (2014) The long noncoding RNA Neat1 is required for mammary gland development and lactation. RNA 20:1844–1849

Steitz J (2015) RNA-RNA base-pairing: theme and variations. RNA 21:476–477

Sunwoo H, Dinger ME, Wilusz JE, Amaral PP, Mattick JS, Spector DL (2009) MEN epsilon/beta nuclear-retained non-coding RNAs are up-regulated upon muscle differentiation and are essential components of paraspeckles. Genome Res 19:347–359

Sunwoo JS, Lee ST, Im W, Lee M, Byun JI, Jung KH, Park KI, Jung KY, Lee SK, Chu K, Kim M (2017) Altered expression of the long noncoding RNA NEAT1 in Huntington's disease. Mol Neurobiol 54:1577–1586

Takahashi H, Carninci P (2014) Widespread genome transcription: new possibilities for RNA therapies. Biochem Biophys Res Commun 452:294–301

Taniue K, Kurimoto A, Sugimasa H, Nasu E, Takeda Y, Iwasaki K, Nagashima T, Okada-Hatakeyama M, Oyama M, Kozuka-Hata H, Hiyoshi M, Kitayama J, Negishi L, Kawasaki Y, Akiyama T (2016) Long noncoding RNA *UPAT* promotes colon tumorigenesis by inhibiting degradation of UHRF1. Proc Natl Acad Sci U S A 113:1273–1278

Taylor JP, Brown RH Jr, Cleveland DW (2016) Decoding ALS: from genes to mechanism. Nature 539:197–206

Ting DT, Lipson D, Paul S, Brannigan BW, Akhavanfard S, Coffman EJ, Contino G, Deshpande V, Iafrate AJ, Letovsky S, Rivera MN, Bardeesy N, Maheswaran S, Haber DA (2011) Aberrant overexpression of satellite repeats in pancreatic and other epithelial cancers. Science 331:593–596

Tollervey JR, Curk T, Rogelj B, Briese M, Cereda M, Kayikci M, König J, Hortobágyi T, Nishimura AL, Zupunski V, Patani R, Chandran S, Rot G, Zupan B, Shaw CE, Ule J (2011) Characterizing the RNA targets and position-dependent splicing regulation by TDP-43. Nat Neurosci 14:452–458

Tsuiji H, Iguchi Y, Furuya A, Kataoka A, Hatsuta H, Atsuta N, Tanaka F, Hashizume Y, Akatsu H, Murayama S, Sobue G, Yamanaka K (2013) Spliceosome integrity is defective in the motor neuron diseases ALS and SMA. EMBO Mol Med 5:221–234

Tycowski KT, Shu MD, Borah S, Shi M, Steitz JA (2012) Conservation of a triple-helix-forming RNA stability element in noncoding and genomic RNAs of diverse viruses. Cell Rep 2:26–32

Tycowski KT, Shu MD, Steitz JA (2016) Myriad triple-helix-forming structures in the transposable element RNAs of plants and fungi. Cell Rep 15:1266–1276

Ulitsky I, Bartel DP (2013) lincRNAs: genomics, evolution, and mechanisms. Cell 154:26–46

Uversky VN (2016) Intrinsically disordered proteins in overcrowded milieu: membrane-less organelles, phase separation, and intrinsic disorder. Curr Opin Struct Biol 44:18–30

Valgardsdottir R, Chiodi I, Giordano M, Rossi A, Bazzini S, Ghigna C, Riva S, Biamonti G (2008) Transcription of Satellite III non-coding RNAs is a general stress response in human cells. Nucleic Acids Res 36:423–434

Wahlestedt C (2013) Targeting long non-coding RNA to therapeutically upregulate gene expression. Nat Rev Drug Discov 12:433–446

West JA, Davis CP, Sunwoo H, Simon MD, Sadreyev RI, Wang PI, Tolstorukov MY, Kingston RE (2014) The long noncoding RNAs NEAT1 and MALAT1 bind active chromatin sites. Mol Cell 55:791–802

West JA, Mito M, Kurosaka S, Takumi T, Tanegashima C, Chujo T, Yanaka K, Kingston RE, Hirose T, Bond C, Fox A, Nakagawa S (2016) Structural, super-resolution microscopy analysis of paraspeckle nuclear body organization. J Cell Biol 214:817–830

Wilusz JE, Freier SM, Spector D (2008) 3′ end processing of a long nuclear-retained noncoding RNA yields a tRNA-like cytoplasmic RNA. Cell 135:919–932

Wilusz JE, JnBaptiste CK, Lu LY, Kuhn CD, Joshua-Tor L, Sharp PA (2012) A triple helix stabilizes the 3′ ends of long noncoding RNAs that lack poly(A) tails. Genes Dev 26:2392–2407

Wojciechowska M, Krzyzosiak WJ (2011) Cellular toxicity of expanded RNA repeats: focus on RNA foci. Hum Mol Genet 20:3811–3821

Wu H (2013) Higher-order assemblies in a new paradigm of signal transduction. Cell 153:287–292

Yamazaki T, Hirose T (2015) The building process of the functional paraspeckle with long non-coding RNAs. Front Biosci (Elite Ed) 7:1–41

Yoon JH, Abdelmohsen K, Kim J, Yang X, Martindale JL, Tominaga-Yamanaka K, White EJ, Orjalo AV, Rinn JL, Kreft SG, Wilson GM, Gorospe M (2013) Scaffold function of long non-coding RNA HOTAIR in protein ubiquitination. Nat Commun 4:2939

Zhang Q, Chen CY, Yedavalli VS, Jeang KT (2013) NEAT1 long noncoding RNA and paraspeckle bodies modulate HIV-1 posttranscriptional expression. MBio 4:e00596–e00512

Zucchelli S, Cotella D, Takahashi H, Carrieri C, Cimatti L, Fasolo F, Jones MH, Sblattero D, Sanges R, Santoro C, Persichetti F, Carninci P, Gustincich S (2015a) SINEUPs: a new class of natural and synthetic antisense long non-coding RNAs that activate translation. RNA Biol 12:771–779

Zucchelli S, Fasolo F, Russo R, Cimatti L, Patrucco L, Takahashi H, Jones MH, Santoro C, Sblattero D, Cotella D, Persichetti F, Carninci P, Gustincich S (2015b) SINEUPs are modular antisense long non-coding RNAs that increase synthesis of target proteins in cells. Front Cell Neurosci 9:174

Zucchelli S, Patrucco L, Persichetti F, Gustincich S, Cotella D (2016) Engineering translation in mammalian cell factories to increase protein yield: the unexpected use of long non-coding SINEUP RNAs. Comput Struct Biotechnol J 14:404–410

Chapter 12
Long Noncoding RNA as Novel Cancer Diagnostic and Effective Therapeutic Targets

Eleonora Leucci

Abstract Recently large-scale genomic analyses have reported that the vast majority of somatic copy number alterations found in human cancers map to transcribed regions lacking protein-coding potential. This is not surprising considering that, although most of the human genome is transcribed into various classes of RNAs, less than 2% of them encodes for proteins. Among them, long noncoding RNAs (lncRNAs) constitute one of the most abundant classes of RNA, defined as RNA longer than 200 nucleotides not coding for proteins. Interestingly, lncRNAs are often expressed in a tissue- and cancer-specific manner, and their expression can be easily manipulated in vivo using antisense oligonucleotides (ASOs) making them attractive cancer-selective markers and therapeutic targets.

Emerging biochemical evidence has revealed an incredible functional diversity for lncRNAs as they can recruit chromatin-modifying proteins, organize nuclear architecture, regulate mRNA stability and translation by competing with microRNA and RNA-binding proteins, and even alter protein localization and function. Accordingly, lncRNAs are emerging as important regulators of cancer initiation and progression. Here we discuss the role of lncRNAs in cancer and their potential as a new promising avenue for the advancement of cancer cell-specific therapeutic design.

Keywords lncRNAs · Cancer · Biomarkers · Oligonucleotides · ASO · siRNAs · Targeted therapy · Small molecules

---

E. Leucci (✉)
Department of Oncology, Laboratory of RNA Cancer Biology, LKI Leuven Cancer Institute, KU Leuven-University of Leuven, Leuven, Belgium
e-mail: eleonora.leucci@KUleuven.be

S. Masuda, S. Izawa (eds.), *Applied RNA Bioscience*,
https://doi.org/10.1007/978-981-10-8372-3_12

12.1 Introduction

The discovery of somatic mutations in protein-coding genes set the basis for targeted therapy and led to major breakthroughs in cancer treatment as exemplified by the introduction of the BRAF inhibitor in melanoma care (Sosman et al. 2012). However, despite the latest advancement in treatment, cancer is still the second leading cause of death worldwide (source: WHO). This is partly due to the fact that protein-coding genes account only for 2% of the human genome, while recent genome-wide studies have revealed that the vast majority of recurrent somatic mutations (Melton et al. 2015), copy number alterations (Beroukhim et al. 2010), and cancer-related SNPs (Cheetham et al. 2013) are found in the so-called gene deserts, namely, regions lacking known protein-coding genes. Interestingly enough, most of the gene desert regions are extensively transcribed into long noncoding RNAs (The FANTOM Consortium 2005).

The term “lncRNAs” embraces a heterogeneous class of transcripts longer than 200 nucleotides and lacking evidence for coding potential (Bánfai et al. 2012; Guttman et al. 2013). Being generally transcribed by RNA polymerase II, lncRNAs are often capped and polyadenylated (Quinn and Chang 2016), and many of them associate with ribosomes (Ingolia et al. 2014; van Heesch et al. 2014); it is therefore still under debate which intrinsic feature and/or transacting factor avoids their translation to proteins.

Analysis of conservation shows that lncRNAs share only short and 5′-biased areas of conserved sequences and that their gene architecture is extensively remodeled during evolution partly by transposable elements (Hezroni et al. 2015). Nevertheless, unexpectedly enough many lncRNAs still have evolutionary conserved functions in spite of their poor sequence conservation (Ulitsky et al. 2011). These findings could be reconciled by the fact that lncRNAs are free from codon preservation constraints, and therefore secondary and tertiary structure may be more important than the primary sequence for their functions (Quinn and Chang 2016). However, in a recent study, Rivas and colleagues tested known lncRNA structures with a new computational tool and found no evidence for structures in lncRNAs (Rivas et al. 2017), thus renewing the controversies about the importance of these transcripts in light of the evolution.

In terms of mechanism of action, lncRNAs are believed to play a large array of molecular roles, ranging from recruitment of chromatin remodeling enzymes (Yap et al. 2010) to sponging of microRNAs and proteins (Poliseno et al. 2010; Kino et al. 2010), and therefore they were implicated in virtually all physiological and pathological processes (Fatica and Bozzoni 2014; Huarte 2015; Lorenzen and Thum 2016).

In addition, considering their tissue- and cancer-specific expression profiles (Cabili et al. 2011; Iyer et al. 2015), long noncoding RNAs are very promising biomarkers and therapeutic targets. Here I discuss the current understanding of the role of lncRNAs in cancer and their potential in the clinic, using the most prominent papers in the field.

12.2 lncRNAs as Products of Aberrant Biology

Due to their cancer-specific expression, lncRNA signatures have better prognostic, diagnostic, and predictive values than protein-coding signatures (Brunner et al. 2012; Derrien et al. 2012; Iyer et al. 2015; Yan et al. 2015). Some lncRNAs (e.g., *SAMMSON*) are not only specifically expressed in one particular cancer, but they are also missing or dispensable in healthy adult tissues. Although it is not possible to exclude that they are expressed in specific stem cell niches or at specific embryonic stages, a possibility is that the expression of these transcripts is simply the by-product of "aberrant biology" caused by genetic (e.g., translocation, amplifications, or mutations of the corresponding DNA) or epigenetic events.

An example of a lncRNA whose overexpression in cancer is driven by an aberrant genetic event is *PVT1*. Located 50 kb downstream of MYC gene, in the region 8q24 frequently amplified in several human cancers, *PVT1* is required for high MYC expression, and its ablation decreases the oncogenic potential of colon cancer cells in xenograft models (Tseng et al. 2014).

Conversely, an epigenetic mechanism leading to aberrant biology is the inappropriate activation of normally dormant, cryptic promoters and/or enhancers derived from ERVs (endogenous retroviral sequences) (Babaian and Mager 2016). Such a mechanism is not uncommon in cancer where hypomethylation of endogenous retroviral sequences has been reported (Szpakowski et al. 2009).

In striking contrast with protein-coding genes (Kelley and Rinn 2012), 97% of lncRNA TSS (transcription start sites) are made of sequences derived from retrotransposons (Göke and Ng 2016). In particular, lncRNA promoters are highly enriched in ERVs and depleted in LINEs (long interspersed nuclear elements) and SINEs (short interspersed nuclear elements) (Kelley and Rinn 2012). In keeping with this, the presence of ERVs at the TSS (transcription start site) characterizes a subclass of lncRNAs with lower and more cell-specific expression (Kelley and Rinn 2012). This phenomenon could therefore be the reason behind lncRNA-specific expression.

A potential example of a lncRNA aberrantly induced by ERV reactivation is *SAMMSON*. Similarly to PVT1, SAMMSON lays 32 kb downstream of an important oncogene (MITF) in a region amplified in 10–15% of melanoma patients. However, in contrast with PVT1, *SAMMSON* does not affect MITF expression and it is virtually undetectable in any adult tissue, although expressed in more than 95% of melanoma patients (Leucci et al. 2016). This last observation suggests that amplification of 3p13–3p14 region is not sufficient to justify the expression of *SAMMSON* in the vast majority of melanoma patients. Other melanoma-specific mechanisms, triggering *SAMMSON* transcription, must exist. In keeping with this, one possibility is that *SAMMSON* expression is driven by aberrant demethylation of an endogenous retroviral sequence in its promoter. Supporting this idea, two different groups recently showed that SAMMSON promoter contains an endogenous retroviral sequence, specifically an LTR1A2 which is primate specific (Babaian and Mager 2016; Göke and Ng 2016). The functionality of LTR1A2 as a promoter seems to be supported by the fact that *SAMMSON* itself is conserved only in primate and could not be detected in mouse melanoma and corresponding normal skin (unpublished data).

The proposed mechanisms are not mutually exclusive and they are equally fascinating. In the next future, a better annotation of lncRNAs and their promoter regions will probably refine our understanding of lncRNA cancer-specific expression.

12.3 Roles of lncRNAs in Cancer

Mounting evidence from in vitro studies supports a link between lncRNA deregulation and cancer development and progression. However only few genetic in vivo models, showing a clear implication of lncRNAs in cancer development and/or progression, exist (Yildirim et al. 2013; Adriaens et al. 2016; Arun et al. 2016; Mello et al. 2017). This is partly due to the tedious procedures needed to obtain appropriate knockout mice, before the advent of the CRISPR-CAS9, and partly to the lack of conservation of a number of lncRNAs. A valid alternative, in case of primate-specific lncRNAs, is the use of patient-derived xenograft models (Leucci et al. 2016), a particularly valuable approach also when the mouse models do not entirely recapitulate the biology of the human tumor.

In spite of the small number of genetic models, a plethora of molecular mechanisms for lncRNAs in cancer have been already described, and the idea that lncRNAs have evolved in highly creative ways to hijack the cellular machinery is emerging from several studies.

Perhaps one of the most interesting raising concepts concerns the idea that noncoding RNAs may contribute to cytoplasmic and nuclear architecture by inducing the phase transition required for the formation of membrane-less bodies (Lin et al. 2015; Zhang et al. 2015). Such a concept is not novel per se since it is well-known that nucleoli are formed around rRNA transcription foci (Berry et al. 2015). It is, however, revolutionizing our understanding of protein-protein and protein-RNA interactions and may help reconcile observations that are otherwise difficult to explain when the interacting molecules are not present in a classical stoichiometric relationship and/or when loss-of-function experiments targeting lowly abundant RNA species still lead to profound phenotypic changes. According to this model, a local increase in a single RNA and/or protein specie would be sufficient to trap surrounding proteins and RNA in a viscous environment where they are forced to interact (Bergeron-Sandoval et al. 2016). Albeit phase transition and granule formation have well-known connections with neurodegeneration (Jain and Vale 2017), they have been largely ignored by the cancer community. Notably, the long noncoding RNA NEAT1, which has recently been unambiguously implicated in cancer (Adriaens et al. 2016; Mello et al. 2017) and neurodegeneration (Nishimoto et al. 2013), is well-known for its ability to induce nuclear membrane-less bodies known as paraspeckles (Hirose et al. 2014). We can speculate that by characterizing more lncRNAs, we will probably find more cancer-related lncRNAs acting through similar mechanism.

Although less revolutionary, other intriguing mechanisms of action exist. NORAD, for instance, is an abundant lncRNA that contains many tandem repeats recognized by RNA-binding proteins of the Pumilio family (PUM1-2). The physiological role of PUM proteins is the posttranscriptional repression of genes implicated in mitosis and DNA replication and repair. By binding PUMILIO proteins, NORAD keeps their activity in check, and therefore loss of this lncRNA in human cancer cells results in dramatic aneuploidy (Lee et al. 2016).

Another ingenious mechanism is adopted by PTENP to ensure proper expression of tumor suppressor PTEN, the negative regulator of AKT/PKB pathway (Ortega-Molina and Serrano 2013). Being a PTEN pseudogene, PTENP shares with it high sequence similarity, and it can therefore bind and sponge PTEN-targeting miRNAs. However, PTENP is often lost in cancer, thus leading to miRNA-dependent decrease in PTEN expression (Poliseno et al. 2010).

One last intriguing example is Zeb2NAT (Zeb2 natural antisense transcript), which is responsible for inclusion of a large intron, containing an internal ribosome entry site (IRES), in the 5'-UTR of Zeb2 mRNA. During the induction of EMT (epithelial-mesenchymal transition), an essential process in metastatic progression of epithelial tumors, Zeb2NAT gets upregulated, thus resulting in increased Zeb2 translation through the IRES and downregulation of E-cadherin (Beltran et al. 2008).

The few examples described above confirm that RNA is the only molecule that can exert also DNA- and protein-associated functions, and as perfect members of this category, lncRNAs are flexible machines that work through a virtually unlimited number of molecular mechanisms. Imagination is therefore the only limit.

12.4 lncRNA as Biomarkers

Early diagnosis and treatment are essential for a positive outcome in the cure of many types of cancers. For this reason, novel biomarkers able to predict responses to therapy and less invasive methods for the detection of malignant lesions are urgently needed.

As biomarkers, lncRNAs can be detected using two different approaches (Fig. 12.1): by RNA FISH or by RT-qPCR/digital PCR-based assays using small amount of material extracted from cancer tissues or liquid biopsies. These methods are very sensitive and they are relatively simple to implement, compared to antibody-based techniques that can be tedious and time-consuming. However, not all tissue-specific transcripts are present and stable in these fluids, and often available material is scarce and does not always allow subsampling for different purposes. In this case, RNA-FISH on sections is a far more convenient alternative, with advantage to combine information derived from the histology and morphology of the tissue with the expression and localization of the lncRNA of interest. Alteration of the cellular

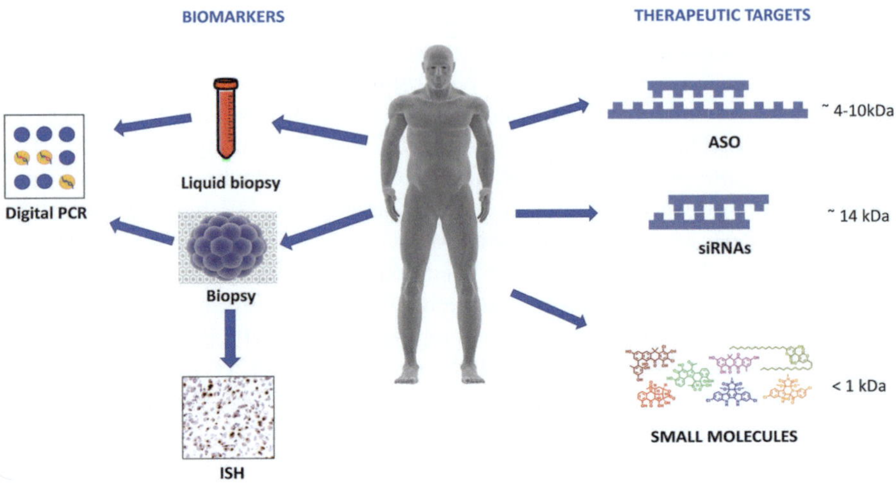

Fig. 12.1 lncRNA as biomarkers and therapeutic targets. lncRNAs show lineage- and disease-specific expression; therefore, they have a great potential as diagnostic markers and therapeutic targets. They can be detected by in situ hybridization (ISH) in regular biopsies or by digital PCR in regular and liquid biopsies. As therapeutic targets, they can be targeted by short interfering RNAs (siRNAs), antisense nucleotides (ASO), or small molecules

localization of a specific lncRNAs might be per se a marker of disease, and therefore its detection by FISH/ISH could be very informative. Lastly, another advantage of the assay is the possibility to combine it with antibody-based methods commonly used for diagnostics. One major limitation of this approach when compared to RT-qPCR is its relatively poor sensitivity, thus decreasing its potential for the detection of many lowly expressed lncRNAs. Nonetheless, methods that amplify the signal are available (e.g., TSA or RNAscope) and relatively easy to implement.

A promising example is *SAMMSON* that is expressed specifically in malignant (vertical growth phase) and not in benign (radial growth phase) lesions in melanoma (Leucci et al. 2016). For this reason, *SAMMSON* could be helpful in the diagnosis of neoplasms that contain lesions with overlapping benign and malignant histopathological features (Shain and Bastian 2016). Another cancer-specific lncRNA is *CCAT1-L* which is exclusively transcribed in colorectal cancer (and not in normal mucosa) from a locus upstream of MYC (Nissan et al. 2012). *CCAT1-L* is expressed already in early lesions, and its levels are constantly high throughout the development of the disease. Moreover, detection of *CCAT1-L* could also be used as a noninvasive diagnostic tool, since it can be retrieved in peripheral blood of 40% of the patients and not in healthy controls (Nissan et al. 2012).

Another example is *PCA3*, a biomarker specifically expressed in prostate cancer and already approved by FDA (Food and Drug Administration) (Bussemakers et al. 1999). *PCA3* can be detected by RT-qPCR in fine needle biopsies and in body fluids, including urine and blood (de Kok et al. 2002).

lncRNA expression has not only diagnostic but also prognostic and predictive value. *HOTAIR*, for instance, was shown to predict poor survival in ovarian cancer patients treated with carboplatin. The same group of patients had then improved responses to cisplatin (Teschendorff et al. 2015). Two more examples are *SChLAP1*, whose high expression is predictive of metastasis within 10 years in prostate cancer (Prensner et al. 2014), and *NEAT1* (long transcript) that reliably predicts response to platinum-based chemotherapy in ovarian cancer (Adriaens et al. 2016).

Considering that lncRNA profiles can classify tumors in clinically relevant groups (Brunner et al. 2012; Iyer et al. 2015; Yan et al. 2015) better than mRNAs, that they can be detected with noninvasive tests, and that they can even be predictive of responses to therapy (Teschendorff et al. 2015; Prensner et al. 2014; Adriaens et al. 2016), lncRNAs could become valuable allies to aid pathologists in issuing an unequivocal diagnosis.

12.5 lncRNA as Therapeutic Targets

Due to their cell type and disease-selective expression profiles, lncRNAs are not only good biomarkers but also attractive targets for the design of anticancer therapies (Fig. 12.1). Two strategies are currently available to interfere with the oncogenic functions of a given lncRNA, direct silencing of lncRNA expression and pharmacological inhibition of its function, both achievable using oligonucleotide-based therapeutics.

Even though they are very versatile, oligonucleotides are often negatively charged and relatively bulky to cross the plasma membrane; therefore, another attractive possibility is the use of small molecules that target RNA-protein interactions, thus directly interfering with specific lncRNA functions. Small molecules are low molecular weight and can be natural or synthetic compounds (including lipids, monosaccharide, and metabolites). Being low molecular weight (less than 1 kDa) and uncharged, many of them can cross the blood-brain barrier, and therefore they could be an attractive option for patients with brain metastasis. To my knowledge, the only example of a study trying to develop a high-throughput method to screen for small molecules is designed to target lncRNA interactions with EZH2 (Pedram Fatemi et al. 2015).

Yet, aside the aforementioned delivery issues, RNA-based therapeutics are still the preferable drugs since they can coevolve easily with their target, are relatively inexpensive, and can be designed to target either lncRNA expression or function.

For depletion of lncRNA expression, an attractive option is the use of aptamer-conjugated small interfering RNAs (siRNA). Aptamers are single-stranded nucleic acid molecules folded into a particular conformation that recognize a specific 3D structure present in their protein or RNA targets (Mayer et al. 2007). They can be designed to recognize specific surface receptors and therefore to be internalized selectively by cancer cells, thus reducing the side effects of non-cancer cell targeting (Sun et al. 2014). Although, currently, not much is known about the structure of

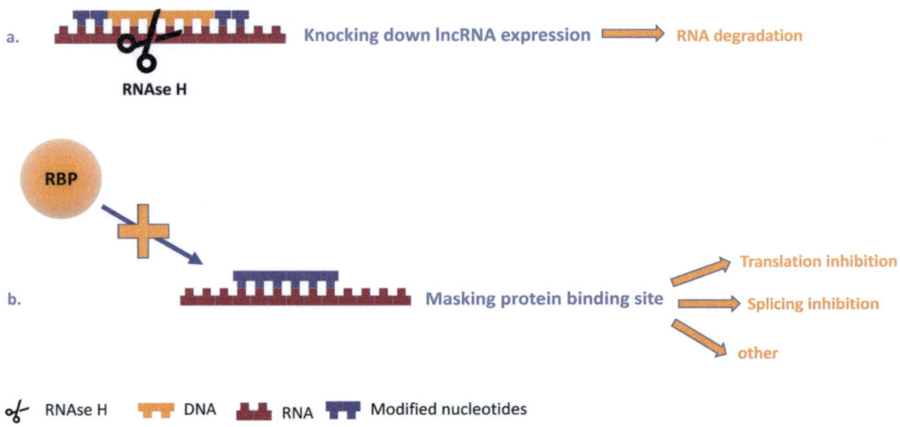

Fig. 12.2 Different strategies to interfere with lncRNA functions. To target lncRNA, two different strategies exist: (**a**) decreasing lncRNA expression (GapmeR approach) or (**b**) interfering with lncRNA-specific functions by impeding the binding of important protein regulators (steric blocker approach)

specific lncRNAs, aptamers could also be used in the future, to target the functions of a lncRNA by hampering the binding of a protein partner (Fig. 12.2).

A caveat of the siRNA approach is their inability to work in certain subcellular compartments. lncRNAs localized into mitochondria, for example, cannot be targeted as mitochondria lack GW182, an important protein needed for the assembly of a functional silencing complexes (Zhang et al. 2014). In addition, siRNAs have numerous hybridization-dependent and hybridization-independent off-target effects. These last ones are being caused by competing with miRNA for RISC (RNA-induced silencing complex) complexes.

An alternative compatible with the clinic and relatively inexpensive is the design of ASO (antisense oligonucleotide)-based drugs. ASOs are single-stranded antisense molecules, usually between 12 and 25 nucleotide in length, which can be used to induce degradation or to block the function of target lncRNAs. Currently, there are 4 FDA-approved ASOs and more than 50 in different phases of clinical trials. Degradation of the target by ASOs involves the recruitment of RNase H by base pairing of the DNA antisense to its RNA target with consequent cleavage of the DNA/RNA hybrid. Compounds triggering RNAse H activity usually contain a stretch of DNA oligonucleotides flanked by modified nucleotides. Since ASO use RNase H for knocking down the target, they have the advantage over siRNAs of not interfering with the RNAi machinery. On the other end ASOs suffer more of hybridization-independent toxicity caused by accumulation of the drug, aptamers binding and pro-inflammatory effects.

Both ASOs and siRNAs can be chemically modified on the sugar moiety or on the backbone to improve the cellular uptake, the stability, and/or the specificity (Fig. 12.3). PS (phosphorothioate) backbone was the first modification to be introduced to improve membrane penetration of the oligonucleotides (Frazier

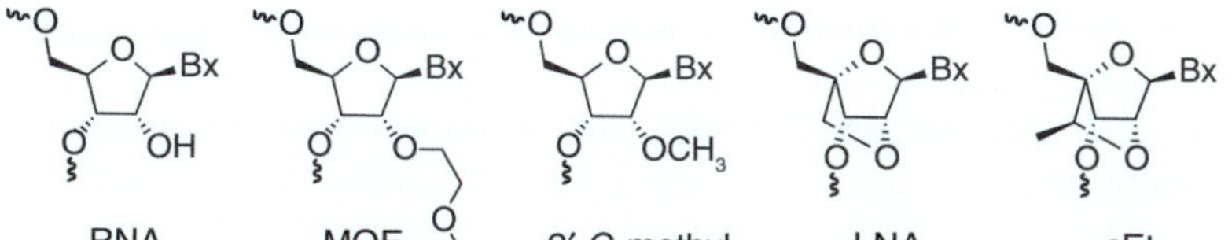

Fig. 12.3 Different ASO modifications. Schematic representation of the different ASO modifications

2015). Oligonucleotides carrying only this modification belong to the first-generation ASO, and only fomivirsen, used to treat cytomegalovirus retinitis, was FDA approved and commercialized (Dias and Stein 2002). The second generation of ASO includes 2′-OMe or 2′-MOE modifications of the backbone to increase stability. Mipomersen, approved by FDA in 2012, is an example. Another example is SPINRAZA, a splice-switching oligonucleotide, which is currently the only approved treatment for spinal muscular atrophy (Hoy 2017).

Since modifications are essential to ensure the stability and delivery of the oligonucleotides, the amount of modifications that are tolerated by a given compound are definitely a yardstick when choosing the right approach. Compared to ASOs, siRNAs do not tolerate many modifications since they impair the binding to RISC. Modifications are better tolerated by the sense strand, while the antisense allows only one or two internal nucleotides to be 2′-OMe (2′-O-methyl) modified (Watts et al. 2008).

Further oligonucleotide modifications to increase the specificity for the target include LNA (locked nucleic acids), BNAs (bicyclic nucleic acids), PNAs (peptide nucleic acids), PMOs (morpholinos), and recently cET (constrained EThyl) from IONIS. PNA and morpholinos are not negatively charged; therefore, compared to classical ASOs, they show stronger binding to nucleic acids, but they are also more difficult to deliver to the cells since they are highly hydrophobic. To improve their intracellular uptake, PMOplus and PPMOs that carry positively charged piperazine residues and arginine-rich peptide residues, respectively, were developed (Kole et al. 2012). PMOs (and derivates) and fully modified oligos (LNA, 2′-OMe, or cET) are normally used as steric blockers, while LNA-containing GapmeRs and other ASOs modified only on the 5′- and 3′-ends are used for RNAse H-mediated degradation of the target. In 2016, a 30-mer PMO antisense causing exon 51 skipping of the dystrophin mRNA (Alter et al. 2006), called Eteplirsen, became the first FDA-approved drug for the treatment of Duchenne muscular dystrophy.

Although the use of oligonucleotide-based therapeutic agents is very promising and dates back to the late 1970s (Stephenson and Zamecnik 1978), its full potential is not yet unleashed mostly because to be really effective the design of these drugs has to overcome a billion years of evolutionary conserved mechanism against foreign nucleic acid invasion, including RNAses and the lipid bilayers (Dowdy 2017).

Despite the recent improvements, homogeneous delivery of ASO into the body is still a challenge for the years to come. To win our battle against cancer, we will need to treat the metastatic disease and therefore to administrate the drug systemically. ASOs as they are described in this review have a clear preference for liver and kidneys, which are also the sites of clearance of these molecules, making the therapy more or less effective depending on the organ that needs to be targeted. Moreover, ASO cannot cross the blood-brain barrier, and therefore the only way to deliver them to the brain is intrathecal injection (Juliano 2016).

The field is going toward the development of conjugates that can deliver siRNAs and ASO directly to specific endocytic receptors on the cell surface to avoid nonproductive uptake (engulfment of the oligonucleotides into lysosomes). A successful example in this sense is the development of GalNAc conjugate that binds the ASGPR (asialoglycoprotein receptor) to specifically target hepatocytes (Prakash et al. 2014).

All the above therapeutic angles have advantages and pitfalls, and they should all be taken into account when envisioning a personalized medicine future.

12.6 Concluding Remarks and Future Perspectives

In spite of the large amount of data implicating them in cancer, the first lncRNA-based anticancer drug is not on the market yet. To go one step further, the field needs definitely to clarify the mechanism of ASO delivery and toxicity. New chemical modifications aiming at ameliorating the biodistribution and/or the toxicity of the ASOs may be needed. Moreover, to design the perfect drug, a more detailed characterization of lncRNA mechanism of action is required. A structural approach aiming at the characterization of RNA structures (Wan et al. 2014) and identification of protein partners of lncRNAs will be essential for the design of aptamers and small molecules against promising candidates.

Although more advanced compared to the therapeutic field, the search for lncRNAs suitable as biomarkers will need more targeted studies, designed to identify the lncRNAs expression profile in specific cell subpopulations, such as drug-resistant cells and/or circulating tumor cells. Single-cell RNA sequencing will be a valuable tool for this purpose.

Concluding, with all the caveat of the case, we can anticipate that lncRNAs will play a central role in the development of personalized cancer therapy, both as biomarkers and therapeutic targets.

Acknowledgments Figures were prepared using somersault libraries available from www.somersault1824.com.

References

Adriaens C, Standaert L, Barra J, Latil M, Verfaillie A, Kalev P, Boeckx B, Wijnhoven PW, Radaelli E, Vermi W, Leucci E, Lapouge G, Beck B, van den Oord J, Nakagawa S, Hirose T, Sablina AA, Lambrechts D, Aerts S, Blanpain C, Marine JC (2016) p53 induces formation of NEAT1 lncRNA-containing paraspeckles that modulate replication stress response and chemosensitivity. Nat Med 22:861–868

Alter J, Lou F, Rabinowitz A, Yin H, Rosenfeld J, Wilton SD, Partridge TA, Lu QL (2006) Systemic delivery of morpholino oligonucleotide restores dystrophin expression bodywide and improves dystrophic pathology. Nat Med 12:175–177

Arun G, Diermeier S, Akerman M, Chang KC, Wilkinson JE, Hearn S, Kim Y, MacLeod AR, Krainer AR, Norton L, Brogi E, Egeblad M, Spector DL (2016) Differentiation of mammary tumors and reduction in metastasis upon Malat1 lncRNA loss. Genes Dev 30:34–51

Babaian A, Mager DL (2016) Endogenous retroviral promoter exaptation in human cancer. Mob DNA 7:24

Bánfai B, Jia H, Khatun J, Wood E, Risk B, Gundling WE Jr, Kundaje A, Gunawardena HP, Yu Y, Xie L, Krajewski K, Strahl BD, Chen X, Bickel P, Giddings MC, Brown JB, Lipovich L (2012) Long noncoding RNAs are rarely translated in two human cell lines. Genome Res 22:1646–1657

Beltran M, Puig I, Peña C, García JM, Alvarez AB, Peña R, Bonilla F, de Herreros AG (2008) A natural antisense transcript regulates Zeb2/Sip1 gene expression during Snail1-induced epithelial-mesenchymal transition. Genes Dev 22:756–769

Bergeron-Sandoval LP, Safaee N, Michnick SW (2016) Mechanisms and consequences of macromolecular phase separation. Cell 165:1067–1079

Beroukhim R, Mermel CH, Porter D, Wei G, Raychaudhuri S, Donovan J, Barretina J, Boehm JS, Dobson J, Urashima M, Mc Henry KT, Pinchback RM, Ligon AH, Cho YJ, Haery L, Greulich H, Reich M, Winckler W, Lawrence MS, Weir BA, Tanaka KE, Chiang DY, Bass AJ, Loo A, Hoffman C, Prensner J, Liefeld T, Gao Q, Yecies D, Signoretti S, Maher E, Kaye FJ, Sasaki H, Tepper JE, Fletcher JA, Tabernero J, Baselga J, Tsao MS, DeMichelis F, Rubin MA, Janne PA, Daly MJ, Nucera C, Levine RL, Ebert BL, Gabriel S, Rustgi AK, Antonescu CR, Ladanyi M, Letai A, Garraway LA, Loda M, Beer DG, True LD, Okamoto A, Pomeroy SL, Singer S, Golub TR, Lander ES, Getz G, Sellers WR, Meyerson M (2010) The landscape of somatic copy-number alteration across human cancers. Nature 463:899–905

Berry J, Weber SC, Vaidya N, Haataja M, Brangwynne CP (2015) RNA transcription modulates phase transition-driven nuclear body assembly. Proc Natl Acad Sci U S A 112:E5237–E5245

Brunner AL, Beck AH, Edris B, Sweeney RT, Zhu SX, Li R, Montgomery K, Varma S, Gilks T, Guo X, Foley JW, Witten DM, Giacomini CP, Flynn RA, Pollack JR, Tibshirani R, Chang HY, van de Rijn M, West RB (2012) Transcriptional profiling of long non-coding RNAs and novel transcribed regions across a diverse panel of archived human cancers. Genome Biol 13:R75

Bussemakers MJ, van Bokhoven A, Verhaegh GW, Smit FP, Karthaus HF, Schalken JA, Debruyne FM, Ru N, Isaacs WB (1999) DD3: a new prostate-specific gene, highly overexpressed in prostate cancer. Cancer Res 59:5975–5979

Cabili MN, Trapnell C, Goff L, Koziol M, Tazon-Vega B, Regev A, Rinn JL (2011) Integrative annotation of human large intergenic noncoding RNAs reveals global properties and specific subclasses. Genes Dev 25:1915–1927

Cheetham SW, Gruhl F, Mattick JS, Dinger ME (2013) Long noncoding RNAs and the genetics of cancer. Br J Cancer 108:2419–2425

de Kok JB, Verhaegh GW, Roelofs RW, Hessels D, Kiemeney LA, Aalders TW, Swinkels DW, Schalken JA (2002) DD3(PCA3), a very sensitive and specific marker to detect prostate tumors. Cancer Res 62:2695–2698

Derrien T, Johnson R, Bussotti G, Tanzer A, Djebali S, Tilgner H, Guernec G, Martin D, Merkel A, Knowles DG, Lagarde J, Veeravalli L, Ruan X, Ruan Y, Lassmann T, Carninci P, Brown JB, Lipovich L, Gonzalez JM, Thomas M, Davis CA, Shiekhattar R, Gingeras TR, Hubbard TJ,

Notredame C, Harrow J, Guigó R (2012) The GENCODE v7 catalog of human long noncoding RNAs: analysis of their gene structure, evolution, and expression. Genome Res 22:1775–1189
Dias N, Stein CA (2002) Antisense oligonucleotides: basic concepts and mechanisms. Mol Cancer Ther 1:347–355
Dowdy SF (2017) Overcoming cellular barriers for RNA therapeutics. Nat Biotechnol 35:222–229
Fatica A, Bozzoni I (2014) Long non-coding RNAs: new players in cell differentiation and development. Nat Rev Genet 15:7–21
Frazier KS (2015) Antisense oligonucleotide therapies: the promise and the challenges from a toxicologic pathologist's perspective. Toxicol Pathol 43:78–89
Göke J, Ng HH (2016) CTRL+INSERT: retrotransposons and their contribution to regulation and innovation of the transcriptome. EMBO Rep 17:1131–1144
Guttman M, Russell P, Ingolia NT, Weissman JS, Lander ES (2013) Ribosome profiling provides evidence that large noncoding RNAs do not encode proteins. Cell 154:240–251
Hezroni H, Koppstein D, Schwartz MG, Avrutin A, Bartel DP, Ulitsky I (2015) Principles of long noncoding RNA evolution derived from direct comparison of transcriptomes in 17 species. Cell Rep 11:1110–1122
Hirose T, Virnicchi G, Tanigawa A, Naganuma T, Li R, Kimura H, Yokoi T, Nakagawa S, Bénard M, Fox AH, Pierron G (2014) NEAT1 long noncoding RNA regulates transcription via protein sequestration within subnuclear bodies. Mol Biol Cell 25(1):169–183
Hoy SM (2017) Nusinersen: first global approval. Drugs 77:473–479
Huarte M (2015) The emerging role of lncRNAs in cancer. Nat Med 21:1253–1261
Ingolia NT, Brar GA, Stern-Ginossar N, Harris MS, Talhouarne GJS, Jackson SE, Wills MR, Weissman JS (2014) Ribosome profiling reveals pervasive translation outside of annotated protein-coding genes. Cell Rep 8:1365–1379
Iyer MK, Niknafs YS, Malik R, Singhal U, Sahu A, Hosono Y, Barrette TR, Prensner JR, Evans JR, Zhao S, Poliakov A, Cao X, Dhanasekaran SM, Wu YM, Robinson DR, Beer DG, Feng FY, Iyer HK, Chinnaiyan AM (2015) The landscape of long noncoding RNAs in the human transcriptome. Nat Publ Group 47(3):199–208
Jain A, Vale RD (2017) RNA phase transitions in repeat expansion disorders. Nature 546(7657):243–247
Juliano RL (2016) The delivery of therapeutic oligonucleotides. Nucleic Acids Res 44(14):6518–6548
Kelley D, Rinn J (2012) Transposable elements reveal a stem cell-specific class of long noncoding RNAs. Genome Biol 13:R107
Kino T, Hurt DE, Ichijo T, Nader N, Chrousos GP (2010) Noncoding RNA gas5 is a growth arrest- and starvation-associated repressor of the glucocorticoid receptor. Sci Signal 3:ra8
Kole R, Krainer AR, Altman S (2012) RNA therapeutics: beyond RNA interference and antisense oligonucleotides. Nat Rev Drug Discov 11:125–140
Lee S, Kopp F, Chang TC, Sataluri A, Chen B, Sivakumar S, Yu H, Xie Y, Mendell JT (2016) Noncoding RNA NORAD regulates genomic stability by sequestering PUMILIO proteins. Cell 164:69–80
Leucci E, Vendramin R, Spinazzi M, Laurette P, Fiers M, Wouters J, Radaelli E, Eyckerman S, Leonelli C, Vanderheyden K, Rogiers A, Hermans E, Baatsen P, Aerts S, Amant F, Van Aelst S, van den Oord J, de Strooper B, Davidson I, Lafontaine DL, Gevaert K, Vandesompele J, Mestdagh P, Marine JC (2016) Melanoma addiction to the long non-coding RNA SAMMSON. Nature 531:518–522
Lin Y, Protter DS, Rosen MK, Parker R (2015) Formation and maturation of phase-separated liquid droplets by RNA-binding proteins. Mol Cell 60:208–219
Lorenzen JM, Thum T (2016) Long noncoding RNAs in kidney and cardiovascular diseases. Nat Rev Nephrol 12:360–373
Mayer G, Raddatz M-SL, Grunwald JD, Famulok M (2007) RNA ligands that distinguish metabolite-induced conformations in the TPP riboswitch. Angew Chem Int Ed Eng 46:557–560

Mello SS, Sinow C, Raj N, Mazur PK, Bieging-Rolett K, Broz DK, Imam JFC, Vogel H, Wood LD, Sage J, Hirose T, Nakagawa S, Rinn J, Attardi LD (2017) Neat1 is a p53-inducible lincRNA essential for transformation suppression. Genes Dev 31:1095–1108

Melton C, Reuter JA, Spacek DV, Snyder M (2015) Recurrent somatic mutations in regulatory regions of human cancer genomes. Nat Genet 47:710–716

Nishimoto Y, Nakagawa S, Hirose T, Okano HJ, Takao M, Shibata S, Suyama S, Kuwako K, Imai T, Murayama S, Suzuki N, Okano H (2013) The long non-coding RNA nuclear-enriched abundant transcript 1_2 induces paraspeckle formation in the motor neuron during the early phase of amyotrophic lateral sclerosis. Mol Brain 6:31

Nissan A, Stojadinovic A, Mitrani-Rosenbaum S, Halle D, Grinbaum R, Roistacher M, Bochem A, Dayanc BE, Ritter G, Gomceli I, Bostanci EB, Akoglu M, Chen YT, Old LJ, Gure AO (2012) Colon cancer associated transcript-1: a novel RNA expressed in malignant and pre-malignant human tissues. Int J Cancer 130:1598–1606

Ortega-Molina A, Serrano M (2013) PTEN in cancer, metabolism, and aging. Trends Endocrinol Metab 24:184–189

Pedram Fatemi R, Salah-Uddin S, Modarresi F, Khoury N, Wahlestedt C, Faghihi MA (2015) Screening for small-molecule modulators of long noncoding RNA-protein interactions using AlphaScreen. J Biomol Screen 20:1132–1141

Poliseno L, Salmena L, Zhang J, Carver B, Haveman WJ, Pandolfi PP (2010) A coding-independent function of gene and pseudogene mRNAs regulates tumour biology. Nature 465:1033–1138

Prakash TP, Graham MJ, Yu J, Carty R, Low A, Chappell A, Schmidt K, Zhao C, Aghajan M, Murray HF, Riney S, Booten SL, Murray SF, Gaus H, Crosby J, Lima WF, Guo S, Monia BP, Swayze EE, Seth PP (2014) Targeted delivery of antisense oligonucleotides to hepatocytes using triantennary *N*-acetyl galactosamine improves potency 10-fold in mice. Nucleic Acids Res 42:8796–8807

Prensner JR, Zhao S, Erho N, Schipper M, Iyer MK, Dhanasekaran SM, Magi-Galluzzi C, Mehra R, Sahu A, Siddiqui J, Davicioni E, Den RB, Dicker AP, Karnes RJ, Wei JT, Klein EA, Jenkins RB, Chinnaiyan AM, Feng FY (2014) RNA biomarkers associated with metastatic progression in prostate cancer: a multi-institutional high-throughput analysis of SChLAP1. Lancet Oncol 15:1469–1480

Quinn JJ, Chang HY (2016) Unique features of long non-coding RNA biogenesis and function. Nat Rev Genet 17:47–62

Rivas E, Clements J, Eddy SR (2017) A statistical test for conserved RNA structure shows lack of evidence for structure in lncRNAs. Nat Methods 14:45–48

Shain AH, Bastian BC (2016) From melanocytes to melanomas. Nat Rev Cancer 16:345–358

Sosman JA, Kim KB, Schuchter L, Gonzalez R, Pavlick AC, Weber JS, McArthur GA, Hutson TE, Moschos SJ, Flaherty KT, Hersey P, Kefford R, Lawrence D, Puzanov I, Lewis KD, Amaravadi RK, Chmielowski B, Lawrence HJ, Shyr Y, Ye F, Li J, Nolop KB, Lee RJ, Joe AK, Ribas A (2012) Survival in BRAF V600-mutant advanced melanoma treated with vemurafenib. N Engl J Med 366:707–714

Stephenson ML, Zamecnik PC (1978) Inhibition of Rous sarcoma viral RNA translation by a specific oligodeoxyribonucleotide. Proc Natl Acad Sci U S A 75(1):285–288

Sun H, Zhu X, Lu PY, Rosato RR, Tan W, Zu Y (2014) Oligonucleotide aptamers: new tools for targeted cancer therapy. Mol Ther Nucleic Acids 3:e182

Szpakowski S, Sun X, Lage JM, Dyer A, Rubinstein J, Kowalski D, Sasaki C, Costa J, Lizardi PM (2009) Loss of epigenetic silencing in tumors preferentially affects primate-specific retroelements. Gene 448:151–167

Teschendorff AE, Lee SH, Jones A, Fiegl H, Kalwa M, Wagner W, Chindera K, Evans I, Dubeau L, Orjalo A, Horlings HM, Niederreiter L, Kaser A, Yang W, Goode EL, Fridley BL, Jenner RG, Berns EM, Wik E, Salvesen HB, Wisman GB, van der Zee AG, Davidson B, Trope CG, Lambrechts S, Vergote I, Calvert H, Jacobs IJ, Widschwendter M (2015) HOTAIR and its surrogate DNA methylation signature indicate carboplatin resistance in ovarian cancer. Genome Med 7:108

The FANTOM Consortium (2005) The transcriptional landscape of the mammalian genome. Science 309(5740):1559–1563
Tseng YY, Moriarity BS, Gong W, Akiyama R, Tiwari A, Kawakami H, Ronning P, Reuland B, Guenther K, Beadnell TC, Essig J, Otto GM, O'Sullivan MG, Largaespada DA, Schwertfeger KL, Marahrens Y, Kawakami Y, Bagchi A (2014) PVT1 dependence in cancer with MYC copy-number increase. Nature 512:82–86
Ulitsky I, Shkumatava A, Jan CH, Sive H, Bartel DP (2011) Conserved function of lincRNAs in vertebrate embryonic development despite rapid sequence evolution. Cell 147:1537–1550
van Heesch S, van Iterson M, Jacobi J, Boymans S, Essers PB, de Bruijn E, Hao W, MacInnes AW, Cuppen E, Simonis M (2014) Extensive localization of long noncoding RNAs to the cytosol and mono- and polyribosomal complexes. Genome Biol 15:R6
Wan Y, Qu K, Zhang QC, Flynn RA, Manor O, Ouyang Z, Zhang J, Spitale RC, Snyder MP, Segal E, Chang HY (2014) Landscape and variation of RNA secondary structure across the human transcriptome. Nature 505:706–709
Watts JK, Deleavey GF, Damha MJ (2008) Chemically modified siRNA: tools and applications. Drug Discov Today 13:842–855
Yan X, Hu Z, Feng Y, Hu X, Yuan J, Zhao SD, Zhang Y, Yang L, Shan W, He Q, Fan L, Kandalaft LE, Tanyi JL, Li C, Yuan CX, Zhang D, Yuan H, Hua K, Lu Y, Katsaros D, Huang Q, Montone K, Fan Y, Coukos G, Boyd J, Sood AK, Rebbeck T, Mills GB, Dang CV, Zhang L (2015) Comprehensive genomic characterization of long non-coding RNAs across human cancers. Cancer Cell 28:529–540
Yap KL, Li S, Muñoz-Cabello AM, Raguz S, Zeng L, Mujtaba S, Gil J, Walsh MJ, Zhou MM (2010) Molecular interplay of the noncoding RNA ANRIL and methylated histone H3 lysine 27 by polycomb CBX7 in transcriptional silencing of INK4a. Mol Cell 38:662–674
Yildirim E, Kirby JE, Brown DE, Mercier FE, Sadreyev RI, Scadden DT, Lee JT (2013) Xist RNA is a potent suppressor of hematologic cancer in mice. Cell 152:727–742
Zhang H, Elbaum-Garfinkle S, Langdon EM, Taylor N, Occhipinti P, Bridges AA, Brangwynne CP, Gladfelter AS (2015) RNA controls PolyQ protein phase transitions. Mol Cell 60(2):220–230
Zhang X, Zuo X, Yang B, Li Z, Xue Y, Zhou Y, Huang J, Zhao X, Zhou J, Yan Y, Zhang H, Guo P, Sun H, Guo L, Zhang Y, Fu XD (2014) MicroRNA directly enhances mitochondrial translation during muscle differentiation. Cell 158:607–619

Chapter 13
Engineered Group I Ribozymes as RNA-Based Modular Tools to Control Gene Expression

Yoshiya Ikawa and Shigeyoshi Matsumura

Abstract Group I ribozymes (RNA enzymes) constitute a class of structural RNAs that promote cleavage and ligation of phosphodiester bonds to conduct RNA splicing. As naturally occurring group I ribozymes are constitutively active in most cases, modular engineering is required to apply them to control gene expression. We first introduce exceptional examples of modified splicing reactions of group I ribozymes. We then summarize strategies to engineer group I ribozymes to develop genetic modular tools and also their application to gene expression control.

Keywords Alternative splicing · Aptamer · Group I intron · Ribozyme · RNA structure · Self-splicing · *Trans*-splicing

13.1 Introduction

RNA splicing is an important process to produce mature RNAs serving as templates for protein synthesis (mRNAs) and as functional RNAs (noncoding RNAs). In eukaryotic cells, introns in primary transcripts produced by RNA polymerase II are excised by *trans*-acting machinery termed the spliceosome, which is a large ribonucleoprotein particle. The catalytic mechanism for excision of introns by the spliceosome relies on its RNA component, suggesting that the spliceosome is an RNA-based enzyme (ribozyme) (Will and Lührmann 2010; Elliott and Ladomery 2016a; Nguyen et al. 2016). This suggests an evolutionary relationship between the spliceosome and self-splicing introns.

Self-splicing introns, which are another class of intervening sequences between exons, form two types of secondary structures designated as group I and group II introns. Typical group I and group II introns exhibit the ability to excise themselves from their precursors by their own activity, the catalytic mechanism of which

Y. Ikawa (✉) · S. Matsumura
Department of Chemistry, Graduate School of Science and Engineering, University of Toyama, Toyama, Japan
e-mail: yikawa@sci.u-toyama.ac.jp

S. Masuda, S. Izawa (eds.), *Applied RNA Bioscience*,
https://doi.org/10.1007/978-981-10-8372-3_13

requires no protein components. Group I and group II introns are classical and major classes of RNA-based catalytic molecules (ribozymes). The similarities in their biochemical mechanisms of catalysis and their core structures suggested an evolutionary relationship between spliceosome-catalyzed excision of pre-mRNA-type introns and self-excision of group II intron, which may be an ancestral form of the spliceosome (Lambowitz and Zimmerly 2010; Nguyen et al. 2016; Zhao and Pyle 2017). Although the evolutionary relationship between the spliceosome and group I introns is less well characterized than that for group II introns, group I introns have also attracted particular attention because some exhibit high catalytic activity and also have defined tertiary structures. Biochemical and structural studies of group I introns have advanced the field of RNA catalysis research. In addition to their contribution to basic research regarding RNA enzymes, group I introns have also been studied as promising tools for biomedical applications.

For biomedical application of self-splicing introns, pre-mRNA-type splicing catalyzed by the spliceosome can be regarded as reference machinery because the functions of the spliceosome, which can conduct several types of regulated or modified splicing, appear to be more sophisticated than those of regular self-splicing introns that splice themselves constitutively. Another significant feature of the spliceosome is to serve as a *trans*-acting catalyst. Alternative splicing is a typical example of regulated spliceosomal splicing, in which multiple exons in a single transcript are excised in two or more different patterns, enabling a single primary transcript to produce several types of mature RNA (Elliott and Ladomery 2016b). *Trans*-splicing is another example in which an intron sequence is separated into two fragments such that the 5′ and 3′ exons in the ligated exons are supplied as two different RNA strands (Elliott and Ladomery 2016a). These regulated or modified splicing mechanisms expand the complexity of genome information without increasing the number of genes encoded in the DNA. Such variations of splicing reactions are also attractive for engineering of self-splicing ribozymes in biotechnology applications, such as RNA-based nanodevices for gene therapy in biomedicine and gene regulation in synthetic biology.

Self-splicing by group I introns seems to be more primitive than pre-mRNA-type splicing catalyzed by the spliceosome and mostly seems to consist of *cis*-splicing and constitutive excision (Fig. 13.1). However, careful and extensive research identified a few interesting examples of regulated or modified splicing of some group I introns (Fig. 13.1b). Most studies of these examples were based on sequence analysis of precursors and ligated exon products, and there have been few biochemical analyses regarding the catalytic efficiencies of the corresponding introns. These observations suggest that engineering of highly active group I introns, such as the *Tetrahymena* intron, is a promising approach to develop regulatory systems for artificial RNA processing in vitro and in vivo. The first part of this review presents a concise summary of regulated or modified versions of group I intron splicing. The later part summarizes artificial engineering of group I intron ribozymes as RNA-based molecular tools for biotechnology applications in gene therapy and synthetic biology.

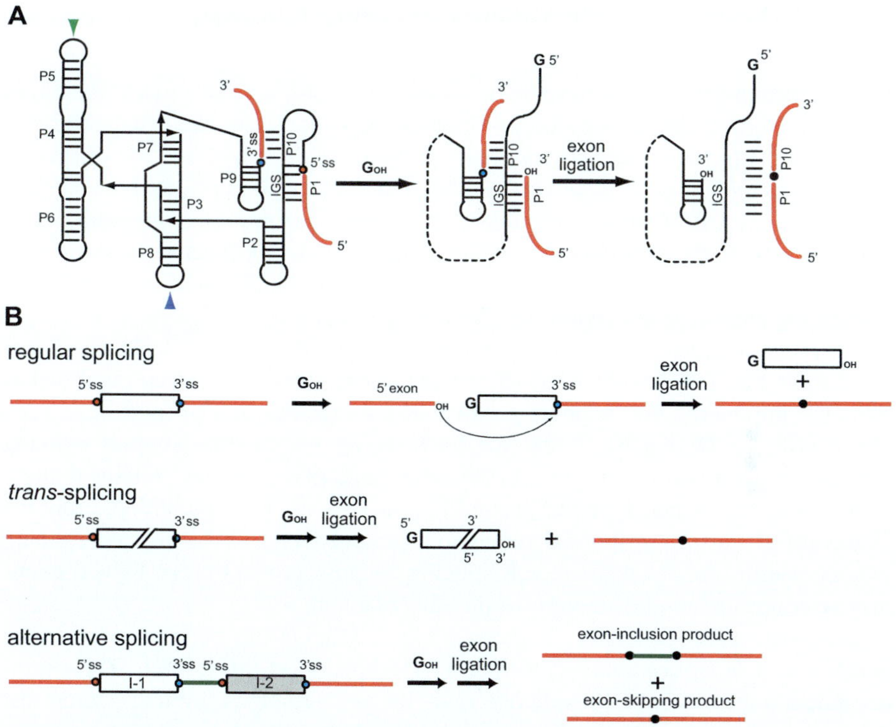

Fig. 13.1 Secondary structure of group I intron ribozyme and its splicing mechanism. (**a**) A commonly conserved secondary structure of group I introns that consist of two helical domains (P5-P4-P6 and P7-P3-P8) as well as splice-site recognition elements (P1, P10, and P9) and its structural changes during self-splicing that is initiated by nucleophilic attack of guanosine at the 5′ splice site. A green arrowhead indicates the position (P5 terminal loop) from which P5abc domain extends. A blue arrowhead indicates the position (P8 terminal loop) at which RNA strand is dissected usually in *trans*-splicing group I introns. (**b**) Regular (*cis*-) splicing, *trans*-splicing, and alternative splicing observed in naturally occurring group I introns

13.2 Naturally Occurring Group I Ribozymes with Regulated Self-Splicing Functions

A limited number of group I intron ribozymes have been used as models in biochemical and structural studies, primarily because of their high catalytic ability in vitro. The *Tetrahymena*, phage T4 td (or phage Twort), and *Azoarcus* introns belonging to the IC1, IA2, and IC3 subgroups, respectively, have been studied in most detail (Woodson 2005; Stahley and Strobel 2006; Vicens and Cech 2006). Among these three introns, the *Tetrahymena* intron has been used extensively in structural and biochemical studies because of its complex and stable 3D structure with modular architecture, as well as robust catalytic activity.

13.2.1 Protein-Assisted Splicing of Group I Introns

While biochemical studies have been carried out using several model ribozymes that show high activities without a protein component, it has been suggested that some group I introns may perform their splicing reactions with the aid of protein components in their host organisms, although their catalysis is primarily dependent on the common RNA-based machinery of group I introns. This suggestion was based mainly on the observation that some intron RNAs are inactive or show poor splicing activity or related reactions in vitro in reaction buffer with a composition resembling physiological conditions (Vicens et al. 2008).

A limited number of group I intron-binding proteins, including CYT-18 (Akins and Lambowitz 1987) and CBP-2 (Shaw and Lewin 1997), have been identified as protein components that support group I intron splicing in vivo and have been characterized with regard to the mechanisms by which they support splicing reactions of their partner introns. On the other hand, many partner proteins thought to assist inactive or poorly active group I introns have not been identified. From the viewpoint of posttranscriptional regulation, protein factors assisting self-splicing may be useful tools to regulate self-splicing because production of ligated exons may be controlled by the amounts of protein cofactors.

Light-dependent activation of group I intron splicing has been reported in the unicellular *Chlamydomonas* chloroplast (Deshpande et al. 1997). This light-dependent enhancement of splicing may be assisted by a protein factor, the production of which is regulated in a light-dependent manner.

13.2.2 Alternative Splicing by a Pair of Group I Introns

A primitive form of alternative splicing is observed in two group IA2 introns in the Orf142 gene in phage Twort and also a pair of IB2 and IA3 introns in 23S rRNA gene in *Coxiella burnetii* (Fig. 13.1b) (Landthaler and Shub 1999; Raghavan et al. 2007). In each host gene, two group I introns were inserted tandemly with a short exon sequence (23 and 34 nucleotides for the Orf142 gene and 23S rRNA gene, respectively). In each case, characterization of splicing products produced from the primary transcript revealed that ligated exons are a mixture of those possessing and lacking the short exon between the two group I introns (Landthaler and Shub 1999; Raghavan et al. 2007). This observation indicates that exon-skipping splicing takes place partially in addition to normal splicing with inclusion of the short exon. In normal self-splicing, the two introns are excised independently. In exon-skipping splicing, however, two introns act cooperatively to remove the short exon between them.

In these cases, it is unclear whether either exon-skipping or exon inclusive splicing is regulated by specific factors. However, these examples indicate that group I introns can be used to produce two distinct RNA products from a single primary transcript.

13.2.3 Trans-Splicing of Group I Introns

A number of *trans*-spliced group II introns, in which the sequences are physically separated into two pieces in the primary transcripts, have been reported since 1988 (Kohchi et al. 1988; Glanz and Kück 2009). On the other hand, *trans*-splicing of group I introns was first described in 2009, and less than 10 examples have been reported to date (Grewe et al. 2009; Pombert et al. 2013; Nadimi et al. 2012).

These exceptionally rare examples of *trans*-spliced group I introns are usually split physically in the P8 terminal loop (see a blue arrowhead in Fig. 13.1a), although few examples are split in the P9 (Fig. 13.2a) or P9.1 terminal loop (Grewe et al. 2009; Nadimi et al. 2012; Pombert et al. 2013). In the *trans*-splicing reaction reported to proceed between two fragments of a group I ribozyme in the fragmented *coxl* gene of *Gigaspora rosea*, the secondary structure of the bimolecular group I intron still lacked the elements between the 3′-half of the P6 region and 5′-half of the P9 region (Nadimi et al. 2012). The missing elements are thought to be supplied as the third RNA strand transcribed from a distinct genome region (intron 3 sequence of the *nad3* gene), and the resulting group I ribozyme may be reconstituted from three distinct RNA fragments to achieve *trans*-exon ligation (Fig. 13.2b) (Nadimi et al. 2012).

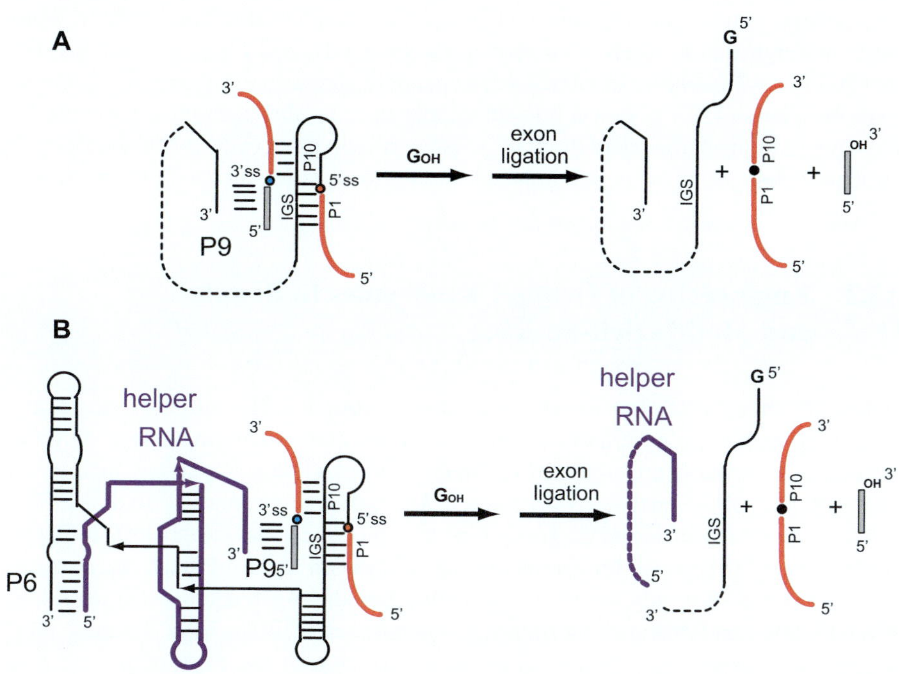

Fig. 13.2 Naturally occurring examples of group I intron *trans*-splicing reactions. (**a**) *Trans*-splicing of bimolecular group I introns whose RNA strand is dissected at P9 terminal loop. (**b**) *Trans*-splicing of trimolecular group I introns involving a helper RNA strand

13.2.4 Conditional Self-Splicing of a Group I Intron Regulated by a Small Molecule

A unique type of regulated splicing has been identified in the pathogenic bacterium, *Clostridium difficile*. Splicing of a group I ribozyme, with insertion in the 5′ untranslated region in its CD3246 ORF such that its ribosome-binding site is disrupted, has been shown to be regulated by a small molecule, c-di-GMP (Lee et al. 2010). This group I ribozyme fused with another class of functional RNA termed a riboswitch, which is an RNA element that specifically captures its target ligand and alters its secondary structure depending on ligand recognition. In this group I ribozyme, the secondary structure of the 5′ splice site is involved in a riboswitch element recognizing c-di-GMP. In the presence of c-di-GMP, this region forms a ligand-bound structure of c-di-GMP that also provides a splicing-proficient P1 structure. The guanosine-dependent 5′ cleavage takes place to conduct the regular self-splicing reaction, and excision of the group I intron results in the formation of a ribosome-binding site to activate protein synthesis. In the absence of c-di-GMP, however, the 5′ splice site is involved in an alternative ligand-free riboswitch structure that cannot serve as a substrate for the group I ribozyme. Thus, the production of the ligated exons by this group I self-splicing depends on the presence of c-di-GMP.

It is also worth noting that conditional splicing of pre-mRNA-type introns by the spliceosome can also be regulated by riboswitch-dependent small molecule recognition, although this is rarely identified in fungi, plants, and algae (Li and Breaker 2013). In a few examples of riboswitch-dependent alternative splicing, the thiamine pyrophosphate (TPP) riboswitch is commonly used to alter the splicing pattern by masking or unmasking one of the 5′ or 3′ splice sites to use its alternative site through structural changes in the region induced by TPP recognition by the TPP riboswitch.

13.3 Engineering of Group I Ribozymes to Regulate and Modify Self-Splicing

Although characterization of group I introns from a wide range of organisms identified both regulated and modified versions of self-splicing, most of these examples have not been subjected to the detailed biochemical analysis that would enable their application directly in research. As an alternative approach to developing RNA molecular tools based on group I intron ribozymes, molecular engineering of model group I introns for biochemical studies has been reported. Such research has been performed not only for their application but also for basic science regarding diversification and evolution of ribozyme functions in laboratories to examine their scopes and limitations. The *Tetrahymena* group I intron has been employed as a model intron for engineering because of its high and robust catalytic activity and also the availability of its structural information, which will facilitate both rational and evolutionary engineering compared with other model group I introns.

13.3.1 Peptide-/Protein-Dependent Splicing of Engineered Tetrahymena *Ribozymes*

As a pioneering work in RNA synthetic biology, semi-rational engineering of the *Tetrahymena* ribozyme was reported (Atsumi et al. 2001), in which the RNA–RNA tertiary interaction in its P4–P6 modular domain was replaced with an artificial RNA-binding peptide. The peptide has two RNA-binding elements (N-peptide and Rev-peptide), designed to bridge two RNA helices through binding to peptide recognition elements (boxB motif and RRE motif) installed in P5b and P6 elements in place of the parent tetraloop and its receptor (Fig. 13.3). The designed peptide bridges P5b and P6 elements to fix the P4–P6 domain into a sharply bent confirmation that is close to the parent conformation held by the tetraloop/receptor interaction in the parent ribozyme. The designed peptide activates the variant ribozyme not only with regard to its splicing in vitro but also in *Escherichia coli*, as demonstrated by regulation of the growth of *E.coli* through production of the reporter lacZ used as a model exon element to which the splicing ribozyme was inserted as an intron (Atsumi et al. 2001).

CYT-18 protein is a naturally occurring protein that is able to bind to and activate group I introns. The naturally occurring partner intron of CYT-18 protein belongs to subgroup IA1, but it can also activate other group IA introns because of their structural similarity. On the other hand, CYT-18 is not capable of binding to the *Tetrahymena* IC1 ribozyme, which possesses a large P5abc peripheral element (Fig. 13.3a, see also a green arrowhead in Fig. 13.1a). The structural region, to which CYT-18 protein binds in group IA introns, is occupied by the P5abc domain in group IC1 introns (Fig. 13.3) (Caprara et al. 1996; Paukstelis et al. 2008). Removal of the P5abc region from the *Tetrahymena* ribozyme significantly diminishes its catalytic activity. The deletion mutant (ΔP5 ribozyme) showed full recovery of its activity when associated with separately prepared P5abc RNA to form a bimolecular version of the full-length ribozyme (Mohr et al. 1994). ΔP5 ribozyme is also activated by CYT-18 protein, indicating that it is a protein-dependent group I ribozyme, splicing of which can be regulated by a nonnatural partner, CYT-18 (Mohr et al. 1994). This finding suggests that intracellular proteins may be used as artificial regulators for engineered *Tetrahymena* ribozymes. Furthermore, this protein-assisted ribozyme demonstrated that the P5abc RNA domain and CYT-18 protein play equivalent roles in activation of the group IC1 intron. These findings are also instructive from the viewpoint of the evolution of group I introns and their complexes with protein components because the P5abc domain may be replaced with CYT-18 protein. Such transition from RNA-only ribozyme to protein-dependent ribozyme may proceed in a gradual manner, in the early stages of which small peptide activators resembling the peptides designed by Atsumi et al. may emerge to trigger the transition (Ikawa et al. 2003).

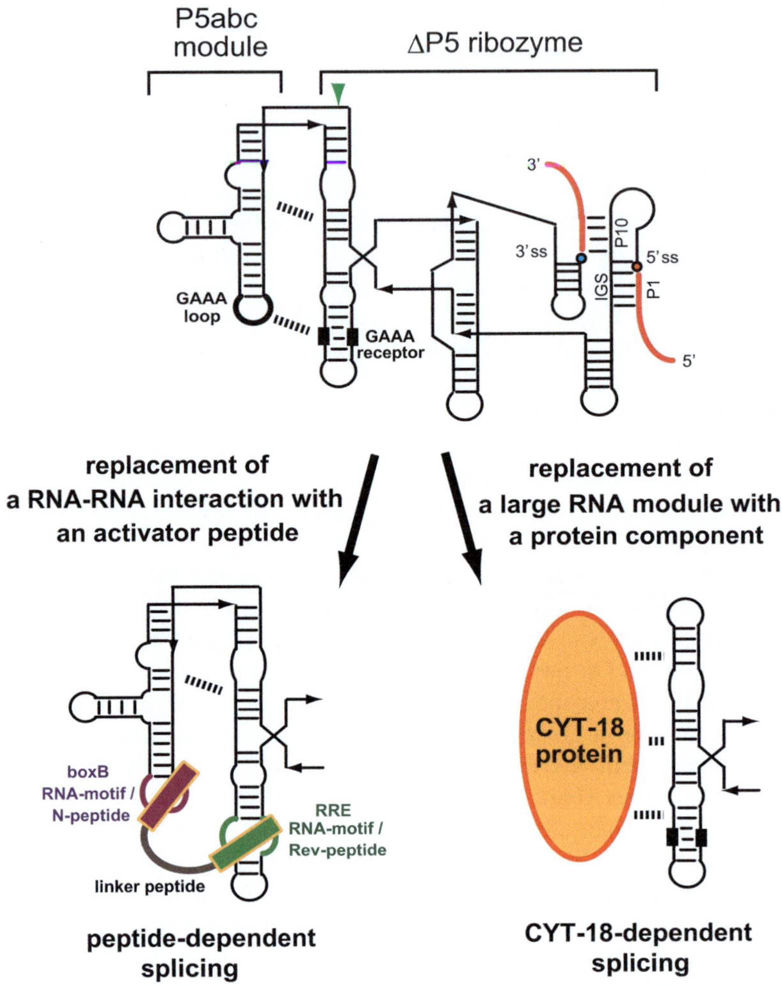

Fig. 13.3 Peptide-/protein-dependent splicing of engineered *Tetrahymena* ribozymes. The *Tetrahymena* group IC1 intron possessing a large P5abc domain (top) has been engineered to be activated by an artificial peptide (bottom-left) or by naturally occurring CYT-18 protein (bottom-right)

13.3.2 Allosteric Group I Ribozymes Regulated by Small Molecule Ligands

Although naturally occurring ribozymes the activity of which is regulated by small molecule ligands are rare, ribozyme engineering to generate allosteric ribozymes (aptazymes) has been performed by connecting ribozyme RNAs and aptamer RNAs in a modular manner via linker elements (Tang and Breaker 1997; Silverman 2003; Felletti and Hartig 2017). Linker elements can be designed rationally or identified by screening from random sequences. This approach has also been applied to group I introns using theophylline and its aptamer RNA as a modular unit to control splicing activity (Kertsburg and Soukup 2002; Kim et al. 2014: Thompson et al. 2002). The *Tetrahymena* ribozyme was engineered to a theophylline-dependent aptazyme by appending the theophylline aptamer with a linker element. Theophylline aptamer unit with the linker was installed by replacing P6 or P8 elements (Fig. 13.1a). The resulting aptazymes showed modestly enhanced (three to fourfold) activity upon addition of theophylline (Kertsburg and Soukup 2002). To improve theophylline-dependent activation, two theophylline aptamers were introduced to both P6 and P8 elements simultaneously, and the activity of the engineered ribozyme showed a greater degree of enhancement (24-fold) by theophylline ligand (Kertsburg and Soukup 2002). These *Tetrahymena* aptazymes have also been used to regulate *trans*-splicing reaction targeting human telomerase reverse transcriptase (hTERT) described in the next section (Kim et al. 2014).

Similar modifications were also examined using the T4 td group IA2 intron, in which the P5 or P6 element of the intron was replaced with the theophylline aptamer (Fig. 13.1a) (Thompson et al. 2002). The modified T4 td introns allowed splicing in *E. coli* to be controlled by theophylline, although the degree of allosteric regulation was also modest.

13.3.3 Engineered Group I Ribozymes for Trans-Splicing Reactions

While naturally occurring examples of group I intron *trans*-splicing are very limited and no biochemical studies have been performed, rational redesign of the *Tetrahymena* group I intron to perform *trans*-splicing reactions has been performed to understand its reaction mechanism, based on which *Tetrahymena* ribozyme-based gene therapy has been performed. Early types of *trans*-splicing ribozymes were developed based on a shortened form of the *Tetrahymena* intron, which can cleave the short substrate RNA catalytically (Inoue et al. 1986). The cleavage reaction mimics the 5′ strand of the P1 element containing the 5′ splice site. The short substrate is recognized by the internal guide sequence (IGS) that corresponds to the 3′ strand of the P1 element containing a sequence complementary to the substrate (Fig. 13.4a, see also Fig. 13.1a). Guanosine-dependent cleavage of the substrate RNA thus produces two RNA fragments, one of which serves as the 5′ exon to attack the 3′ splice site of the ribozyme possessing the 3′ exon. This reaction yields the product of *trans*-exon ligation and also the free ribozyme (Fig. 13.4a). This *trans*-splicing reaction differs from naturally occurring reactions in the structure of the catalytically active group I ribozyme, which is provided by one RNA fragment in artificial reactions but is reconstituted from two fragments in natural examples in which the core catalytic elements are usually disconnected into two fragments at P8 region.

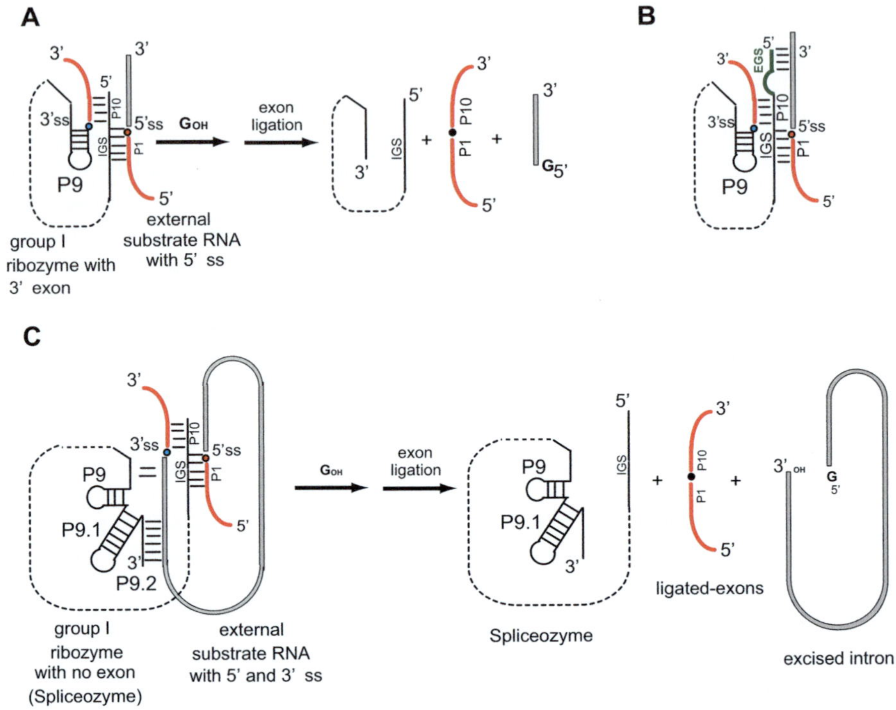

Fig. 13.4 Engineered group I ribozymes for *trans*-splicing reactions. (**a**) *Trans*-splicing reaction at the 5′ splice site. (**b**) Extended guide sequence (EGS) developed to improve the efficiency of *trans*-splicing reactions. (**c**) Spliceozyme that catalyzes an intact group I-type splicing reaction. Spliceozyme is shown in black line. In substrate RNA, two (5-′ and 3′-) exons and an intron are colored with red and gray, respectively

To apply the artificial *trans*-splicing reaction at the 5′ splice site to long RNA substrates, such as mRNAs with aberrant mutations, the IGS element to recognize *trans*-acting substrate and P10 element guiding the 3′ exon in the *Tetrahymena* ribozyme must be engineered to adapt to given target sequences (Sullenger and Cech 1994; Fiskaa and Birgisdottir 2010; Müller 2017). The resulting ribozyme variant has customized IGS and P10 elements as well as the 3′ exon that serves as the sequence to replace the aberrant portion of the target mRNA. The recognition of target RNA can be further improved by addition of base pairs between the target RNA and the ribozyme. These base pairs form between the antisense region and extended guide sequence (EGS) installed upstream of the IGS region in the ribozyme and its complementary region in the substrate (Fig. 13.4b) (Köhler et al. 1999). *Trans*-splicing reaction at the 5′ splice site was first tested in *E. coli* and mammalian cells to repair *lacZ* mRNA bearing a mutation (Sullenger and Cech 1994). This approach has subsequently been applied to repair or reprogram several gene transcripts, protein products of which are relevant to diseases, such as β-globin (Lan et al. 1998), myotonic dystrophy protein kinase (MDPK) (Phylactou et al. 1998), hTERT (Hong et al. 2008; Kim et al. 2014), and oncogenic KRAS (Kim et al. 2017).

An aberrant element located upstream of the target site processed by the intron ribozyme can also be replaced using an engineered ribozyme possessing the 5′ exon but lacking its 3′ exon and the 3′ splice site (Alexander et al. 2005). The aberrant mRNA serves as a *trans*-acting 3′ element bearing the 3′ exon and 3′ splice site. To accept external RNA sequences as the 3′ element bearing the 3′ splice site, rational engineering is required according to the ribozymes used for this type of *trans*-splicing. This engineering resembles a naturally occurring *trans*-splicing introns whose strand is disconnected at P9 or P9.1 element (Fig. 13.2a). In contrast to the P1 element, the structural design of which is common between the *Tetrahymen*a intron and its relatives, structural design of 3′ elements (downstream of P9 elements) is varied even between the *Tetrahymena* ribozyme and its relatives (Alexander et al. 2005; Amini et al. 2014). *Trans*-splicing reaction at the 3′ splice site has been applied to the *Tetrahymena* ribozyme and the closely related *Pneumocystis carinii* ribozyme.

13.3.4 Spliceozymes to Catalyze Intact Splicing of External Substrates

A catalytically active ribozyme unit covalently possessing a 3′ exon or 5′ exon can conduct *trans*-splicing reactions with external RNA substrates serving as the 5′ or 3′ splice site, respectively. Improvement of rational design enables integration of *trans*-splicing at 5′ and 3′ sites together to develop intact splicing reactions of external RNA, in which both 5′ and 3′ splice site elements are present without the ribozyme unit (Bell et al. 2002; Amini et al. 2014). To this substrate RNA, an engineered group I ribozyme can serve as a *trans*-acting enzyme to catalyze intact splicing reactions, named *trans*-excision splicing (Fig. 13.4c) (Bell et al. 2002). Such group I ribozymes, designated as group I spliceozymes (Amini et al. 2014), have been constructed based on the *Tetrahymena* ribozyme and the *P. carinii* ribozyme. Excision of short (single nucleotide) and long (100 nucleotides) intervening sequences (introns) has been achieved using this type of splicing reaction (Bell et al. 2002; Baum and Testa 2005; Amini et al. 2014; Amini and Müller 2015).

13.3.5 Optimization of Trans-Splicing Reactions by Group I Ribozymes

Although rational prediction is the first choice to identify a target site for *trans*-splicing reactions, the efficiency of the reactions is often insufficient. Screening of the most accessible site in a given target RNA is useful to improve the reaction efficiency. For such optimization, an evolutionary approach is used using libraries of ribozyme variants bearing randomized IGS (Jones and Sullenger 1997). Through *trans*-splicing between the target mRNA and the ribozyme library, active variants that should bear IGS sequences recognizing more accessible site in the target RNA can be enriched. Their IGS can be identified through selective RT-PCR of the *trans*-splicing product. This approach allows the mapping of sites on the target RNAs accessible to the *trans*-splicing ribozymes. This approach can also be employed to optimize the sequence of the antisense region (or EGS) element developed for *trans*-splicing (Jones and Sullenger 1997; Ryu and Lee 2003; Olson and Müller 2012).

13.3.6 Rational Dimerization of the Tetrahymena *Ribozyme to Develop Pairs of Cooperative Splicing Reactions*

Both naturally occurring and artificially engineered examples of regulated or modified splicing ribozymes introduced functions depending on one ribozyme unit, which is composed of either a single RNA strand or reconstituted from two or more strands. Regulated splicing or related reactions by group I ribozymes may also be designed by the cooperative actions of two or more ribozymes. Such engineered ribozymes may be attractive not only as gene regulatory tools in gene therapy or synthetic biology but also as building blocks for RNA-based nanostructures. We engineered the *Tetrahymena* group I ribozyme to design a splicing system in which two group I ribozyme molecules were activated cooperatively to conduct a pair of either *trans*- or regular (*cis*-) splicing reactions.

A paired *trans*-splicing reaction system was developed by designing a pair of heterodimerizing variants of the *Tetrahymena* intron through two-step redesign of its secondary structure (Fig. 13.5) (Tanaka et al. 2017). In the first step, mutations were introduced into the primary sequence of the ribozyme without disruption of its secondary and tertiary structures supporting the self-splicing activity (Che and Knight Jr. 2010). The three resulting variants (M1, M2, and M3) with distinct primary sequences from the parent wild-type retained catalytic activity comparable to that of the wild-type ribozyme in vitro and also in *E. coli* cells. In the second step, two pairs of chimeric ribozymes were designed based on the three variants as well as the wild-type. In each pair of chimeric ribozymes, a pair of ribozymes (M1+M2 or M3+WT) was dissected at a particular position (the terminal loop of P6 for M1+M2 or the terminal loop of P8 for M3+WT), and the resulting fragments were then swapped and joined, producing two pairs of chimeric sequences (M1/M2+M2/M1 and M3/WT+WT/M3). Each variant with a chimeric sequence showed complete or near loss of self-splicing activity due to disruptive changes in the secondary and tertiary structures. Assembly of the pair of ribozymes, however, enabled them to form the correct secondary structures of two ribozyme units. In the resulting dimers, two catalytically active ribozyme units were formed, each of which was composed of two RNA fragments corresponding to the 5′- and 3′-portions of the full-length ribozyme.

One ribozyme dimer, in which two fragments from two ribozymes (M1 and M2) were joined at the terminal loop of the P6 element, had a 3D structure with two active ribozymes that were connected in a tail-to-tail manner at their P6 elements (Fig. 13.5). The other dimer, in which two fragments from two ribozymes (M3+WT) were joined at the terminal loop of the P8 element, had a 3D structure with two ribozymes connected in a tail-to-tail manner at their P8 elements.

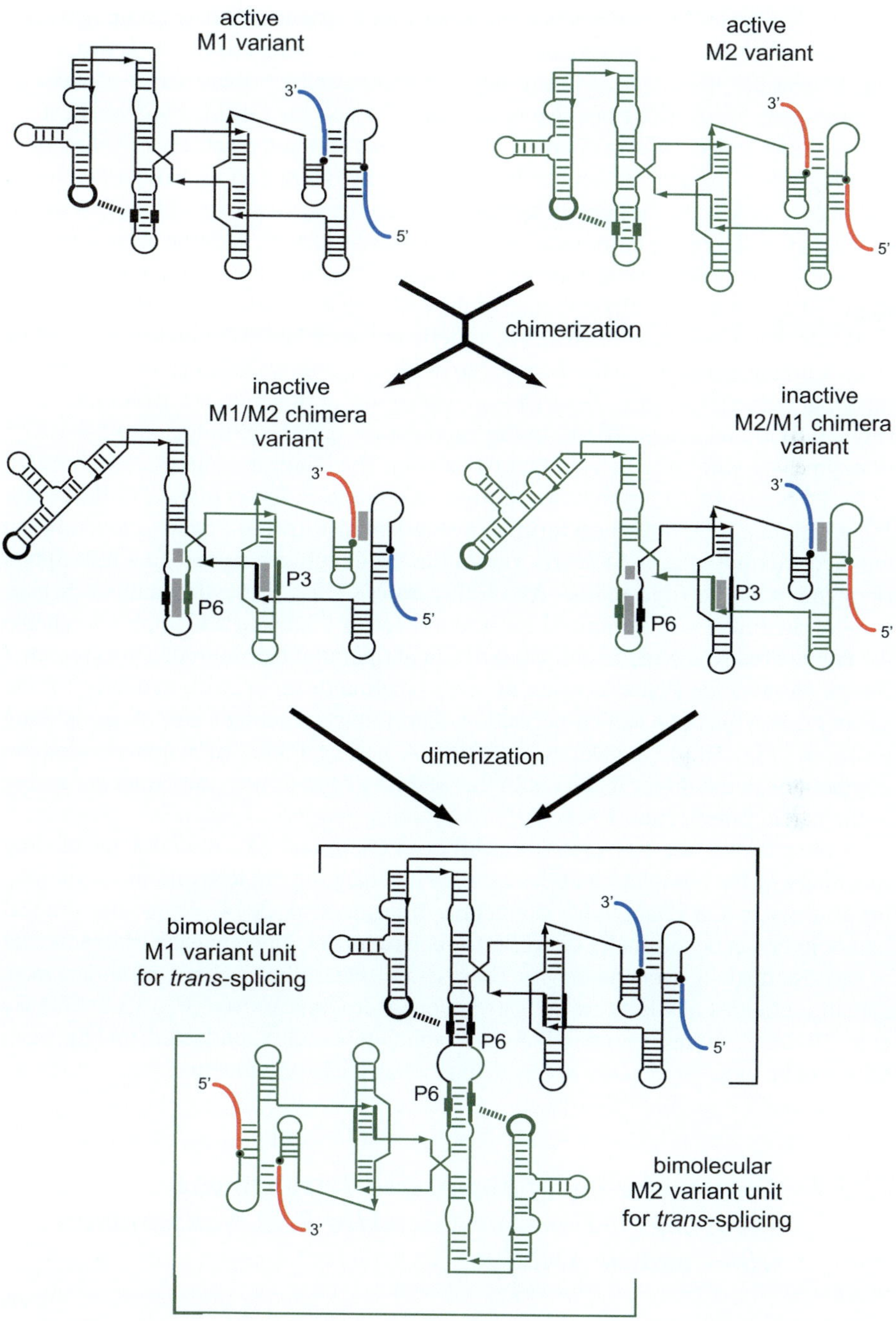

Fig. 13.5 Redesign of the *Tetrahymena* ribozyme for a pair of *trans*-splicing reactions that depends on heterodimerization of the variant ribozymes. In secondary structures of the ribozyme variants, black and green lines indicate differences in their primary sequences. The two sequences were designed to form common secondary and tertiary structures. In two chimera variants (M1/M2 and M2/M1), gray boxes mean that base-paired regions are disrupted due to different primary sequences between M1 and M2 variants

In the substrate cleavage reaction mimicking the first step of self-splicing, two ribozyme dimers showed dimerization-dependent activation of their ribozyme activity, with which two ribozymes cleaved two distinct substrates, each of which served as the cognate substrate for the two ribozyme units, each of which has a distinct IGS element.

Two ribozymes can also conduct a pair of *trans*-splicing reactions in a dimerization-dependent manner, which can be monitored by co-transcriptional *trans*-splicing reactions to produce a fluorescent aptamer (Spinach) as ligated exons (Furukawa et al. 2016). In this fluorescence-reporter system, however, it is not possible to monitor two *trans*-splicing reactions simultaneously because of the lack of a second reporter RNA that can be monitored in the presence of the first reporter product (Spinach RNA).

While dimerization of the pair of chimeric introns enables the design of a pair of *trans*-splicing reactions, paired cooperative *cis*-splicing was also achieved through modular redesign of the *Tetrahymena* ribozyme at the tertiary structural level (Fig. 13.6) (Tanaka et al. 2016). In this modular design, the bimolecular P5abc/ΔP5 ribozyme was employed as a structural platform. The P5abc domain was reconnected to the P5 region of ΔP5 ribozyme through a rigid duplex linker instead of the parent J5/5a element, a flexible loop forming a sharp bend allowing P5abc to assemble to the ΔP5 module intramolecularly within the full-length ribozyme. In a redesigned ribozyme with the rigid P5abc–ΔP5 linker duplex (Fig. 13.6), the number of base pairs of the linker was optimized for intermolecular P5abc/ΔP5 assembly to enable the engineered ribozyme to self-dimerize in antiparallel orientation. This structural design allowed the P5abc domain in one ribozyme to serve as an activator for the partner ribozyme upon self-dimerization, therefore conducting a pair of *cis*-splicing reactions (Fig. 13.6) (Tanaka et al. 2016). A pair of P5abc/ΔP5 domain–domain interactions in the dimer results in higher stability of the ribozyme dimer compared to the parent bimolecular P5abc/ΔP5 ribozyme.

In the pair of RNA–RNA interfaces between P5abc and ΔP5 modules, the binding specificity of the modules can be controlled by changing the RNA motifs participating directly in the P5abc/ΔP5 interaction (Tanaka et al. 2014). Thus, the original homodimer can be rationally converted to heterodimers allowing each ribozyme unit to possess distinct exon sequences. Two sets of heterodimers have been designed, splicing activities of which are activated upon heterodimerization (Fig. 13.6) (Tanaka et al. 2016). Co-transcriptional double *cis*-splicing was also conducted, and the reactions can be monitored using Spinach aptamer as the exon sequence.

13.3.7 Oligomerization of Engineered Tetrahymena *Ribozymes to Generate Assembled RNA Nanostructures with Catalytic Abilities*

Modular engineering of the *Tetrahymena* ribozyme to design dimeric derivatives for double *cis*-splicing reaction can be applied to RNA nanostructure design. The modular design of dimeric ribozymes, in which two ribozyme units dimerize in

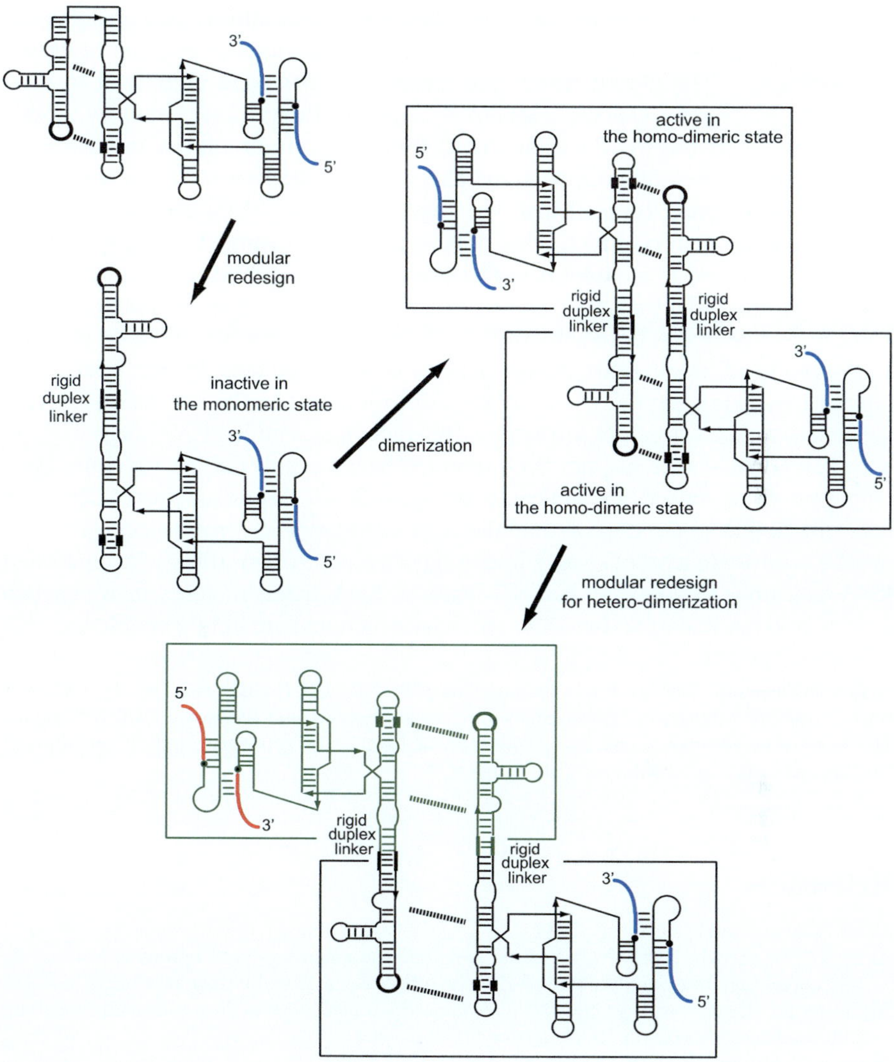

Fig. 13.6 Modular redesign of the *Tetrahymena* ribozyme to develop dimeric variants for a pair of *cis*-splicing reactions. In secondary structures of the ribozyme variants, black and green lines indicate differences in their primary sequences. The two sequences were designed to form common secondary and tertiary structures

antiparallel orientation, can be expanded to directional oligomerization of engineered ribozymes, assembly of which is also mediated by P5abc/ΔP5 interaction. To expand the number of ribozyme units to be assembled, the P5abc domain was connected to the P8 element of ΔP5 ribozyme. The resulting design with a linker duplex between P5 and P8 elements predicts that the ribozyme unit can homotrimerize to form a closed trimer with a triangular shape (Oi et al. 2017). Biochemical analysis of the engineered ribozyme by native gel electrophoresis confirmed the predominant formation of a closed homotrimer but also revealed a minor population of closed homotetramers. The closed trimer and closed tetramer were confirmed to have triangular and square shapes, respectively, through direct observation by atomic force microscopy (AFM) (Oi et al. 2017). Homooligomerization of the ribozyme unit produces closed homotrimers and closed homotetramers as major and minor assembly structures, respectively. By engineering the P5abc/ΔP5 interface to generate a pair or a trio of interfaces, each of which has distinct binding specificity, selective formation of a closed heterotetramer as a dimer of dimers of two distinct ribozyme units and also a closed heterotrimer consisting of three distinct ribozyme units, respectively, has been achieved.

Assembly of these RNA nanostructures is also mediated by the P5abc/ΔP5 interface, meaning that the activity of each ribozyme unit can be activated positively depending on nanostructure formation. This type of assembled RNA nanostructure with polygonal shapes has not been reported because no catalytic RNA has been employed as a unit RNA molecule for such RNA nanostructures. Thus, this nanostructure design based on modular engineering of the *Tetrahymena* ribozyme may be regarded as a proof of principle study of a novel design strategy for controlled RNA assembly to construct a novel class of RNA nanostructures in which two aspects of RNA (catalytic functions and nanostructures) are tightly coupled.

Acknowledgments This work was supported by JSPS KAKENHI Grant Number JP 15 K05561 (to Y.I.) from the Ministry of Education, Culture, Sports, Science and Technology (MEXT), Japan. This work was also supported partly by University of Toyama Discretionary Funds of the President "Toyama RNA Research Alliance" (to Y.I. and S.M.).

References

Akins RA, Lambowitz AM (1987) A protein required for splicing group I introns in Neurospora mitochondria is mitochondrial tyrosyl-tRNA synthetase or a derivative thereof. Cell 50:331–345

Alexander RC, Baum DA, Testa SM (2005) 5′ transcript replacement in vitro catalyzed by a group I intron-derived ribozyme. Biochemistry 44:7796–7804

Amini ZN, Müller UF (2015) Increased efficiency of evolved group I intron spliceozymes by decreased side product formation. RNA 21:1480–1489

Amini ZN, Olson KE, Müller UF (2014) Spliceozymes: ribozymes that remove introns from pre-mRNAs in trans. PLoS One 9:e101932. https://doi.org/10.1371/journal.pone.0101932

Atsumi S, Ikawa Y, Shiraishi IT (2001) Design and development of a catalytic ribonucleoprotein. EMBO J 20:5453–5460

Baum DA, Testa SM (2005) In vivo excision of a single targeted nucleotide from an mRNA by a trans excision-splicing ribozyme. RNA 11:897–905

Bell MA, Johnson AK, Testa SM (2002) Ribozyme-catalyzed excision of targeted sequences from within RNAs. Biochemistry 41:15327–15333

Caprara MG, Lehnert V, Lambowitz AM, Westhof E (1996) A tyrosyl-tRNA synthetase recognizes a conserved tRNA-like structural motif in the group I intron catalytic core. Cell 87:1135–1145

Che AJ, Knight TF Jr (2010) Engineering a family of synthetic splicing ribozymes. Nucleic Acids Res 38:2748–2755

Deshpande NN, Bao Y, Herrin DL (1997) Evidence for light/redox-regulated splicing of psbA pre-RNAs in Chlamydomonas chloroplasts. RNA 3:37–48
Elliott D, Ladomery M (2016a) Pre-mRNA splicing by the spliceosome. In: Elliott D, Ladomery M (eds) Molecular biology of RNA, 2nd edn. Oxford University Press, Oxford, pp 84–110
Elliott D, Ladomery M (2016b) Regulated alternative splicing. In: Elliott D, Ladomery M (eds) Molecular biology of RNA, 2nd edn. Oxford University Press, Oxford, pp 111–137
Felletti M, Hartig JS (2017) Ligand-dependent ribozymes. Wiley Interdiscip Rev RNA 8. https://doi.org/10.1002/wrna.1395
Fiskaa T, Birgisdottir AB (2010) RNA reprogramming and repair based on trans-splicing group I ribozymes. New Biotechnol 27:194–203
Furukawa A, Tanaka T, Furuta H, Matsumura S, Ikawa Y (2016) Use of a fluorescent aptamer RNA as an exonic sequence to analyze self-splicing ability of a group I Intron from structured RNAs. Biology 5:pii: E43. https://doi.org/10.3390/biology5040043
Glanz S, Kück U (2009) Trans-splicing of organelle introns – a detour to continuous RNAs. Bioessays 31:921–934
Grewe F, Viehoever P, Weisshaar B, Knoop V (2009) A trans-splicing group I intron and tRNA-hyperediting in the mitochondrial genome of the lycophyte Isoetes engelmannii. Nucleic Acids Res 37:5093–5104
Hong SH, Jeong JS, Lee YJ, Jung HI, Cho KS, Kim CM, Kwon BS, Sullenger BA, Lee SW, Kim IH (2008) In vivo reprogramming of hTERT by trans-splicing ribozyme to target tumor cells. Mol Ther 16:74–80
Ikawa Y, Tsuda K, Matsumura S, Atsumi S, Inoue T (2003) Putative intermediary stages for the molecular evolution from a ribozyme to a catalytic RNP. Nucleic Acids Res 31:1488–1496
Inoue T, Sullivan FX, Cech TR (1986) New reactions of the ribosomal RNA precursor of Tetrahymena and the mechanism of self-splicing. J Mol Biol 189:143–165
Jones JT, Sullenger BA (1997) Evaluating and enhancing ribozyme reaction efficiency in mammalian cells. Nat Biotechnol 15:902–905
Kertsburg K, Soukup GA (2002) A versatile communication module for controlling RNA folding and catalysis. Nucleic Acids Res 30:4599–4606
Kim J, Jeong S, Kertsburg A, Soukup GA, Lee SW (2014) Conditional and target-specific transgene induction through RNA replacement using an allosteric trans-splicing ribozyme. ACS Chem Biol 9:2491–2495
Kim SJ, Kim JH, Yang B, Jeong JS, Lee SW (2017) Specific and efficient regression of cancers harboring KRAS mutation by targeted RNA replacement. Mol Ther 25:356–367
Kohchi T, Umesono K, Ogura Y, Komine Y, Nakahigashi K, Komano T, Yamada Y, Ozeki H, Ohyama K (1988) A nicked group II intron and trans-splicing in liverwort, Marchantia polymorpha, chloroplasts. Nucleic Acids Res 16:10025–10036
Köhler U, Ayre BG, Goodman HM, Haseloff J (1999) Trans-splicing ribozymes for targeted gene delivery. J Mol Biol 285:1935–1950
Lambowitz AM, Zimmerly S (2010) Group II introns: mobile ribozymes that invade DNA. In: Atkins JF, Gesteland RF, Cech TR (eds) RNA worlds: from life's origins to diversity in gene regulation. CSHL Press, New York, pp 103–121
Lan N, Howrey RP, Lee SW, Smith CA, Sullenger BA (1998) Ribozyme-mediated repair of sickle beta-globin mRNAs in erythrocyte precursors. Science 280:1593–1596
Landthaler M, Shub DA (1999) Unexpected abundance of self-splicing introns in the genome of bacteriophage Twort: introns in multiple genes, a single gene with three introns, and exon skipping by group I ribozymes. Proc Natl Acad Sci U S A 96:7005–7010
Lee ER, Baker JL, Weinberg Z, Sudarsan N, Breaker RR (2010) An allosteric self-splicing ribozyme triggered by a bacterial second messenger. Science 329:845–848
Li S, Breaker RR (2013) Eukaryotic TPP riboswitch regulation of alternative splicing involving long-distance base pairing. Nucleic Acids Res 41:3022–3031
Mohr G, Caprara MG, Guo Q, Lambowitz AM (1994) A tyrosyl-tRNA synthetase can function similarly to an RNA structure in the Tetrahymena ribozyme. Nature 370:147–150

Müller UF (2017) Design and experimental evolution of trans-splicing group I intron ribozymes. Molecules 22:pii: E75. https://doi.org/10.3390/molecules22010075

Nadimi M, Beaudet D, Forget L, Hijri M, Lang BF (2012) Group I intron-mediated trans-splicing in mitochondria of Gigaspora rosea and a robust phylogenetic affiliation of arbuscular mycorrhizal fungi with Mortierellales. Mol Biol Evol 29:2199–2210

Nguyen TH, Galej WP, Fica SM, Lin PC, Newman AJ, Nagai K (2016) CryoEM structures of two spliceosomal complexes: starter and dessert at the spliceosome feast. Curr Opin Struct Biol 36:48–57

Oi H, Fujita D, Suzuki Y, Sugiyama H, Endo M, Matsumura S, Ikawa Y (2017) Programmable formation of catalytic RNA triangles and squares by assembling modular RNA enzymes. J Biochem 61:451–462

Olson KE, Müller UF (2012) An in vivo selection method to optimize trans-splicing ribozymes. RNA 18:581–589

Paukstelis PJ, Chen JH, Chase E, Lambowitz AM, Golden BL (2008) Structure of a tyrosyl-tRNA synthetase splicing factor bound to a group I intron RNA. Nature 451:94–97

Phylactou LA, Darrah C, Wood MJ (1998) Ribozyme-mediated trans-splicing of a trinucleotide repeat. Nat Genet 18:378–381

Pombert JF, Otis C, Turmel M, Lemieux C (2013) The mitochondrial genome of the prasinophyte Prasinoderma coloniale reveals two trans-spliced group I introns in the large subunit rRNA gene. PLoS One 8:e84325. https://doi.org/10.1371/journal.pone.0084325

Raghavan R, Miller SR, Hicks LD, Minnick MF (2007) The unusual 23S rRNA gene of Coxiella burnetii: two self-splicing group I introns flank a 34-base-pair exon, and one element lacks the canonical omegaG. J Bacteriol 189:6572–6579

Ryu KJ, Lee SW (2003) Identification of the most accessible sites to ribozymes on the hepatitis C virus internal ribosome entry site. J Biochem Mol Biol 36:538–544

Shaw LC, Lewin AS (1997) The Cbp2 protein stimulates the splicing of the omega intron of yeast mitochondria. Nucleic Acids Res 25:1597–1604

Silverman SK (2003) Rube Goldberg goes (ribo)nuclear? Molecular switches and sensors made from RNA. RNA 9:377–383

Stahley MR, Strobel SA (2006) RNA splicing: group I intron crystal structures reveal the basis of splice site selection and metal ion catalysis. Curr Opin Struct Biol 16:319–326

Sullenger BA, Cech TR (1994) Ribozyme-mediated repair of defective mRNA by targeted, trans-splicing. Nature 371:619–622

Tanaka T, Furuta H, Ikawa Y (2014) Installation of orthogonality to the interface that assembles two modular domains in the Tetrahymena group I ribozyme. J Biosci Bioeng 117:407–412

Tanaka T, Matsumura S, Furuta H, Ikawa Y (2016) Tecto-GIRz: engineered group I ribozyme the catalytic ability of which can be controlled by self-dimerization. ChemBioChem 17:1448–1455

Tanaka T, Hirata Y, Tominaga Y, Furuta H, Matsumura S, Ikawa Y (2017) Heterodimerization of group I ribozymes enabling exon recombination through pairs of cooperative trans-splicing reactions. ChemBioChem 18:1659–1667

Tang J, Breaker RR (1997) Rational design of allosteric ribozymes. Chem Biol 4:453–459

Thompson KM, Syrett HA, Knudsen SM, Ellington AD (2002) Group I aptazymes as genetic regulatory switches. BMC Biotechnol 2. https://doi.org/10.1186/1472-6750-2-21

Vicens Q, Cech TR (2006) Atomic level architecture of group I introns revealed. Trends Biochem Sci 31:41–51

Vicens Q, Paukstelis PJ, Westhof E, Lambowitz AM, Cech TR (2008) Toward predicting self-splicing and protein-facilitated splicing of group I introns. RNA 14:2013–2029

Will CL, Lührmann R (2010) Spliceosome structure and function. In: Atkins JF, Gesteland RF, Cech TR (eds) RNA worlds: from life's origins to diversity in gene regulation. CSHL Press, New York, pp 181–203

Woodson SA (2005) Structure and assembly of group I introns. Curr Opin Struct Biol 15:324–330

Zhao C, Pyle AM (2017) Structural insights into the mechanism of group II intron splicing. Trends Biochem Sci 42:470–482

Chapter 14
Chemical and CRISPR/Cas9 Tools for Functional Characterization of RNA Helicases

Jennifer Chu and Jerry Pelletier

Abstract RNA helicases are remodeling proteins implicated in all aspects of RNA biology. Several gene expression pathways, such as ribosome biogenesis, splicing, and translation, are mis-regulated in cancer, making the helicases involved in maintaining expression flux through these networks putative drug targets. It is thus important to better understand the normal regulatory constraints of these proteins, as well as develop small molecule tools that can be used to probe their function. CRISPR/Cas9 can make a significant impact by helping to probe RNA helicase function, determining druggability, and linking the biological activity of small molecules to perturbation of RNA helicase activity.

Keywords CRISPR/Cas9 · Chemical biology · RNA helicases · Translation · Target validation · Cancer

14.1 A Brief Introduction to Eukaryotic RNA Helicases

DNA/RNA helicases are categorized into six superfamilies (SF) based on the presence and organization of signature motifs. SF1 and SF2 helicases are characterized by a common monomeric helicase core that is formed by two tandem RecA domains, a structural motif that was first observed in the *E. coli* RecA protein (Story et al. 1992). In contrast, SF3-6 helicases oligomerize into hexameric rings. All RNA helicases in eukaryotes fall into SF1 or SF2 and are further categorized into five different subfamilies according to their level of shared sequence conservation, substrate specificity, and mechanistic features (Jankowsky 2010) (Fig. 14.1). RNA helicases by and large operate via a similar mechanism. When

J. Chu · J. Pelletier (✉)
Department of Biochemistry, McGill University, Montréal, Canada
e-mail: jerry.pelletier@mcgill.ca

S. Masuda, S. Izawa (eds.), *Applied RNA Bioscience*,
https://doi.org/10.1007/978-981-10-8372-3_14

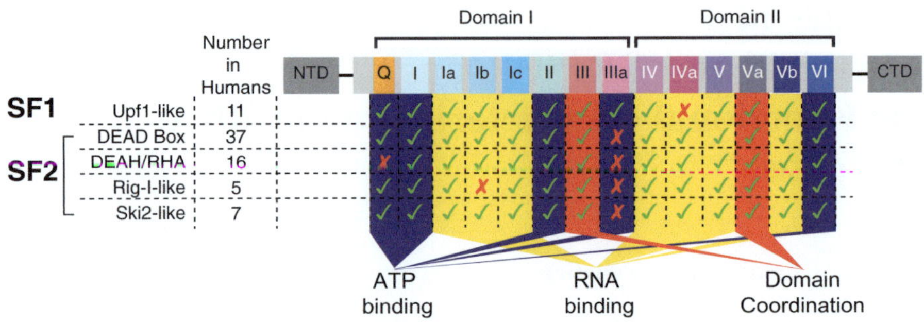

Fig. 14.1 Schematic showing the structural organization of the different RNA helicase subfamilies. While certain characteristic motifs are found within all subfamilies, there is a level of sequence variability in the motifs between different subfamilies. Sequence conservation of structural motifs is much higher when comparing helicases within the same subfamily

bound to ATP, the helicase transitions from an open to closed conformation, in which the two RecA domains are brought into close proximity. In this state, the RNA affinity of the helicase is increased. This is then followed by ATP hydrolysis and P_i release, which restores the helicase back to its "open", low RNA affinity state. Although ATPase activity is not required for RNA binding per se, hydrolysis of ATP to ADP decreases the helicase's RNA-binding affinity, leading to substrate release and enzyme turnover. It should be noted, however, that in spite of their namesake, the primary roles of many RNA helicases are not necessarily dependent on their ability to unwind RNA. For instance, DDX48 (aka eukaryotic initiation factor [eIF] 4A3) is a component of the exon junction complex that acts as an anchor to facilitate the attachment of additional RNA-binding factors – a function that does not require ATPase activity (Shibuya et al. 2004, 2006). In addition to the common conserved core, several RNA helicases are flanked by N- and C- terminal extensions that confer RNA substrate specificity and/or additional catalytic properties. These ancillary domains diversify the range of cellular processes in which RNA helicases partake – consequently, RNA helicases are involved in nearly every gene expression pathway involving RNA.

It is perhaps not surprising that several RNA helicases have been implicated in tumor initiation and progression. However, the precise contributory roles of RNA helicases in cancer are generally not well defined, and this is further complicated by the fact that several RNA helicases are multifunctional proteins reported to possess both tumor-promoting and tumor-suppressing properties depending on the context. Moreover, a number of RNA helicases have also been reported to be promising therapeutic targets, emphasizing the necessity for a comprehensive understanding of their role in neoplasia. In this chapter, we discuss recent developments in the pursuit of small molecules that can target RNA helicases as well as the use of CRISPR/Cas9 as a general tool for functional characterization of RNA helicases.

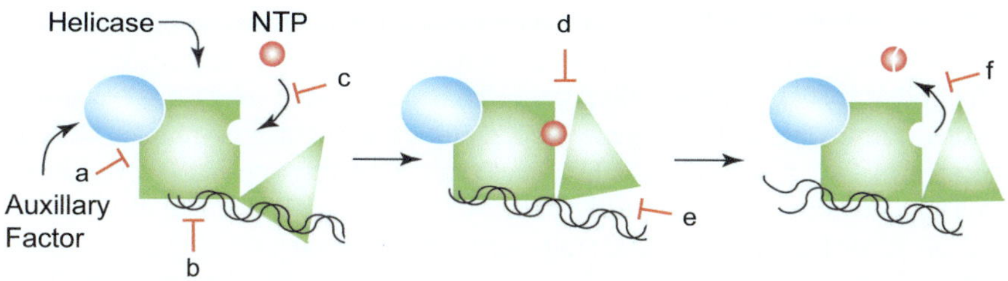

Fig. 14.2 Points of interdiction available to block RNA helicase activity. The RNA helicase is represented as two domains (square and triangular) that undergo conformational changes as a consequence of NTP hydrolysis. An auxiliary protein is depicted as a blue-shaded circle and could function to anchor or modulate helicase activity. Small molecule inhibitors could inhibit (*a*) ancillary factor/helicase interaction, (*b*) helicase/RNA binding, (*c*) NTP binding, (*d*) conformational changes during helicase domain movement, (*e*) helicase activity, and (*f*) NTP hydrolysis or release

14.2 Recent Advances in Targeting RNA Helicases

RNA helicases can potentially be targeted by small molecules through a wide variety of mechanisms. These include preventing ATP and RNA binding, blocking cofactor interaction, and inhibition of ATP hydrolysis and RNA unwinding. Since each conformation change undertaken by the helicase during these events represents a different pose, there are potentially multiple druggable opportunities (Fig. 14.2).

Several RNA helicases have been reported to promote tumor onset and progression. Suppression of these using RNAi has demonstrated antitumor effects. These results bode well for the discovery of small molecules to block RNA helicases as having significant potential to the field of cancer therapeutics. However, there are certain challenges toward identifying such inhibitors. For one, many compounds found in chemical libraries possess inherent affinity toward RNA or are contaminated with low levels of RNAses – both of which can lead to artifacts in screens that utilize RNA binding or unwinding as a readout. The development ATP-competitive inhibitors of RNA helicases faces the obstacle that the ATP-binding pocket is highly conserved between RNA helicases, thus making it challenging to identify a selective inhibitor. These are indeed early days with respect to identifying and crafting selective helicase inhibitors, and there remains much to be learned.

14.2.1 DDX3 Inhibitors

DDX3X is a multifunctional RNA helicase implicated in multiple processes, including transcription regulation, mRNA splicing, mRNA nucleocytoplasmic transport, and translation (Sharma and Jankowsky 2014). There is much interest in finding small molecule inhibitors of DDX3X, especially within the field of virology as DDX3X has been identified as a crucial host factor imperative for the replication of several viruses of medical importance, including HCV, HIV, and HBV (Fullam and Schroder 2013). Additionally, genomic studies have identified recurrent mutations within the DDX3X gene in a significant number of human cancers (Stransky et al. 2011; Wang et al. 2011; Jones et al. 2012; Pugh et al. 2012; Robinson et al. 2012; Brandimarte et al. 2013, 2014; Kandoth et al. 2013; Bol et al. 2015; Jiang et al. 2015; Landau et al. 2015; Ojha et al. 2015). These mutations are thought to lead to a loss of function (Epling et al. 2015) and in medullablastoma have been associated with increased stress granule formation and reduced global translation (Valentin-Vega et al. 2016). DDX3X has also been described to be oncogenic in other settings, with its suppression by RNAi leading to cell cycle blockade (Lai et al. 2010), reduced proliferation in cell culture models of lung (Bol et al. 2015) and breast cancer (Xie et al. 2015), and metastatic capacity in in vivo mouse models (Chen et al. 2015a; Xie et al. 2015). High levels of DDX3X in human tumors have also been associated with poor prognosis (Miao et al. 2013; Heerma van Voss et al. 2015). Therefore, the activity of DDX3X (oncogenic or tumor suppressive) may very well be context-dependent. Germline mutations in DDX3X have also been linked to intellectual disabilities, highlighting a critical role for DDX3X in development (Blok et al. 2015).

Currently there are a number of small molecules that have been proposed to target DDX3 (Fig. 14.3a). Whereas the original intent leading to their development was to assess their ability at reducing viral replication, there is also much interest in assessing their potential as antineoplastics. The majority of DDX3 inhibitors at the moment were designed to target the DDX3 ATP-binding pocket. Rhodanine derivatives were among the first ATP-competitive inhibitors toward DDX3 discovered and were identified through pharmacophore modeling using an X-ray crystal structure of human DDX3 bound to AMP (Maga et al. 2008). Similarly, triazine derivatives were identified as ATP-competitive inhibitors using in silico docking simulations (Maga et al. 2008). Both rhodanine- and triazine-based inhibitors were originally developed with the goal of inhibiting HIV-1 replication and, accordingly, are effective in reducing HIV-1 viral titers at micromolar concentrations (Maga et al. 2008, 2011). However, the effects of these compounds toward malignant cells have yet to be investigated.

Ring-expanded nucleoside (REN) analogues were also found to be effective in inhibiting DDX3 activity based on in vitro dsRNA binding and helicase assays (Yedavalli et al. 2008). Like the rhodanine and triazine analogues, REN derivatives were also able to reduce viral replication and, moreover, were well tolerated in mice (Yedavalli et al. 2008). With respect to cancer therapy, several REN analogues were shown to exhibit cytotoxic activity in cell culture within micromolar or submicromolar concentrations (Kondaskar et al. 2013). One REN analogue in particular, NZ51, was found to induce cell cycle arrest, inhibit proliferation, and impede cellular migration of MCF-7 and MDA-MB-231 breast cancer cell lines (Xie et al. 2015). However, despite its efficacy in cell culture, NZ51 ultimately had no effect on tumor growth when used in a breast cancer xenograft model in mice (Xie et al. 2015). It is possible that the absence of in vivo activity of NZ51 is a con-

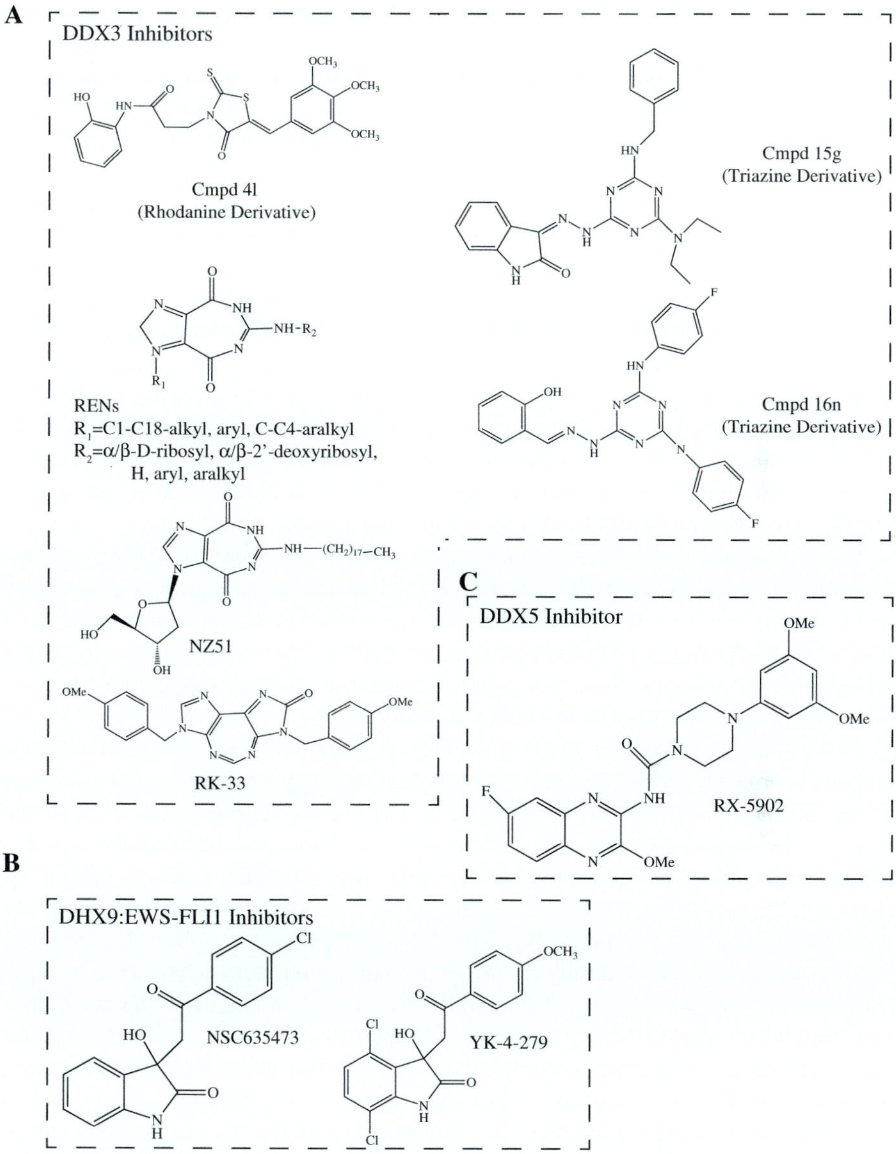

Fig. 14.3 Chemical structures of small molecule inhibitors of DDX3X (**a**), DHX:EWS-FLi1 interaction (**b**), and DDX5 (**c**)

sequence of poor pharmacokinetic properties, although this aspect has yet to be investigated. RK-33, another REN derivative, has also demonstrated promising activity as an anticancer agent (Bol et al. 2015; Xie et al. 2016). As a single agent, RK-33 is cytotoxic toward numerous lung and prostate cancer cell lines within low micromolar concentrations and is also capable of sensitizing the cells to γ-radiation

(Bol et al. 2015; Xie et al. 2016). Unlike NZ51, *in vivo* activity was observed with RK-33, as it was able to enhance radiation-induced tumor regression in an orthotopic lung cancer model as well as a xenograft model of prostate cancer (Bol et al. 2015; Xie et al. 2016). Synergy between RK-33 and the PARP inhibitor olaparib in BRCA1-proficient breast cancer cells was also recently reported (Heerma van Voss et al. 2017).

An alternative strategy to inhibit RNA helicases is to target the RNA-binding site of the helicase. While the ATP-binding pocket is highly similar between RNA helicases (and can potentially offer a challenge when it comes to developing a selective inhibitor), there is relatively more variability in the RNA-binding domains. With this in mind, a small molecule that targets the DDX3 RNA-binding site was recently discovered by in silico docking (Radi et al. 2012). The lead compound identified in this study, Cmpd 16d, inhibits DDX3 helicase activity without affecting its ATPase activity by competitively binding to its RNA-binding site (Fig. 14.3). This compound is cytotoxic toward several strains of HIV-1 known to be resistant to treatments that are currently used in the clinic (Brai et al. 2016). It is not known whether this class of DDX3 inhibitors is effective against cancer cells.

Small molecules can also exert their mechanism of action by abrogating protein-protein interaction. NSC305787 was found to block the interaction with DDX3 and ezrin, a membrane protein that is integral for proper cell adhesion and motility (Çelik et al. 2015). Long-term exposure to NSC305787 led to reduced DDX3 protein levels without affecting DDX3 transcript levels, suggesting that its interaction with ezrin may modulate its stability.

Overall, there are several small molecules targeting DDX3 that exhibit antiviral and antineoplastic properties. The development of many of these compounds was driven by rational drug design strategies and given the reported efficacies of the compounds identified by this approach; in silico molecule docking may be a promising avenue that can be applied to the discovery of other helicase inhibitors. However, there is much work that remains to be undertaken in the characterization of DDX3 inhibitors. The majority of DDX3 inhibitors reported thus far have not been tested against other family members to define their selectivity. This is especially concerning for the competitive ATP inhibitors given the high degree of conservation within the ATP-binding domain. In the case of RK-33, some attempt has been undertaken to evaluate compound specificity (Bol et al. 2015). Biotinylated derivatives of RK-33 were used to demonstrate that RK-33 interacts with DDX3 in cells, but not with DDX5 or DDX17. The efficacy of RK-33 also appears to be dependent on cellular levels of DDX3, with cells possessing high DDX3 levels being more sensitive compared to cells expressing low levels of DDX3. Additionally, microarray expression profiling indicated that perturbations in gene expression resulting from RK-33 exposure largely overlapped with depletion of DDX3 using shRNAs (Bol et al. 2015). While these experiments are suggestive that DDX3 is the relevant therapeutic target, it is now critical to utilize genome engineering tools to link the in vivo biological effects of these inhibitors to perturbation of DDX3X activity.

14.2.2 DDX23

DDX23 is a DEAD-box helicase that plays a role in pre-mRNA splicing through mediation of spliceosome assembly (Teigelkamp et al. 1997; Boesler et al. 2016). There is also evidence suggesting that DDX23 associates with Drosha and promotes biogenesis of mir-21, a micro-RNA whose overexpression is associated with poor prognosis in gliomas (Yin et al. 2015). Upregulation of DDX23 appears to correlate with malignancy in gliomas, and depletion of DDX23 using shRNAs reduced glioma invasiveness and proliferation in cell culture and in orthotopic brain cancer models in mice (Yin et al. 2015). In the quest for a compound that targets DDX23, Yin et al. (2015) assayed a number of viral helicase inhibitors for cross-reactivity toward DDX23. Ivermectin, an anti-parasitic drug that also inhibits HCV NS3 RNA helicase, was found to repress DDX23-mediated processing of pri-mir-21 (Mastrangelo et al. 2012; Yin et al. 2015). Exposure to ivermectin was also able to phenocopy the effects of RNAi-mediated DDX23 depletion (Yin et al. 2015). However, the evidence indicating that the activity of ivermectin is mediated through inhibition of DDX23 is correlative as there is no biochemical data reporting the direct effects of this compound on DDX23 enzymatic activity.

14.2.3 DHX9

DHX9 (also known as RNA helicase A (RHA)) is a multifaceted DNA/RNA helicase that is implicated in several cellular processes including DNA replication, transcription, miRNA processing, and translation (Lee and Pelletier 2016). Like DDX3, DHX9 is also co-opted by several viruses for their own replication (Fullam and Schroder 2013). As DHX9 possesses multiple roles in gene expression and regulation, deregulation of DHX9 has been shown to lead to aberrant cellular activity. DHX9 has been found to be elevated in lung cancer (Sun et al. 2014) and missense mutations have been identified in familial breast cancer settings (Guénard et al. 2009). Moreover, overexpression of DHX9 was shown to increase expression of the multi-drug resistance protein (MDR1), thus establishing a link between DHX9 and chemoresistance (Zhong and Safa 2004). DHX9 suppression exerts different biological consequences depending on context. Prolonged systemic DHX9 suppression in a conditional shDHX9 mouse is tolerated at the organismal level (Lee et al. 2016), and DHX9 knockdown in primary fibroblasts leads to p53-dependent growth arrest and premature senescence (Lee et al. 2014). However, in cancer cells, DHX9 suppression is cytotoxic and can resensitize drug-resistant lymphomas to ABT-737 (Mills et al. 2013; Lee et al. 2016). The fact that systemic DHX9 suppression is well-tolerated in mice, yet cytotoxic to malignant cells suggests that inhibitors targeting DHX9 may offer an attractive therapeutic window for cancer therapy.

There are several reports indicating that DHX9 plays a maintenance role in the Ewing's sarcoma family of tumors (ESFT). The vast majority of ESFT cases are driven by the EWS/FLI1 gene fusion that produces a potent transcription activator that induces malignant transformation (May et al. 1993a, b). DHX9 was shown to interact with the EWS/FLI1 fusion protein, and this interaction appears to enhance the oncogenic activity of EWS/FLI1 (Toretsky et al. 2006). Following screening of a small collection of 3000 compounds, NSC635437 was identified as an inhibitor of DHX9:EWS-FLI1 interaction, and this later led to the development of a more potent derivative, YK-4-27 (Erkizan et al. 2009) (Fig. 14.3b). YK-4-279 induced apoptosis in ESFT cells and blocked tumor growth in xenograft models (Erkizan et al. 2009). DHX9 is also overexpressed in lung cancer, and here, another potential inhibitor – the antibiotic enoxacin – leads to reduced DHX9 protein levels and a modest effect on A549 tumor cell survival (Cao et al. 2017). The mechanism of action of enoxacin vis-à-vis the reported effects on DHX9 has not been elucidated.

14.2.4 DDX5

DDX5, also known as p68, is a DEAD-box helicase that has been strongly associated with tumorigenesis and is overexpressed in several cancer types, including colon (Causevic et al. 2001; Shin et al. 2007; Sarkar et al. 2015), prostate (Clark et al. 2008), breast (Wang et al. 2012a), stomach (Du et al. 2017), and high-grade gliomas (Wang et al. 2012b). In terms of cellular function, DDX5 is developmentally regulated (Stevenson et al. 1998) and is reported to possess both nuclear- and cytoplasmic-specific functions. Notably, DDX5 plays an important role in transcription regulation and can act as an activator by directly interacting with a variety factors such as β-catenin (Shin et al. 2007; Yang et al. 2006), p53 (Bates et al. 2005), and p300/CBP (Rossow and Janknecht 2003) and in certain circumstances may also repress transcription (Wilson et al. 2004). DDX5 has also been implicated in miRNA processing by associating with the Drosha/DGCR8 complex (Suzuki et al. 2009), ribosome biogenesis (Saporita et al. 2011), and regulation of alternative splicing (Liu 2002; Dardenne et al. 2012, 2014).

In preclinical studies, ectopic overexpression of DDX5 has been shown to stimulate tumor cell proliferation, increase tumor size in xenograft models of gastric cancer (Du et al. 2017) and colon cancer (Sarkar et al. 2015), and increase metastatic rates (Sarkar et al. 2015). Due to its ability to cooperate with the onco-protein β-catenin, RNAi depletion of DDX5 blocked nuclear translocation of β-catenin, reduced expression of downstream transcriptional targets (such as c-Myc and c-Jun), inhibited epithelial-to-mesenchymal transition in cell culture, and attenuated tumor growth in vivo (Yang et al. 2006; Shin et al. 2007). Overall, DDX5 is a well-documented promoter of neoplastic growth, and the preclinical data from mice is suggestive that its suppression may attenuate multiple aspects of tumor progression.

In many instances, the oncogenic properties of DDX5 appear to be dependent on its phosphorylation at Y593 (Yang et al. 2006). Consequently, Wang et al. (2013) designed peptides spanning, and proximal to, the phosphorylation site, which were then assessed for antineoplastic activity in a colon cancer xenograft. Of note, one of these, PepIQ (which corresponds to amino acids spanning 549–568), significantly reduced tumor metastasis without affecting the size or proliferation rate of the primary tumor. This result appears to be attributed to DDX5 promoting the localization of calmodulin (CaM) to the leading edge of migrating cells. Interestingly, Wang et al. (2013) also showed that DDX5 exhibits CaM-dependent microtubule motor activity and that DDX5 ATPase activity can be stimulated by microtubules instead of RNA. Although there are a number of issues associated with peptide-based therapies (for instance, induction of an immune response), this work illustrates the therapeutic potential of targeting DDX5.

Through the drug affinity responsive target stability (DARTS) method coupled with LC-MS/MS, DDX5 was also identified to be the molecular target of RX-5902 (Fig. 14.3c), a quinoxalinyl-piperazine derivative that was previously shown to possess growth inhibitory effects against several human tumor cell lines with nanomolar potency (Lee et al. 2010;Kost et al. 2015). While phospho-DDX5 (Y593) was confirmed to bind to RX-5902 using filter-binding assays, the compound was unable to interact with unphosphorylated DDX5. Interestingly, in vitro biochemical assays showed that although RX-5902 has little effect on the RNA-dependent ATPase activity of DDX5, it was able to inhibit its β-catenin-/microtubule-dependent ATPase activity. Exposure of cancer cell lines to RX-5902 resulted in lowered expression of many β-catenin transcription targets. This compound was also shown to display favorable pharmacokinetic properties in rats (Lee et al. 2012), but its efficacy as an antineoplastic agent has yet to be tested in vivo.

14.2.5 DDX2A (eIF4A1) and DDX2B (eIF4A2)

EIF4F has recently emerged as an anticancer target (Bhat et al. 2015). This complex consists of (I) eIF4E, the cap-binding protein responsible for directing eIF4F to mRNA cap structures; (II) eIF4A, a DEAD-box RNA helicase required to unwind local secondary structure to facilitate access of the 43S preinitiation complex (PIC) to the mRNA template; and (III) eIF4G, a large scaffolding protein that recruits the 43S PIC to the mRNA template. Efforts to drug this complex lead to the identification of three natural products that appear to be quite selective inhibitors of eIF4A – pateamine A (PatA), hippuristanol, and rocaglates (Fig. 14.4a).

The properties of these three inhibitors have been extensively reviewed, and only the important pertinent properties of these compounds will be discussed herein (Bhat et al. 2015; Chu and Pelletier 2015). All three molecules have different mechanisms of action while appearing quite selective for eIF4A. (I) Pat A was initially isolated from the marine sponge, *Mycale* sp., and is an irreversible inhibitor of eIF4A (Northcote et al. 1991). This compound causes eIF4A to clamp down onto RNA and leads to depletion of eIF4A from the eIF4F complex. Affinity chromatography experiments with immobilized PatA by two independent groups

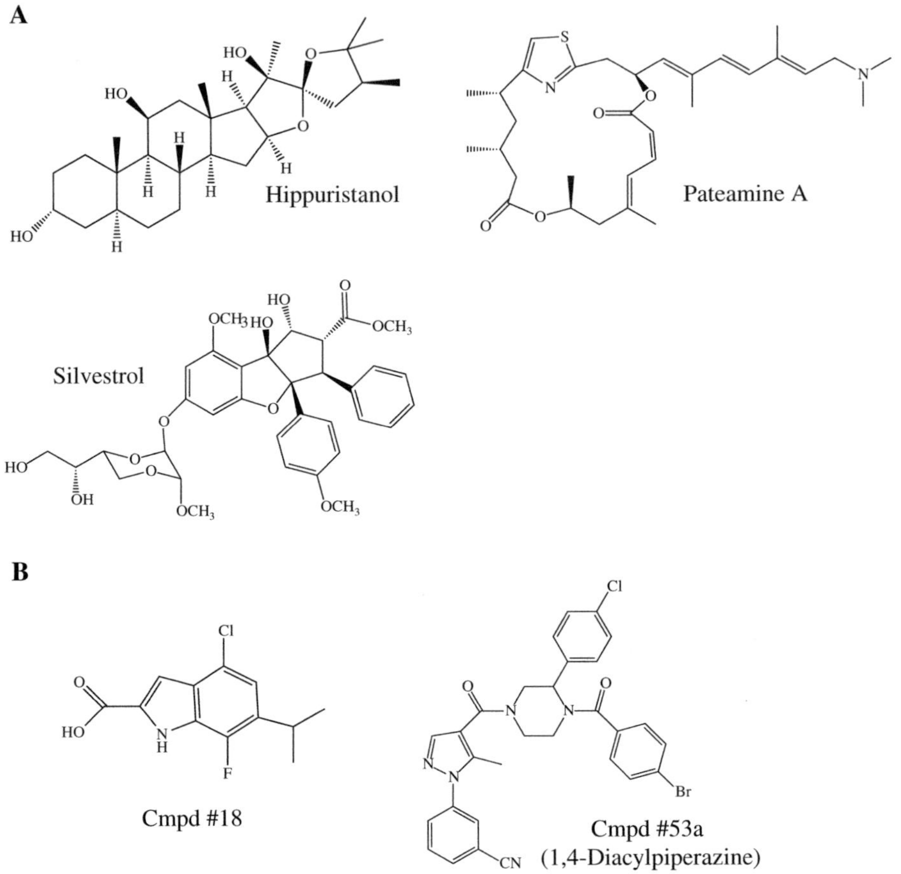

Fig. 14.4 Chemical structures of eIF4A1/2 (**a**) and eIF4A3 (**b**) inhibitors

has shown that PatA is not a pan-helicase inhibitor and recognizes a handful of cellular targets – of which eIF4A is the only translation factor directly bound by the compound (Bordeleau et al. 2005; Low et al. 2005). (II) Hippuristanol is a polyoxygenated steroid first isolated from the gorgonian *Isis hippuris* (Higa et al. 1981). This compound inhibits eIF4A RNA binding and ribosome recruitment to mRNA templates. It binds to the C-terminal domain of eIF4A, in a pocket that encompasses regions of conserved motifs V and VI (Fig. 14.1a) and adjacent non-conserved amino acids (Lindqvist et al. 2008). The hippuristanol-binding site is present in the yeast eIF4A orthologs, TIF1 and TIF2, explaining its inhibitory activity in yeast translation extracts (Lindqvist et al. 2008). DDX48 contains three critical amino acid differences in its hippuristanol-binding site compared to DDX2, and indeed tenfold more hippuristanol is required to inhibit DDX48 ATPase activity compared to DDX2 (Lindqvist et al. 2008). Hippuristanol-binding sites are not present within other DEAD-box RNA helicases rationalizing the selectivity of this compound for DDX2. (III) Rocaglates were first identified from the roots and stems of the *Aglaia elliptifolia* plant (King et al. 1982). This class of compounds causes eIF4A to clamp onto polypurine RNA sequences and also leads to depletion of eIF4A from the eIF4F complex (Bordeleau et al. 2008; Iwasaki et al. 2016). Affinity chromatography experiments have shown that only eIF4A is bound by these compounds, and genetic approaches (described below) have shown that eIF4A is the primary target that can explain the biological activity of these compounds (Chambers et al. 2013; Chu et al. 2016).

14.2.6 DDX48 (eIF4A3)

EIF4A3 is highly related to eIF4A1 and eIF4A2 (sharing ~68% identity) but has predominantly a nuclear function. It is one of four core components (along with Y14, MAGOH, and MLN51) that form the exon junction complex (EJC) (Chan et al. 2004; Shibuya et al. 2004; Palacios et al. 2004; Le Hir et al. 2016). As a result, eIF4A3 is a modulator of pre-mRNA splicing, nucleocytoplasmic shuttling of spliced mRNAs, translation, and nonsense-mediated decay (NMD) (Shibuya et al. 2004; Palacios et al. 2004; Le Hir et al. 2016). eIF4A3 is first recruited in its "open" conformation (i.e., low RNA and ATP affinity state) to the pre-mRNA through its interaction with the essential splicing factor, CWC22 (Barbosa et al. 2012; Steckelberg et al. 2012). As splicing proceeds, eIF4A3 is dissociated from CWC22 and transitions into its "closed" conformation (i.e., high RNA and ATP affinity state) and binds onto the EJC site, which, based on CLIP (cross-linking and immunoprecipitation) experiments and in vitro biochemical data, is primarily ~24 nucleotides upstream from exon-exon junctions (Shibuya et al. 2004; Saulière et al. 2012). The MAGOH:Y14 heterodimer assembles onto eIF4A3 and locks it in a state that prevents eIF4A3 from hydrolyzing ATP – this consequently locks eIF4A3 in its closed/high RNA affinity conformation and allows it to act as a stable RNA clamp (Ballut et al. 2005; Nielsen et al. 2009). Accordingly, eIF4A3 mutants that are unable to hydrolyze ATP are still capable of promoting EJC assembly and activating NMD (Shibuya et al. 2006). Beyond the EJC, eIF4A3 is also a regulator of rRNA biogenesis (Alexandrov et al. 2011) as well as a repressor of selenoprotein translation (Budiman et al. 2009).

As one of the core components of the EJC, deregulated expression of eIF4A3 leads to considerable changes in global gene expression. Suppression of eIF4A3 using shRNAs or in a haploinsufficient mouse model was shown to cause alternative splicing of several genes (Wang et al. 2014; Mao et al. 2016; Miller et al. 2017). Furthermore, copy number variations of eIF4A3 (as well as other factors involved in EJC formation and NMD) have been linked to neurodevelopmental disorders (Nguyen et al. 2013). Families with Richieri-Costa-Pereira syndrome (an autosomal recessive disorder that is often characterized by neurological defects, and abnormal craniofacial and limb development) have been found to carry either a mutation within eIF4A3 that would prevent its interaction with CWC22 or an expansion within the eIF4A3 5′ leader region that reduces eIF4A3 mRNA levels (Favaro et al. 2014). In line with this, suppression of eIF4A3 levels in zebrafish using morpholino antisense oligonucleotides leads to defective craniofacial cartilage development and abnormal bone formation (Favaro et al. 2014). Furthermore, eIF4A3 haploinsufficient mice also exhibited malformed mandibles, delayed limb development, microencephaly, and reduced body size (Mao et al. 2016; Miller et al. 2017).

Systemic suppression of eIF4A3 has revealed its importance in development. Nonetheless, the transient suppression of eIF4A3 post-development is currently not well characterized. HTS efforts have been undertaken to identify eIF4A3 inhibitors, from which was identified an indole-2-carboxylic acid derivative (Ito et al. 2017a). Further chemical optimization of this derivative resulted in the synthesis of compound 18 (4-chloro-7-fluoro-6-isopropyl-1H-indole-2-carboxylic acid) which was found to act as a ATP-competitive inhibitor of eIF4A3 (Ito et al. 2017a). Importantly, one of the compounds (cmpd#18) exhibited high selectively toward eIF4A3 over other helicases, including eIF4A1, eIFA2, and DHX29 (Fig. 14.4b) (Ito et al. 2017a). However, cmpd#18 was reported to show little effects when tested in cell culture setting (Ito et al. 2017b). From the same HTS campaign, 1,4-diacylpiperazine derivatives were also identified to be inhibitors of eIF4A3 and were later shown to be able to suppress a reported based NMD assay in cell culture at micromolar concentrations (Ito et al. 2017b). These compounds were also found to bind directly to eIF4A3 using SPR and were able to selectively inhibit eIF4A3 ATPase activity (Ito et al. 2017b). In addition, affinity and kinetic assays indicate that this class of compounds are not ATP-competitive inhibitors and may exert their effects through an allosteric mechanism. It will be of interest to map the binding site of these compounds on eIF4A3.

14.3 CRISPR/Cas9 to Investigate the Role of RNA Helicases in Tumorigenesis and for Drug Target Validation

Since its advent as a gene-editing tool in 2012, CRISPR has been a powerful technique that has allowed researchers to introduce modifications to their genomic locus of choice with relative ease (Jinek et al. 2012). In brief, this technique usually involves the co-introduction of the bacterial endonuclease Cas9 (or, alternatively, Cpf1 from *Francisella novicida*) and a single-guide RNA (sgRNA) that directs Cas9 to the specific genomic site of interest. Cas9 (or Cpf1) then generates a double-stranded break at the targeted DNA, which, in the absence of a suitable repair template, is then mended through nonhomologous end joining (NHEJ). As this repair pathway is error-prone, often introducing random insertions and deletions (indels) at the cleavage site, Cas9 can be easily harnessed to knockout any gene by simply changing the sequence of the sgRNA. Alternatively, in the presence of a DNA repair template with sufficient homology to the targeted DNA, the double-stranded break can also be repaired through homology-directed recombination (HDR). As a result, specific modifications can be introduced into a locus of interest by HDR with high specificity, although the efficiency of achieving this is much lower compared to the generation of indels by NHEJ (Cong et al. 2013; Jinek et al. 2013; Mali et al. 2013; Malina et al. 2013).

Due to the relative ease of introducing frameshifts via NHEJ, early studies primarily implemented CRISPR in loss of function/knockout studies. Since then Cas9 has been repurposed to alter gene expression through other means. For instance, CRISPR interference (CRISPRi) utilizes a catalytically inactive Cas9 (dCas9) to repress transcription of specific genes by acting as a steric barrier to RNA polymerase or to transcription factor binding (Qi et al. 2013). Moreover, dCas9 can also be coupled to a variety of effector domains, including, but not limited to, transcription activation domains (for gene overexpression), DNA methyltransferases and acetyltransferases (for targeted epigenetic modification), fluorescent proteins (for localization studies), and DNA base editing enzymes (for targeted cytosine hypermutagenesis) (Gilbert et al. 2013; Hess et al. 2016; Komor et al. 2016; Ma et al. 2016; Vojta et al. 2016). In summary, CRISPR can be applied to manipulate gene expression through a wide variety of mechanisms, and one can surmise the potential of applying this technology to investigate RNA helicases with respect to cellular function, neoplasia, and therapeutic applications.

Within the field of cancer research, a number of groups have employed genome-wide CRISPR/Cas9 knockout screens to identify important modulators of tumor onset, progression (Chen et al. 2015b), metastasis (Chen et al. 2015b), drug resistance (Kurata et al. 2016), as well as novel synergistic drug pairs (Han et al. 2017). As critical regulators of several essential cellular processes, sgRNAs targeting RNA helicases tend to be early drop-offs in these knockout screens (Chen et al. 2015b). Accordingly, roughly half of all eukaryotic RNA helicases were identified as being essential in a CRISPR-mediated whole-genome screen using a variety of in vitro cell lines (Wang et al. 2015). This work not only highlights the importance of RNA helicases for normal cellular function but also suggests that functional redundancies may be shared between certain helicases.

14.4 Probing Vasa/DDX4 Function

In addition to genome-wide screens, researchers have implemented CRISPR/Cas9 to specifically knockout RNA helicase expression or to generate precise mutations into endogenous loci. Cas9 has been used to examine the unique role of Vasa (also known as DDX4) in drosophila germ cell development. DDX4 was not categorized as an essential gene according to the data from the aforementioned genome-wide CRISPR screen (Wang et al. 2015). Orthologs of DDX4/Vasa possess a conserved tryptophan residue in its C-terminal domain (W660 in *Drosophila melanogaster*) that is not found in other DEAD-box helicases. Introduction of a W660E point mutation in DDX4 using Cas9 partially mimicked the Vasa-null phenotype, in which germ cell development and piRNA biogenesis were severely impaired (Dehghani and Lasko 2016). However, unlike vas-null females, vas^{W660E} females did not show defects regarding egg production, suggesting that this point mutation does not entirely abrogate Vasa function (Dehghani and Lasko 2016). Nonetheless this work illustrates that this unique yet highly conserved W660 residue is essential for the role of Vasa in germ cell development.

14.5 Probing DHX33 Function

The DHX33 helicase is found in both the cytoplasm and the nucleus and has been implicated in rRNA biogenesis, in cytosolic double-stranded RNA sensing (Mitoma et al. 2013; Zhang et al. 2013; Liu et al. 2014) and in regulating translation. RNAi-mediated suppression of DHX33 induces a reduction in the polysome/monosome ratio in cells and decreases incorporation of ^{35}S-methionine into newly synthesized proteins. Immunoprecipitation studies also suggest that DHX33 associates with several components of the translation apparatus (Zhang et al. 2015). DHX33 has also been implicated in transcriptional regulation of cell cycle genes by regulating loading of RNA polymerase onto the promoters of target genes (Yuan et al. 2016). Acute depletion of DHX33 by RNAi in H1299 and Calu-1 cells leads to increased cell death (Yuan et al. 2016). An 11 bp deletion in exon 33 of the zebra fish DHX33 has been engineered using CRISPR/Cas9, and homozygote mutants show a developmental delay visible at 48 h post fertilization and ultimately die – possibly due to heart edema (Yuan et al. 2016). Although this is a very interesting phenotype, further experiments are required to ensure that these are not the consequences of an off-target activity of Cas9.

14.6 Probing eIF4A/DDX2 Function

Mammals encode two highly related eIF4A homologs with the human proteins showing 90% amino acid identity (Conroy et al. 1990; Yoder-Hill et al. 1993). In general, eIF4A1 is the more abundantly expressed gene, with notable exceptions being fetal brain, fetal heart, adult brain, and ovaries (Nielsen et al. 1985; Galicia-Vázquez et al. 2012). The majority (~90%) of eIF4A exists as a free form (eIF4A$_f$) with only a small proportion present in the eIF4F complex (eIF4A$_c$) (Thomas et al. 1979; Edery et al. 1983; Grifo et al. 1983). Although both helicases can assemble into the eIF4F complex, most studies on translation have focused on the more abundant eIF4A1 protein (Yoder-Hill et al. 1993). There is also evidence suggesting that eIF4A1 and eIF4A2 are not redundant proteins and may possess unique biological properties (Li et al. 2001; Galicia-Vázquez et al. 2012). For instance, it was reported that eIF4A2, but not eIF4A1, was critical for miRNA-mediated silencing (Meijer et al. 2013). In order to investigate potential functional differences between eIF4A1 and eIF4A2, we targeted eIF4A2 exon 5 for inactivation in NIH/3 T3 cells using CRISPR/Cas9 (Galicia-Vázquez et al. 2015). Selection of a desired cell clone was achieved by screening colonies established by limiting dilution using a PCR-based assay. Isolation of RNA from one selected clone showed the expected deletions in exon 5, but subsequent sequencing of an eIF4A2 cDNA generated from primers spanning a larger region of the mRNA (from exons 4–9) indicated that one of the alleles had undergone a large internal deletion, possibly due to partial homology of the sgRNA with a target sequence in exon 8. These results serve as an example that genetic engineering products arising from CRISPR/Cas9 can be quite complex, and resulting cells lines should be cautiously used with validation experiments of key results being a critical component of any study using them. Notwithstanding these complexities, the absence of eIF4A2 in the isolated cell line was confirmed by Western blotting, and we observed no deleterious effects on cell proliferation or viability (Galicia-Vázquez et al. 2015). This is in stark contrast to experiments in which cells are infected with retroviruses expressing eIF4A1-directed shRNAs where cellular proliferation and viability are dramatically impaired (Galicia-Vázquez et al. 2012). These results indicate that (at least in NIH/3T3 cells) eIF4A2 is dispensable, whereas eIF4A1 is essential. Importantly, we also did not observe any defects in miRNA-mediated gene repression in our eIF4A2 KO cells, suggesting that eIF4A2 is not an essential factor for this process.

14.7 Using CRISPR/Cas9 for Drug Target Validation

One important issue that arises during drug development programs is whether observed biological effects of compounds in vivo are the consequence of on-target versus off-target inhibition. The failure to correctly address this early on in research programs can lead to faulty conclusions and in the drug development field to a significant waste of resources (Cook et al. 2014). One of the aforementioned eIF4A inhibitors, rocaglates, displays significant anticancer activity in a large number of preclinical models (Bordeleau et al. 2008; Cencic et al. 2009; Lucas et al. 2009; Alachkar et al. 2013; Santagata et al. 2013; Wolfe et al. 2014). Although in vitro biochemistry data had linked eIF4A inhibition to rocaglates (Bordeleau et al. 2008), it was critical to demonstrate via genetics that the in vivo properties were a consequence of perturbing eIF4A activity. To this end, a specific mutation, F163L, was engineered into NIH/3T3 cells (Chu et al. 2016). The nature of the F163L mutation was based on experiments that had been previously performed in yeast in which rocaglate-resistant cells had been isolated and among which one of the lines harboring the corresponding F151L TIF1 (the yeast eIF4A ortholog) mutation had been characterized (Sadlish et al. 2013). Critical to undertaking this endeavor was reassurance that the F163L did not in any manner grossly affect eIF4A activity. Indeed, assessment of V_{max} and K_m values for ATP hydrolysis indicated these to be quite similar to values obtained for the recombinant wild-type protein (Chu et al. 2016). As well, differential scanning fluorimetry provided evidence that whereas wild-type recombinant eIF4A1 could bind to rocaglates, the eIF4A1(F163L) mutant did not (Chu et al. 2016). These results prompted us to generate an NIH/3T3 cell line using

Cas9 to stimulate HDR within eIF4A1 exon 5 with an oligonucleotide donor harboring the desired F163L nucleotide change. A cell line in which all wild-type eIF4A1 alleles had been modified to F163L was found to be resistant to translation inhibition by rocaglates. To demonstrate that this resistance was an on-target consequence of the intended modification and not an off-target Cas9 byproduct, a genetic rescue experiment was undertaken in which the eIF4A1(F163L) mRNA was suppressed using RNAi and full-length shRNA-resistant wild-type eIF4A1 cDNA delivered to the engineered cells. This resensitized the cells to rocaglates, thus confirming that the observed resistance with the parental eIF4A1(F163L) line was due to on-target modification of the eIF4A1 allele. The parental and eIF4A1(F163L) lines were transformed with the potent myr-Akt oncogene and used to generate xenograft tumors in nude mice, which turned out to be, respectively, responsive and non-responsive to rocaglate treatment (Chu et al. 2016). In sum, these experiments outline a clear strategy that can be followed to validate drug-target relationships in vitro and in vivo and extended to study any small molecule-helicase relationship.

14.8 Future Perspectives

Currently, there is a major need to develop more selective inhibitors to a larger set of RNA helicases. While approaches like CRISPR knockouts and RNAi-based strategies are capable of mimicking (to some extent) an inhibited RNA helicase activity, there is a significant delay in kinetics of onset (relative to a small molecule inhibitor). This makes interpretation of subsequent downstream phenotypes complicated due to the potential for secondary effects to confound resulting analyses. As well, small molecule inhibitors do not necessarily lead to target depletion, which is quite distinct than what is achieved by CRISPR or RNAi, and may thus have a different impact on helicase-dependent ancillary factors.

One strategy by which some of these issues can be addressed is to borrow the concept of "gatekeeper" residue mutations in kinases that enable the design of analog-sensitive proteins (Lopez et al. 2014). Applying this to helicases by engineering the ATP-binding site with an appropriately designed novel space or cavity that can accommodate a small molecule, the activity of the altered helicase can be placed under regulation of a unique small molecule inhibitor. Indeed, this approach has been implemented by Floor et al. (2016) for DDX3. Here, an F182A change placed DDX3^{F182A} under regulation of an ATP (GXJ1-76) and anilinoquinazoline analogues in vitro (Floor et al. 2016). Given how the development of CRISPR has markedly enhanced the efficiency of introducing specific point mutations into the genome, it is reasonable to expect that expanded active-site mutations in RNA helicases can be generated in cell lines or in organisms and in combination with appropriate selective inhibitors be used to probe for in vivo function.

References

Alachkar H, Santhanam R, Harb JG, Lucas DM, Oaks JJ, Hickey CJ, Pan L, Kinghorn AD, Caligiuri MA, Perrotti D, Byrd JC, Garzon R, Grever MR, Marcucci G (2013) Silvestrol exhibits significant in vivo and in vitro antileukemic activities and inhibits FLT3 and miR-155 expressions in acute myeloid leukemia. J Hemat Oncol 6:21

Alexandrov A, Colognori D, Steitz JA (2011) Human eIF4AIII interacts with an eIF4G-like partner, NOM1, revealing an evolutionarily conserved function outside the exon junction complex. Genes Dev 25:1078–1090

Ballut L, Marchadier B, Baguet A, Tomasetto C, Séraphin B, Le Hir H (2005) The exon junction core complex is locked onto RNA by inhibition of eIF4AIII ATPase activity. Nat Struct Mol Biol 12:861–869

Barbosa I, Haque N, Fiorini F, Barrandon C, Tomasetto C, Blanchette M, Le Hir H (2012) Human CWC22 escorts the helicase eIF4AIII to spliceosomes and promotes exon junction complex assembly. Nat Struct Mol Biol 19:983–990

Bates GJ, Nicol SM, Wilson BJ, Jacobs AM, Bourdon JC, Wardrop J, Gregory DJ, Lane DP, Perkins ND, Fuller-Pace FV (2005) The DEAD box protein p68: a novel transcriptional coactivator of the p53 tumour suppressor. EMBO J 24:543–553

Bhat M, Robichaud N, Hulea L, Sonenberg N, Pelletier J, Topisirovic I (2015) Targeting the translation machinery in cancer. Nat Rev Drug Discov 14:261–278

Blok LS, Madsen E, Juusola J, Gilissen C, Baralle D, Reijnders MRF, Venselaar H, Helsmoortel C, Cho MT, Hoischen A, Vissers LELM, Koemans TS, Wissink-Lindhout W, Eichler EE, Romano C, Esch HV, Stumpel C, Vreeburg M, Smeets E, Oberndorff K, van Bon BWM, Shaw M, Gecz J, Haan E, Bienek M, Jensen C, Loeys BL, Dijck AV, Innes AM, Racher H, Vermeer S, Donato ND, Rump A, Tatton-Brown K, Parker MJ, Henderson A, Lynch SA, Fryer A, Ross A, Vasudevan P, Kini U, Newbury-Ecob R, Chandler K, Male A, the DDD study, Dijkstra S, Schieving J, Giltay J, van Gassen KLI, Schuurs-Hoeijmakers J, Tan PL, Pediaditakis I, Haas SA, Retterer K, Reed P, Monaghan KG, Haverfield E, Natowicz M, Myers A, Kruer MC, Stein Q, Strauss KA, Brigatti KW, Keating K, Burton BK, Kim KH, Charrow J, Norman J, Foster-Barber A, Kline AD, Kimball A, Zackai E, Harr M, Fox J, McLaughlin J, Lindstrom K, Haude KM, van Roozendaal K, Brunner H, Chung WK, Kooy RF, Pfundt R, Kalscheuer V, Mehta SG, Katsanis N, Kleefstra T (2015) Mutations in DDX3X are a common cause of unexplained intellectual disability with gender-specific effects on Wnt signaling. Am J Hum Genet 97:343–352

Boesler C, Rigo N, Anokhina MM, Tauchert MJ, Agafonov DE, Kastner B, Urlaub H, Ficner R, Will CL, Lührmann R (2016) A spliceosome intermediate with loosely associated tri-snRNP accumulates in the absence of Prp28 ATPase activity. Nat Commun 7:11997

Bol GM, Vesuna F, Xie M, Zeng J, Aziz K, Gandhi N, Levine A, Irving A, Korz D, Tantravedi S, van Voss MRH, Gabrielson K, Bordt EA, Polster BM, Cope L, van der Groep P, Kondaskar A, Rudek MA, Hosmane RS, van der Wall E, van Diest PJ, Tran PT, Raman V (2015) Targeting DDX3 with a small molecule inhibitor for lung cancer therapy. EMBO Mol Med 7:648–669

Bordeleau ME, Matthews J, Wojnar JM, Lindqvist L, Novac O, Jankowsky E, Sonenberg N, Northcote P, Teesdale-Spittle P, Pelletier J (2005) Stimulation of mammalian translation initiation factor eIF4A activity by a small molecule inhibitor of eukaryotic translation. Proc Natl Acad Sci U S A 102:10460–10465

Bordeleau ME, Robert F, Gerard B, Lindqvist L, Chen SM, Wendel HG, Brem B, Greger H, Lowe SW, Porco JA Jr, Pelletier J (2008) Therapeutic suppression of translation initiation modulates chemosensitivity in a mouse lymphoma model. J Clin Invest 118:2651–2660

Brai A, Fazi R, Tintori C, Zamperini C, Bugli F, Sanguinetti M, Stigliano E, Esté J, Badia R, Franco S, Martinez MA, Martinez JP, Meyerhans A, Saladini F, Zazzi M, Garbelli A, Maga G, Botta M (2016) Human DDX3 protein is a valuable target to develop broad spectrum antiviral agents. Proc Natl Acad Sci U S A 113:5388–5393

Brandimarte L, Pierini V, Giacomo DD, Borga C, Nozza F, Gorello P, Giordan M, Cazzaniga G, Kronnie G, Starza RL, Mecucci C (2013) New MLLT10 gene recombinations in pediatric T-acute lymphoblastic leukemia. Blood 121:5064–5067

Brandimarte L, Starza RL, Gianfelici V, Barba G, Pierini V, Giacomo DD, Cools J, Elia L, Vitale A, Luciano L, Bardi A, Chiaretti S, Matteucci C, Specchia G, Mecucci C (2014) DDX3X-MLLT10 fusion in adults with NOTCH1 positive T-cell acute lymphoblastic leukemia. Haematologica 99:64–66

Budiman ME, Bubenik JL, Miniard AC, Middleton LM, Gerber CA, Cash A, Driscoll DM (2009) Eukaryotic initiation factor 4a3 is a selenium-regulated RNA-binding protein that selectively inhibits selenocysteine incorporation. Mol Cell 35:479–489

Cao S, Sun R, Wang W, Meng X, Zhang Y, Zhang N, Yang S (2017) RNA helicase DHX9 may be a therapeutic target in lung cancer and inhibited by enoxacin. Am J Transl Res 9:674–682

Causevic M, Hislop RG, Kernohan NM, Carey FA, Kay RA, Steele RJ, Fuller-Pace FV (2001) Overexpression and poly-ubiquitylation of the DEAD-box RNA helicase p68 in colorectal tumours. Oncogene 20:7734–7743

Çelik H, Sajwan KP, Selvanathan SP, Marsh BJ, Pai AV, Kont YS, Han J, Minas TZ, Rahim S, Erkizan HV, Toretsky JA, Üren A (2015) Ezrin binds to DEAD-box RNA helicase DDX3 and regulates its function and protein level. Mol Cell Biol 35:3145–3162

Cencic R, Carrier M, Galicia-Vázquez G, Bordeleau ME, Sukarieh R, Bourdeau A, Brem B, Teodoro JG, Greger H, Tremblay ML, Porco JA Jr, Pelletier J (2009) Antitumor activity and mechanism of action of the cyclopenta[b]benzofuran, silvestrol. PLoS One 4:e5223

Chambers JM, Lindqvist LM, Webb A, Huang DC, Savage GP, Rizzacasa MA (2013) Synthesis of biotinylated episilvestrol: highly selective targeting of the translation factors eIF4AI/II. Org Lett 15:1406–1409

Chan CC, Dostie J, Diem MD, Feng W, Mann M, Rappsilber J, Dreyfuss G (2004) eIF4A3 is a novel component of the exon junction complex. RNA 10:200–209

Chen HH, Yu HI, Cho WC, Tarn WY (2015a) DDX3 modulates cell adhesion and motility and cancer cell metastasis via Rac1-mediated signaling pathway. Oncogene 34:2790–2800

Chen S, Sanjana NE, Zheng K, Shalem O, Lee K, Shi X, Scott DA, Song J, Pan JQ, Weissleder R, Lee H, Zhang F, Sharp PA (2015b) Genome-wide CRISPR screen in a mouse model of tumor growth and metastasis. Cell 160:1246–1260

Chu J, Pelletier J (2015) Targeting the eIF4A RNA helicase as an anti-neoplastic approach. Biochim Biophys Acta 1849:781–791

Chu J, Galicia-Vázquez G, Cencic R, Mills JR, Katigbak A, Porco JA Jr, Pelletier J (2016) CRISPR-mediated drug-target validation reveals selective pharmacological inhibition of the RNA helicase, eIF4A. Cell Rep 15:2340–2347

Clark EL, Coulson A, Dalgliesh C, Rajan P, Nicol SM, Fleming S, Heer R, Gaughan L, Leung HY, Elliott DJ, Fuller-Pace FV, Robson CN (2008) The RNA helicase p68 is a novel androgen receptor coactivator involved in splicing and is overexpressed in prostate cancer. Cancer Res 68:7938–7946

Cong L, Ran FA, Cox D, Lin S, Barretto R, Habib N, Hsu PD, Wu X, Jiang W, Marraffini LA, Zhang F (2013) Multiplex genome engineering using CRISPR/Cas systems. Science 339:819–823

Conroy SC, Dever TE, Owens CL, Merrick WC (1990) Characterization of the 46,000-dalton subunit of eIF-4F. Arch Biochem Biophys 282:363–371

Cook D, Brown D, Alexander R, March R, Morgan P, Satterthwaite G, Pangalos MN (2014) Lessons learned from the fate of AstraZeneca's drug pipeline: a five-dimensional framework. Nat Rev Drug Discov 13:419–431

Dardenne E, Pierredon S, Driouch K, Gratadou L, Lacroix-Triki M, Espinoza MP, Zonta E, Germann S, Mortada H, Villemin JP, Dutertre M, Lidereau R, Vagner S, Auboeuf D (2012) Splicing switch of an epigenetic regulator by RNA helicases promotes tumor-cell invasiveness. Nat Struct Mol Biol 19:1139–1146

Dardenne E, Polay Espinoza M, Fattet L, Germann S, Lambert MP, Neil H, Zonta E, Mortada H, Gratadou L, Deygas M, Chakrama FZ, Samaan S, Desmet FO, Tranchevent LC, Dutertre M,

Rimokh R, Bourgeois CF, Auboeuf D (2014) RNA helicases DDX5 and DDX17 dynamically orchestrate transcription, miRNA, and splicing programs in cell differentiation. Cell Rep 7:1900–1913

Dehghani M, Lasko P (2016) C-terminal residues specific to Vasa among DEAD-box helicases are required for its functions in piRNA biogenesis and embryonic patterning. Dev Genes Evol 226:401–412

Du C, Li DQ, Li N, Chen L, Li SS, Yang Y, Hou MX, Xie MJ, Zheng ZD (2017) DDX5 promotes gastric cancer cell proliferation in vitro and in vivo through mTOR signaling pathway. Sci Rep 7:42876

Edery I, Hümbelin M, Darveau A, Lee KA, Milburn S, Hershey JW, Trachsel H, Sonenberg N (1983) Involvement of eukaryotic initiation factor 4A in the cap recognition process. J Biol Chem 258:11398–11403

Epling LB, Grace CR, Lowe BR, Partidge JF, Enemark EJ (2015) Cancer-associated mutants of RNA helicase DDX3X are defective in RNA-stimulated ATP hydrolysis. J Mol Biol 427:1779–1796

Erkizan HV, Kong Y, Merchant M, Schlottmann S, Barber-Rotenberg JS, Yuan L, Abaan OD, Chou TH, Dakshanamurthy S, Brown ML, Uren A, Toretsky JA (2009) A small molecule blocking oncogenic protein EWS-FLI1 interaction with RNA helicase A inhibits growth of Ewing's sarcoma. Nat Med 15:750–756

Favaro FP, Alvizi L, Zechi-Ceide RM, Bertola D, Felix TM, de Souza J, Raskin S, Twigg SR, Weiner AM, Armas P, Margarit E, Calcaterra NB, Andersen GR, McGowan SJ, Wilkie AO, Richieri-Costa A, de Almeida ML, Passos-Bueno MR (2014) A noncoding expansion in EIF4A3 causes Richieri-Costa-Pereira syndrome, a craniofacial disorder associated with limb defects. Am J Hum Genets 94:120–128

Floor SN, Barkovich KJ, Condon KJ, Shokat KM, Doudna JA (2016) Analog sensitive chemical inhibition of the DEAD-box protein DDX3. Protein Sci 25:638–649

Fullam A, Schroder M (2013) DExD/H-box RNA helicases as mediators of anti-viral innate immunity and essential host factors for viral replication. Biochim Biophys Acta 1829:854–865

Galicia-Vázquez G, Cencic R, Robert F, Agenor AQ, Pelletier J (2012) A cellular response linking eIF4AI activity to eIF4AII transcription. RNA 18:1373–1384

Galicia-Vázquez G, Chu J, Pelletier J (2015) eIF4AII is dispensable for miRNA-mediated gene silencing. RNA 21:1826–1833

Gilbert LA, Larson MH, Morsut L, Liu Z, Brar GA, Torres SE, Stern-Ginossar N, Brandman O, Whitehead EH, Doudna JA, Lim WA, Weissman JS, Qi LS (2013) CRISPR-mediated modular RNA-guided regulation of transcription in eukaryotes. Cell 154:442–451

Grifo JA, Tahara SM, Morgan MA, Shatkin AJ, Merrick WC (1983) New initiation factor activity required for globin mRNA translation. J Biol Chem 258:5804–5810

Guénard F, Labrie Y, Ouellette G, Beauparlant CJ, Durocher F, INHERIT BRCAs (2009) Genetic sequence variations of BRCA1-interacting genes AURKA, BAP1, BARD1 and DHX9 in French Canadian families with high risk of breast cancer. J Hum Genet 54:152–161

Han K, Jeng EE, Hess GT, Morgens DW, Li A, Bassik MC (2017) Synergistic drug combinations for cancer identified in a CRISPR screen for pairwise genetic interactions. Nat Biotechnol 35:463–474

Heerma van Voss MR, van Kempen PM, Noorlag R, van Diest PJ, Willems SM, Raman V (2015) DDX3 has divergent roles in head and neck squamous cell carcinomas in smoking versus non-smoking patients. Oral Dis 21:270–271

Heerma van Voss MR, Brilliant JD, Vesuna F, Bol GM, van der Wall E, van Diest PJ, Raman V (2017) Combination treatment using DDX3 and PARP inhibitors induces synthetic lethality in BRCA1-proficient breast cancer. Med Oncol 34:33

Hess GT, Frésard L, Han K, Lee CH, Li A, Cimprich KA, Montgomery SB, Bassik MC (2016) Directed evolution using dCas9-targeted somatic hypermutation in mammalian cells. Nat Methods 13:1036–1042

Higa T, Tanaka J, Tsukitani Y, Kikuchi H (1981) Hippuristanols, cytotoxic polyoxygenated steroids from the gorgonian *Isis hippuris*. Chem Lett 10:1647–1650
Ito M, Iwatani M, Kamada Y, Sogabe S, Nakao S, Tanaka T, Kawamoto T, Aparicio S, Nakanishi A, Imaeda Y (2017a) Discovery of selective ATP-competitive eIF4A3 inhibitors. Bioorg Med Chem 25:2200–2209
Ito M, Tanaka T, Cary DR, Iwatani-Yoshihara M, Kamada Y, Kawamoto T, Aparicio S, Nakanishi A, Imaeda Y (2017b) Discovery of novel 1,4-diacylpiperazines as selective and cell-active eIF4A3 inhibitors. J Med Chem 60:3335–3351
Iwasaki S, Floor SN, Ingolia NT (2016) Rocaglates convert DEAD-box protein eIF4A into a sequence-selective translational repressor. Nature 534:558–561
Jankowsky E (2010) RNA helicases. RSC Publishing, London
Jiang L, Gu ZH, Yan ZX, Zhao X, Xie YY, Zhang ZG, Pan CM, Hu Y, cai CP, Dong Y, Huang JY, Wang L, Shen Y, Meng G, Zhou JF, Hu JD, Wang JF, Liu YH, Yang LH, Zhang F, Wang JM, Wang Z, Peng ZG, Chen FY, Sun ZM, Ding H, Shi JM, Hou J, Yan JS, Shi JY, Xu L, Li Y, Lu J, Zheng Z, Xue W, Zhao WL, Chen Z, Chen SJ (2015) Exome sequencing identifies somatic mutations of DDX3X in natural killer/T-cell lymphoma. Nat Genet 47:1061–1066
Jinek M, Chylinski K, Fonfara I, Hauer M, Doudna JA, Charpentier E (2012) A programmable dual-RNA-guided DNA endonuclease in adaptive bacterial immunity. Science 337:816–821
Jinek M, East A, Cheng A, Lin S, Ma E, Doudna J (2013) RNA-programmed genome editing in human cells. eLife 2:e00471
Jones DTW, Jäger N, Kool M, Zichner T, Hutter B, Sultan M, Cho YJ, Pugh TJ, Hovestadt V, Stütz AM, Rausch T, Warnatz HJ, Ryzhova M, Bender S, Sturm D, Pleier S, Cin H, Pfaff E, Sieber L, Wittmann A, Remke M, Witt H, Hutter S, Tzaridis T, Weischenfeldt J, Raeder B, Avci M, Amstislavskiy V, Zapatka M, Weber UD, Wang Q, Lastischka B, Bartholomae CCSchmidt M, von Kalle C, Ast V, lawerenz C, Eils J, Kabbe R, Benes V, van Sluis P, Koster J, Volckmann R, Shih D, Betts MJ, Russell RB, Çoco S, Tonini GP, Schüller U, Hans V, Graf N, Kim YJ, Monoranu C, Roggendorf W, Unterberg A, Herold-Mende C, Milde T, Kulozik AE, von Deimling A, Witt O, Maass E, Rössler J, Ebinger M, Schuhmann MU, Frühwald MC, Hasselbatt M, Jabado N, Rutkowski S, von Bueren AO, Williamson D, Clifford SC, McCabe MG, Collins VP, Wolf S, Wiemann S, Lehrach H, Brors B, Scheurlen W, Felsberg J, Reifenberger G, Northcott PA, Taylor MD, Meyerson M, Pomeroy SL, Yaspo ML, Korbel JO, Korshunov A, Eils R, Pfister SM, Lichter P (2012) Dissecting the genomic complexity underlying medulloblastoma. Nature 488:100–105
Kandoth C, McLellan MD, Vandin F, Ye K, Niu B, Lu C, Xie M, Zhang Q, McMichael JF, Wyczalkowski MA, Leiserson MDM, Miller CA, Welch JS, Walter MJ, Wendl MC, Ley TJ, Wilson RK, Raphael BJ, Ding L (2013) Mutational landscape and significance across 12 major cancer types. Nature 502:333–339
King ML, Chiang CC, Ling HC, Fujita E, Ochiai M, McPhail AT (1982) X-ray crystal structure of rocaglamide, a novel antileukemic 1H-cyclopenta[b]benzofuran from *Aglaia elliptifolia*. J Chem Soc Chem Commun 20:1150–1151
Komor AC, Kim YB, Packer MS, Zuris JA, Liu DR (2016) Programmable editing of a target base in genomic DNA without double-stranded DNA cleavage. Nature 533:420–424
Kondaskar A, Kondaskar S, Fishbein JC, Carter-Cooper BA, Lapidus RG, Sadowska M, Edelman MJ, Hosmane RS (2013) Structure-based drug design and potent anti-cancer activity of tricyclic 5:7:5-fused diimidazo[4,5-d:4′,5′-f][1,3]diazepines. Bioorg Med Chem 21:618–631
Kost GC, Yang MY, Li L, Zhang Y, Liu CY, Kim DJ, Ahn CH, Lee YB, Liu ZR (2015) A novel anti-cancer agent, 1-(3,5-dimethoxyphenyl)-4-[(6-fluoro-2-methoxyquinoxalin-3-yl) aminocarbonyl] piperazine (RX-5902), interferes with beta-catenin function through Y593 phospho-p68 RNA helicase. J Cell Biochem 116:1595–1601
Kurata M, Rathe SK, Bailey NJ, Aumann NK, Jones JM, Veldhuijzen GW, Moriarity BS, Largaespada DA (2016) Using genome-wide CRISPR library screening with library resistant DCK to find new sources of Ara-C drug resistance in AML. Sci Rep 6:36199

Lai MC, Chang WC, Shieh SY, Tarn WY (2010) DDX3 regulates cell growth through translational control of cyclin E1. Mol Cell Biol 30:5444–5453

Landau DA, Tausch E, Taylor-Weiner AN, Stewart C, Reiter JG, Bahlo J, Kluth S, Bozic I, Lawrence M, Böttcher S, Carter SL, Cibulskis K, Mertens D, Sougnez CL, Rosenberg M, Hess JM, Edelmann J, Kless S, Kneba M, Ritgen M, Fink A, Fischer K, Gabriel S, Lander ES, Nowak MA, Dohner H, Hallek M, Neuberg D, Getz G, Stilgenbauer S, Wu CJ (2015) Mutations driving CLL and their evolution in progression and relapse. Nature 526:525–530

Le Hir H, Saulière J, Wang Z (2016) The exon junction complex as a node of post-transcriptional networks. Nat Rev Mol Cell Biol 17:41–54

Lee T, Pelletier J (2016) The biology of DHX9 and its potential as a therapeutic target. Oncotarget 7:42716–42739

Lee YB, Gong YD, Yoon H, Ahn CH, Jeon MK, Kong JY (2010) Synthesis and anticancer activity of new 1-[(5 or 6-substituted 2-alkoxyquinoxalin-3-yl)aminocarbonyl]-4-(hetero) arylpiperazine derivatives. Bioorg Med Chem 18:7966–7974

Lee YB, Gong YD, Kim DJ, Ahn CH, Kong JY, Kang NS (2012) Synthesis, anticancer activity and pharmacokinetic analysis of 1-[(substituted 2-alkoxyquinoxalin-3-yl)aminocarbonyl]-4-(hetero)arylpiperazine derivatives. Bioorg Med Chem 20:1303–1309

Lee T, Di Paola D, Malina A, Mills JR, Kreps A, Grosse F, Tang H, Zannis-Hadjopoulos M, Larsson O, Pelletier J (2014) Suppression of the DHX9 helicase induces premature senescence in human diploid fibroblasts in a p53-dependent manner. J Biol Chem 289:22798–22814

Lee T, Paquet M, Larsson O, Pelletier J (2016) Tumor cell survival dependence on the DHX9 DExH-box helicase. Oncogene 35:5093–5105

Li W, Ross-Smith N, Proud CG, Belsham GJ (2001) Cleavage of translation initiation factor 4AI (eIF4AI) but not eIF4AII by foot-and-mouth disease virus 3C protease: identification of the eIF4AI cleavage site. FEBS Lett 507:1–5

Lindqvist L, Oberer M, Reibarkh M, Cencic R, Bordeleau ME, Vogt E, Marintchev A, Tanaka J, Fagotto F, Altmann M, Wagner G, Pelletier J (2008) Selective pharmacological targeting of a DEAD box RNA helicase. PLoS One 3:e1583

Liu ZR (2002) p68 RNA helicase is an essential human splicing factor that acts at the U1 snRNA-5′ splice site duplex. Mol Cell Biol 22:5443–5450

Liu Y, Lu N, Yuan B, Weng L, Wang F, Liu YJ, Zhang Z (2014) The interaction between the helicase DHX33 and IPS-1 as a novel pathway to sense double-stranded RNA and RNA viruses in myeloid dendritic cells. Cell Mol Immunol 11:49–57

Lopez MS, Kliegman JI, Shokat KM (2014) The logic and design of analog-sensitive kinases and their small molecule inhibitors. Methods Enzymol 548:189–213

Low WK, Dang Y, Schneider-Poetsch T, Shi Z, Choi NS, Merrick WC, Romo D, Liu JO (2005) Inhibition of eukaryotic translation initiation by the marine natural product pateamine A. Mol Cell 20:709–722

Lucas DM, Edwards RB, Lozanski G, West DA, Shin JD, Vargo MA, Davis ME, Rozewski DM, Johnson AJ, Su BN, Goettl VM, Heerema NA, Lin TS, Lehman A, Zhang X, Jarjoura D, Newman DJ, Byrd JC, Kinghorn AD, Grever MR (2009) The novel plant-derived agent silvestrol has B-cell selective activity in chronic lymphocytic leukemia and acute lymphoblastic leukemia in vitro and in vivo. Blood 113:4656–4666

Ma Y, Zhang J, Yin W, Zhang Z, Song Y, Chang X (2016) Targeted AID-mediated mutagenesis (TAM) enables efficient genomic diversification in mammalian cells. Nat Methods 13:1029–1035

Maga G, Falchi F, Garbelli A, Belfiore A, Witvrouw M, Manetti F, Botta M (2008) Pharmacophore modeling and molecular docking led to the discovery of inhibitors of human immunodeficiency virus-1 replication targeting the human cellular aspartic acid-glutamic acid-alanine-aspartic acid box polypeptide 3. J Med Chem 51:6635–6638

Maga G, Falchi F, Radi M, Botta L, Casaluce G, Bernardini M, Irannejad H, Manetti F, Garbelli A, Samuele Z, Zanoli S, Esté JA, Gonzalez E, Zucca E, Paolucci S, Baldanti F, De Rijck J, Debyser Z, Botta M (2011) Toward the discovery of novel anti-HIV drugs. Second-generation inhibitors

of the cellular ATPase DDX3 with improved anti-HIV activity: synthesis, structure-activity relationship analysis, cytotoxicity studies, and target validation. ChemMedChem 6:1371–1389
Mali P, Yang L, Esvelt KM, Aach J, Guell M, DiCarlo JE, Norville JE, Church GM (2013) RNA-guided human genome engineering via Cas9. Science 339:823–826
Malina A, Mills JR, Cencic R, Yan Y, Fraser J, Schippers LM, Paquet M, Dostie J, Pelletier J (2013) Repurposing CRISPR/Cas9 for in situ functional assays. Genes Dev 27:2602–2614
Mao H, McMahon JJ, Tsai YH, Wang Z, Silver DL (2016) Haploinsufficiency for core exon junction complex components disrupts embryonic neurogenesis and causes p53-mediated microcephaly. PLoS Genet 12:e1006282
Mastrangelo E, Pezzullo M, De Burghgraeve T, Kaptein S, Pastrino B, Dallmeier K, de Lamballerie X, Neyts J, Hanson AM, Frick DN, Bolognesi M, Milani M (2012) Ivermectin is a potent inhibitor of flavivirus replication specifically targeting NS3 helicase activity: new prospects for an old drug. J Antimicrob Chemother 67:1884–1894
May WA, Gishizky ML, Lessnick SL, Lunsford LB, Lewis BC, Delattre O, Zucman J, Thomas G, Denny CT (1993a) Ewing sarcoma 11;22 translocation produces a chimeric transcription factor that requires the DNA-binding domain encoded by FLI1 for transformation. Proc Natl Acad Sci U S A 90:5752–5756
May WA, Lessnick SL, Braun BS, Klemsz M, Lewis BC, Lunsford LB, Hromas R, Denny CT (1993b) The Ewing's sarcoma EWS/FLI-1 fusion gene encodes a more potent transcriptional activator and is a more powerful transforming gene than FLI-1. Mol Cell Biol 13:7393–7398
Meijer HA, Kong YW, Lu WT, Wilczynska A, Spriggs RV, Robinson SW, Godfrey JD, Willis AE, Bushell M (2013) Translational repression and eIF4A2 activity are critical for microRNA-mediated gene regulation. Science 340:82–85
Miao X, Yang ZL, Xiong L, Zou Q, Yuan Y, Li J, Liang L, Chen M, Chen S (2013) Nectin-2 and DDX3 are biomarkers for metastasis and poor prognosis of squamous cell/adenosquamous carcinomas and adenocarcinoma of gallbladder. Int J Clin Exp Pathol 6:179–190
Miller EE, Kobayashi GS, Musso CM, Allen M, Ishiy FAA, de Caires LC Jr, Goulart E, Griesi-Oliveira K, Zechi-Ceide RM, Richieri-Costa A, Bertola DR, Passos-Bueno MR, Silver DL (2017) EIF4A3 deficient human iPSCs and mouse models demonstrate neural crest defects that underlie Richieri-Costa-Pereira syndrome. Hum Mol Genet 26:2177–2191
Mills JR, Malina A, Lee T, Di Paola D, Larsson O, Miething C, Grosse F, Tang H, Zannis-Hadjopoulos M, Lowe SW, Pelletier J (2013) RNAi screening uncovers Dhx9 as a modifier of ABT-737 resistance in an Emu-myc/Bcl-2 mouse model. Blood 121:3402–3412
Mitoma H, Hanabuchi S, Kim T, Bao M, Zhang Z, Sugimoto N, Liu YJ (2013) The DHX33 RNA helicase senses cytosolic RNA and activates the NLRP3 inflammasome. Immunity 39:123–135
Nguyen LS, Kim HG, Rosenfeld JA, Shen Y, Gusella JF, Lacassie Y, Layman LC, Shaffer LG, Gécz J (2013) Contribution of copy number variants involving nonsense-mediated mRNA decay pathway genes to neuro-developmental disorders. Hum Mol Genet 22:1816–1825
Nielsen PJ, McMaster GK, Trachsel H (1985) Cloning of eukaryotic protein synthesis initiation factor genes: isolation and characterization of cDNA clones encoding factor eIF-4A. Nucleic Acids Res 13:6867–6880
Nielsen KH, Chamieh H, Andersen CB, Fredslund F, Hamborg K, Le Hir H, Andersen GR (2009) Mechanism of ATP turnover inhibition in the EJC. RNA 15:67–75
Northcote PT, Blunt JW, Munro MHG (1991) Pateamine – a potent cytotoxin from the New-Zealand marine sponge, Mycale Sp. Tetrahedron Lett 32:6411–6414
Ojha J, Secreto CR, Rabe KG, Van Dyke DL, Kortum KM, Slager SL, Shanafelt TD, Fonseca R, Kay NE, Braggio E (2015) Identification of recurrent truncated DDX3X mutations in chronic lymphocytic leukaemia. Br J Haematol 169:445–448
Palacios IM, Gatfield D, St Johnston D, Izaurralde E (2004) An eIF4AIII-containing complex required for mRNA localization and nonsense-mediated mRNA decay. Nature 427:753–757
Pugh TJ, Weeraratne SD, Archer TC, Krummel DAP, Auclair D, Bochicchio J, Careiro MO, Carter SL, Cibulskis K, Erlich RL, Greulich H, Lawrence MS, Lennon NJ, McKenna A, Meldrim J, Ramos AH, Ross MG, Russ C, Shefler E, Sivachenko A, Sogoloff B, Stojanov P, Tamayo

P, Mesirov JP, Amani V, Teider N, Sengupta S, Francois JP, Northcott PA, Taylor MD, Yu F, Crabtree GR, Kautzman AG, Gabriel SB, Getz G, Jäger N, Jones DTW, Lichter P, Pfister SM, Roberts TM, Meyerson M, Pomeroy SL, Cho YJ (2012) Medulloblastoma exome sequencing uncovers subtype-specific somatic mutations. Nature 488:106–110

Qi LS, Larson MH, Gilbert LA, Doudna JA, Weissman JS, Arkin AP, Lim WA (2013) Repurposing CRISPR as an RNA-guided platform for sequence-specific control of gene expression. Cell 152:1173–1183

Radi M, Falchi F, Garbelli A, Samuele A, Bernardo V, Paolucci S, Baldanti F, Schenone S, Manetti F, Maga G, Botta M (2012) Discovery of the first small molecule inhibitor of human DDX3 specifically designed to target the RNA binding site: towards the next generation HIV-1 inhibitors. Bioorg Med Chem Lett 22:2094–2098

Robinson G, Parker M, Kranenburg TA, Lu C, Chen X, Ding L, Phoenix TN, Hedlund E, Wei L, Zhu X, Chalhoub N, Baker SJ, Huether R, Kriwacki R, Curley N, Thiruvenkatam R, Wang J, Wu G, Rusch M, Hong X, Becksfort J, Gupta P, Ma J, Easton J, Vadodaria B, Onar-Thomas A, Lin T, Li S, Pounds S, Paugh S, Zhao D, Kawauchi D, Roussel MF, Finkelstein D, Ellison DW, Lau CC, Bouffet E, Hassall T, Gururangana S, Cohn R, Fulton RS, Fulton LL, Dooling DJ, Ochoa K, Gajjar A, Mardis ER, Wilson RK, Downing JR, Zhang J, Gilbertson RJ (2012) Novel mutations target distinct subgroups of medulloblastoma. Nature 488:43–48

Rossow KL, Janknecht R (2003) Synergism between p68 RNA helicase and the transcriptional coactivators CBP and p300. Oncogene 22:151–156

Sadlish H, Galicia-Vazquez G, Paris CG, Aust T, Bhullar B, Chang L, Helliwell SB, Hoepfner D, Knapp B, Riedl R, Roggo S, Schuierer S, Studer C, Porco JA Jr, Pelletier J, Movva NR (2013) Evidence for a functionally relevant rocaglamide binding site on the eIF4A-RNA complex. ACS Chem Biol 8:1519–1527

Santagata S, Mendillo ML, Tang YC, Subramanian A, Perley CC, Roche SP, Wong B, Narayan R, Kwon H, Koeva M, Amon A, Golub TR, Porco JA Jr, Whitesell L, Lindquist S (2013) Tight coordination of protein translation and HSF1 activation supports the anabolic malignant state. Science 341:1238303

Saporita AJ, Chang HC, Winkeler CL, Apicelli AJ, Kladney RD, Wang J, Townsend RR, Michel LS, Weber JD (2011) RNA helicase DDX5 is a p53-independent target of ARF that participates in ribosome biogenesis. Cancer Res 71:6708–6717

Sarkar M, Khare V, Guturi KK, Das N, Ghosh MK (2015) The DEAD box protein p68: a crucial regulator of AKT/FOXO3a signaling axis in oncogenesis. Oncogene 34:5843–5856

Saulière J, Murigneux V, Wang Z, Marquenet E, Barbosa I, Le Tonquèze O, Audic Y, Paillard L, Roest Crollius H, Le Hir H (2012) CLIP-seq of eIF4AIII reveals transcriptome-wide mapping of the human exon junction complex. Nat Struct Mol Biol 19:1124–1131

Sharma D, Jankowsky E (2014) The Ded1/DDX3 subfamily of DEAD-box RNA helicases. Crit Rev Biochem Mol Biol 49:343–360

Shibuya T, Tango TØ, Sonenberg N, Moore MJ (2004) eIF4AIII binds spliced mRNA in the exon junction complex and is essential for nonsense-mediated decay. Nat Struct Mol Biol 11:346–351

Shibuya T, Tango TØ, Stroupe ME, Moore MJ (2006) Mutational analysis of human eIF4AIII identifies regions necessary for exon junction complex formation and nonsense-mediated mRNA decay. RNA 12:360–374

Shin S, Rossow KL, Grande JP, Janknecht R (2007) Involvement of RNA helicases p68 and p72 in colon cancer. Cancer Res 67:7572–7578

Steckelberg AL, Boehm V, Gromadzka AM, Gehring NH (2012) CWC22 connects pre-mRNA splicing and exon junction complex assembly. Cell Rep 2:454–461

Stevenson RJ, Hamilton SJ, MacCallum DE, Hall PA, Fuller-Pace FV (1998) Expression of the 'dead box' RNA helicase p68 is developmentally and growth regulated and correlates with organ differentiation/maturation in the fetus. J Pathol 184:351–359

Story RM, Weber IT, Steitz TA (1992) The structure of the *E. coli* recA protein monomer and polymer. Nature 355:318–325

Stransky N, Egloff AM, Tward AD, Kostic AD, Cibulskis K, Sivachenko A, Kryukov GV, Lawrence MS, Sougnez C, McKenna A, Shefler E, Ramos AH, Stojanov P, Carter SL, Voet D, Cortés ML, Auclair D, Berger MF, Saksena G, Guiducci C, Onofrio RC, Parkin M, Romkes M, Weissfeld JL, Seethala RR, Wang L, Rangel-Escareño C, Fernandez-Lopez JC, Hidalgo-Miranda A, Melendez-Zajgla J, Winckler W, Ardlie K, Gabriel SB, Meyerson M, Lander ES, Getz G, Golub TR, Garraway LA, Grandis JR (2011) The mutational landscape of head and neck squamous cell carcinoma. Science 333:1157–1160

Sun Z, Wang L, Eckloff BW, Deng B, Wang Y, Wampfler JA, Jang J, Wieben ED, Jen J, You M, Yang P (2014) Conserved recurrent gene mutations correlate with pathway deregulation and clinical outcomes of lung adenocarcinoma in never-smokers. BMC Med Genet 7:32

Suzuki HI, Yamagata K, Sugimoto K, Iwamoto T, Kato S, Miyazono K (2009) Modulation of microRNA processing by p53. Nature 460:529–533

Teigelkamp S, Mundt C, Achsel T, Will CL, Lührmann R (1997) The human U5 snRNP-specific 100-kD protein is an RS domain-containing, putative RNA helicase with significant homology to the yeast splicing factor Prp28p. RNA 3:1313–1326

Thomas A, Goumans H, Amesz H, Benne R, Voorma HO (1979) A comparison of the initiation factors of eukaryotic protein synthesis from ribosomes and from the postribosomal supernatant. Eur J Biochem 98:329–337

Toretsky JA, Erkizan V, Levenson A, Abaan OD, Parvin JD, Cripe TP, Rice AM, Lee SB, Uren A (2006) Oncoprotein EWS-FLI1 activity is enhanced by RNA helicase A. Cancer Res 66:5574–5581

Valentin-Vega YA, Wang YD, Parker M, Patmore DM, Kanagaraj A, Moore J, Rusch M, Finkelstein D, Ellison DW, Gilbertson RJ, Zhang J, Kim HJ, Taylor P (2016) Cancer-associated DDX3X mutations drive stress granule assembly and impair global translation. Sci Rep 6:25996

Vojta A, Dobrinić P, Tadić V, Bočkor L, Korać P, Julg B, Klasić M, Zoldoš V (2016) Repurposing the CRISPR-Cas9 system for targeted DNA methylation. Nucleic Acids Res 44:5615–5628

Wang L, Lawrence MS, Wan Y, Stojanov P, Sougnez C, Stevenson K, Werner L, Sivachenko A, DeLuca DS, Zhang L, Zhang W, Vartanov AR, Fernandes SM, Goldstein NR, Folco EG, Cibulskis K, Tesar B, Sievers QL, Shefler E, Gabriel S, Hacohen N, Reed R, Meyerson M, Golub TR, Lander ES, Neuberg D, Brown JR, Getz G, Wu CJ (2011) SF3B1 and other novel cancer genes in chronic lymphocytic leukemia. N Engl J Med 365:2497–2506

Wang D, Huang J, Hu Z (2012a) RNA helicase DDX5 regulates microRNA expression and contributes to cytoskeletal reorganization in basal breast cancer cells. Mol Cell Proteomics 11:M111.011932

Wang R, Jiao Z, Li R, Yue H, Chen L (2012b) p68 RNA helicase promotes glioma cell proliferation in vitro and in vivo via direct regulation of NF-kappaB transcription factor p50. Neuro-Oncology 14:1116–1124

Wang H, Gao X, Yang JJ, Liu ZR (2013) Interaction between p68 RNA helicase and Ca^{2+}-calmodulin promotes cell migration and metastasis. Nat Commun 4:1354

Wang Z, Murigneux V, Le Hir H (2014) Transcriptome-wide modulation of splicing by the exon junction complex. Genome Biol 15:551

Wang T, Birsoy K, Hughes NW, Krupczak KM, Post Y, Wei JJ, Lander ES, Sabatini DM (2015) Identification and characterization of essential genes in the human genome. Science 350:1096–1101

Wilson BJ, Bates GJ, Nicol SM, Gregory DJ, Perkins ND, Fuller-Pace FV (2004) The p68 and p72 DEAD box RNA helicases interact with HDAC1 and repress transcription in a promoter-specific manner. BMC Mol Biol 5:11

Wolfe AL, Singh K, Zhong Y, Drewe P, Rajasekhar VK, Sanghvi VR, Mavrakis KJ, Jiang M, Roderick JE, Van der Meulen J, Schatz JH, Rodrigo CM, Zhao C, Rondou P, de Stanchina E, Teruya-Feldstein J, Kelliher MA, Speleman F, Porco JA Jr, Pelletier J, Rätsch G, Wendel HG (2014) RNA G-quadruplexes cause eIF4A-dependent oncogene translation in cancer. Nature 513:65–70

Xie M, Vesuna F, Botlagunta M, Bol GM, Irving A, Bergman Y, Hosmane RS, Kato Y, Winnard PT Jr, Raman V (2015) NZ51, a ring-expanded nucleoside analog, inhibits motility and viability of breast cancer cells by targeting the RNA helicase DDX3. Oncotarget 6:29901–29913

Xie M, Vesuna F, Tantravedi S, Bol GM, Heerma van Voss MR, Nugent K, Malek R, Gabrielson K, van Diest PJ, Tran PT, Raman V (2016) RK-33 Radiosensitizes prostate cancer cells by blocking the RNA helicase DDX3. Cancer Res 76:6340–6350

Yang L, Lin C, Liu ZR (2006) P68 RNA helicase mediates PDGF-induced epithelial mesenchymal transition by displacing Axin from beta-catenin. Cell 127:139–155

Yedavalli VS, Zhang N, Cai H, Zhang P, Starost MF, Hosmane RS, Jeang KT (2008) Ring expanded nucleoside analogues inhibit RNA helicase and intracellular human immunodeficiency virus type 1 replication. J Med Chem 51:5043–5051

Yin J, Park G, Lee JE, Choi EY, Park JY, Kim TH, Park N, Jin X, Jung JE, Shin D, Hong JH, Kim H, Yoo H, Lee SH, Kim YJ, Park JB, Kim JH (2015) DEAD-box RNA helicase DDX23 modulates glioma malignancy via elevating miR-21 biogenesis. Brain 138:2553–2570

Yoder-Hill J, Pause A, Sonenberg N, Merrick WC (1993) The p46 subunit of eukaryotic initiation factor (eIF)-4F exchanges with eIF-4A. J Biol Chem 268:5566–5573

Yuan B, Wang X, Fan C, You J, Liu Y, Weber JD, Zhong H, Zhang Y (2016) DHX33 transcriptionally controls genes involved in the cell cycle. Mol Cell Biol 36:2903–2917

Zhang Y, Saporita AJ, Weber JD (2013) P19ARF and RasV(1)(2) offer opposing regulation of DHX33 translation to dictate tumor cell fate. Mol Cell Biol 33:1594–1607

Zhang Y, You J, Wang X, Weber J (2015) The DHX33 RNA helicase promotes mRNA translation initiation. Mol Cell Biol 35:2918–2931

Zhong X, Safa AR (2004) RNA helicase A in the MEF1 transcription factor complex up-regulates the MDR1 gene in multidrug-resistant cancer cells. J Biol Chem 279:17134–17141

Chapter 15
Application of Systemic Transcriptional Gene Silencing for Plant Breeding

Songling Bai, Takeo Harada, and Atsushi Kasai

Abstract Small interfering RNA (siRNA)-mediated gene silencing has been observed in eukaryotes across all kingdoms from fungi to mammals. In plants, this phenomenon influences resistance to pathogenic viruses, suppression of transgene expression, and the inactivation of transposable elements. Recent studies have revealed that double-stranded RNA-derived siRNAs are able to induce systemic transcriptional gene silencing (TGS) in graft partners. In particular, when the scion is used as the siRNA donor, the roots exhibit strong systemic TGS, especially the lateral roots. Such gene silencing can be maintained through in vitro regeneration and is heritable. We developed a novel method for transforming plants using this process. The expression of the target gene can be arrested without inserting exogenous DNA into the genome of the target organism. We herein review the recent advances in research related to systemic TGS. We also describe the potential utility of systemic TGS for plant breeding.

Keywords Transcriptional gene silencing · RNA-directed DNA methylation · Epigenetics · Grafting · Genetically modified organism · Plant breeding technique

15.1 Introduction

Genome-editing systems involving zinc finger nucleases, transcription activator-like effector nucleases, and clustered regulatory interspaced short palindromic repeat/Cas-based RNA-guided DNA endonucleases have recently been extensively used by plant molecular biologists and plant breeders. These systems rely on highly efficient stable or transient genetic transformations to introduce the genome-editing

S. Bai (✉)
College of Agriculture & Biotechnology, Zhejiang University, Zhejiang, China
e-mail: songlingbai@zju.edu.cn

T. Harada · A. Kasai
Department of Agriculture and Life Science, Hirosaki University, Hirosaki, Japan
e-mail: tharada@hirosaki-u.ac.jp; kasaia@hirosaki-u.ac.jp

S. Masuda, S. Izawa (eds.), *Applied RNA Bioscience*,
https://doi.org/10.1007/978-981-10-8372-3_15

tools into living cells and lines. The transgenic lines exhibiting the desired genetic edit are then selected from multiple regenerated lines. The exogenous loci that are inserted into the genome during stable transformations are eliminated during self-fertilizations. Thus, the first generation plants are considered to be genetically modified organisms, but there is still some controversy regarding whether the genome-edited second generation plants, which lack a transgene, can be considered the same as wild-type plants or plants with induced mutations. Additionally, for many vegetatively propagated horticultural crops, such as fruit trees, a similar genome-editing strategy is not applicable because of the high heterogeneity among genomes and potential self-incompatibilities.

Phenotypic variations in plants are commonly the result of epigenetic alterations. Recent studies have suggested that somaclonal variation (Miguel and Marum 2011; Stroud et al. 2013), heterosis (Chen 2013), sex determination (Martin et al. 2009), and graft hybrid (Wu et al. 2013) phenomena involve epigenetic changes. Therefore, heritable phenotypic changes are not only due to altered nucleotide sequences but are also caused by epigenetic alterations. Consequently, artificially changing the epigenetic state (i.e., chromosome structure) represents a promising method for modifying the function of specific genes without altering the DNA sequence.

In this chapter, we describe a novel breeding system called grafting-induced epigenetic modification (GrIEM). This breeding system is able to induce epigenetic alterations at target loci using a grafting technique and transcriptional gene silencing (TGS) mediated by small interfering RNAs (siRNAs) transported over long distances to generate heritable epimutations.

15.2 What Is Grafting?

Most fruit tree species cannot be propagated clonally from seeds because of the high genomic heterogeneity in seeds. Therefore, grafting is widely used to maintain the favorable characteristics of the original cultivar. Grafting involves the natural or deliberate fusion of plant parts so that vascular continuity is established and the resulting genetically composite organism functions as a single plant (Pina and Errea 2005; Mudge et al. 2009; Goldschmidt 2014). Generally, the shoot piece or bud cut from a donor plant that will grow into the upper portion of the grafted plant is called the scion, while the plant that fuses with the scion and functions as the root system is called the stock or rootstock (Fig. 15.1) (Mudge et al. 2009). Grafting is considered to have originated 2000 years ago in China when fruit growers attached the shoot of a desired citrus tree to the roots of another tree (Mudge et al. 2009). Therefore, plants produced by grafting consist of a scion that produces fruits with the desired quality and a rootstock that confers favorable characteristics to the scion, such as vigor, climatic adaptability, and resistance to biotic and abiotic stresses (Fig. 15.1). Despite the considerable advances in plant propagation techniques, grafting is still important for the production of fruits and vegetables, including apple (*Malus domestica*), pear (*Pyrus communis*), grape (*Vitis vinifera*), sweet cherry (*Prunus avium*), tomato (*Lycopersicon esculentum*), eggplant (*Solanum*

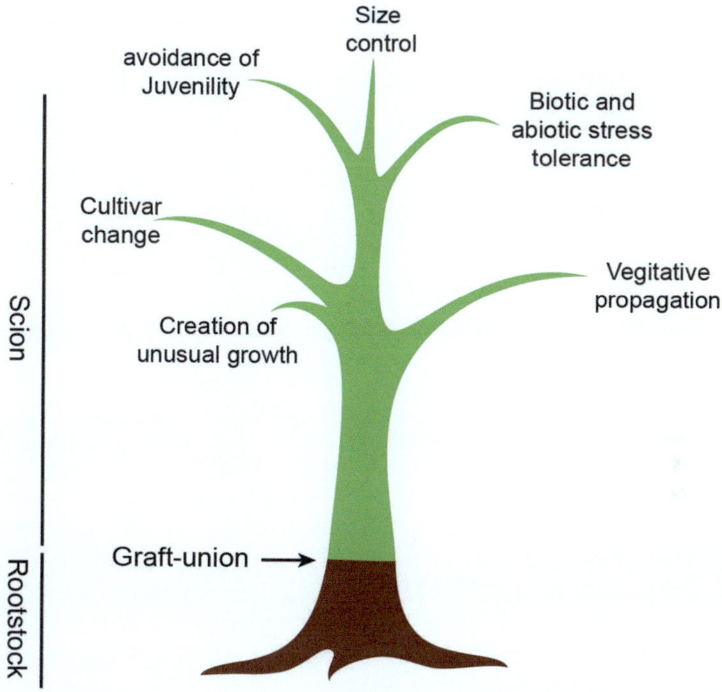

Fig. 15.1 Illustration of grafting technique used in the cultivation of horticultural crops

Table 15.1 Transportation of molecules between sicon and rootstock

Type of molecules	Direction
Water	Upward
Mineral nutrients	Upward
Photosynthate	Downward
Phytohormones	Upward or downward (depends on the type of phytohormones)
RNAs	Upward and downward
Proteins	Upward and downward
Other biochemical compounds	Upward and downward

melongena), watermelon (*Citrullus lanatus*), and cucumber (*Cucumis sativus*) (Mudge et al. 2009; Harada 2010).

In grafted plants, interactions between the rootstock and scion are essential for obtaining the desired trait. Although the mechanism regulating this interaction has not been fully characterized, there is increasing evidence that the upward transport of water and mineral nutrients and the downward flow of photosynthates are modified by grafting, as is the root–top interchange of hormonal signals. These changes may account for the many well-described grafting effects (Table 15.1) (Cutting and Lyne 1993; Aloni et al. 2008; Goldschmidt 2014). In the recent years, interest in the long-distance transport of macromolecules (e.g., proteins and RNAs) through phloem and their potential roles in inter-organ signaling has increased considerably

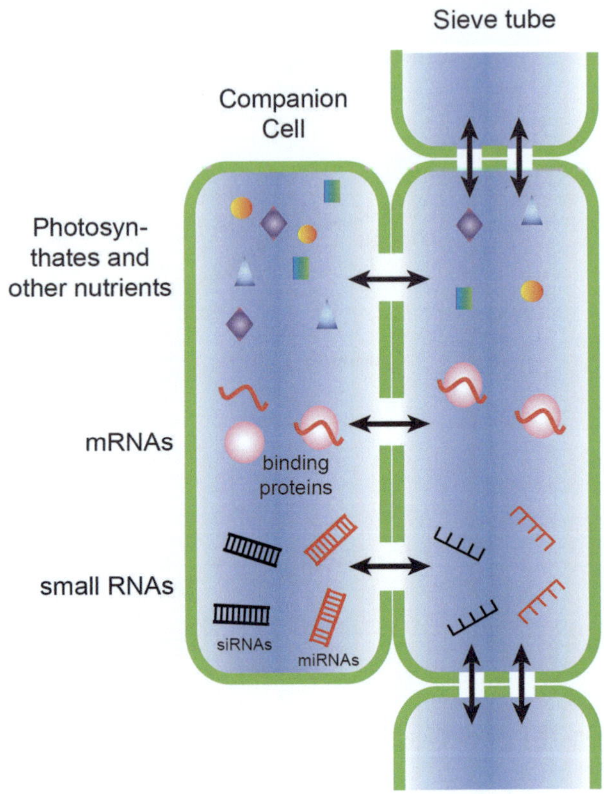

Fig. 15.2 Transportation of macromolecules between the rootstock and scion. Macromolecules, including photosynthates, proteins, and RNAs, can be transported through phloem in both directions along with phloem sap

(Lough and Lucas 2006; Kalantidis et al. 2008; Harada 2010; Mermigka et al. 2016) (Fig. 15.2). It should be noted that merely identifying a mobile macromolecule does not necessarily provide information regarding its physiological function(s) (Goldschmidt 2014). Thus, identified macromolecules should be functionally characterized.

15.3 Transcriptional Gene Silencing

15.3.1 Small RNA and Gene Silencing

In 1990, Napoli et al. (1990) and van der Krol et al. (1990) independently reported the RNA interference (RNAi) phenomenon for the first time. When additional genes encoding chalcone synthase (*CHS*) and dihydoflavonol-4-reductase (*DFR*), which are the key enzymes in flavonoid biosynthesis, were overexpressed in petunia to generate darker violet flowers, white- or mosaic-colored flowers were unexpectedly produced because the exogenous *CHS*/*DFR* silenced the endogenous *CHS*/*DFR* genes. Since then, a similar phenomenon was observed in a fungus (*Neurospora crassa*) as well as animal species, and it became clear that introducing sense or

Table 15.2 Classification of small RNAs in plant

Primary classification	Secondary classification	Biogenesis	Mechanism of action
microRNA (miRNA)	miRNAs	Cleavage of stem-loop precursors by DCL1	mRNA degradation Translational repression
Small interfering RNA (siRNA)	Heterochromatic siRNA (hc-siRNA)	Cleavage from dsRNA by DCL3	Mediate RNA-dependent DNA methylation (RdDM)
	Phased siRNA (phasiRNA)	Cleavage from dsRNA whose synthesis depends on an upstream small RNA trigger and subsequent RDR activity	RNA cleavage
	NAT-siRNA	Cleavage from dsRNA that formed by the hybridization of complementary and independently transcribed RNAs	RNA cleavage

antisense RNA into cells results in the degradation of the mRNA for the targeted gene (Guo and Kemphues 1995). Furthermore, Fire et al. (1998) clarified the RNAi mechanism by revealing that double-stranded RNA (dsRNA) produce small RNAs (sRNAs) and further degrade endogenous mRNA. Andrew Fire and Craig Mello were awarded the Nobel Prize in 2006 for this discovery.

In eukaryotes, small noncoding RNAs [i.e., 20–30 nucleotides (nt) long] have emerged as essential molecules for regulating various biological processes. Plant small noncoding RNAs, which have been relatively well studied, are important for regulating diverse biological processes, including resistance to viruses, suppression of transgenes, and inactivation of transposable elements. According to their origin and biogenesis, endogenous sRNAs in plants are categorized into two major classes, namely, microRNAs (miRNAs) and siRNAs, which are further subclassified as heterochromatic siRNAs (hc-siRNAs), phased-secondary siRNAs (phasiRNAs), and natural antisense transcript siRNAs (NAT-siRNAs) (Table 15.2) (Axtell and Merchant 2013). These sRNAs are produced from dsRNAs by Dicer-like (DCL) family members and are subsequently incorporated into Argonaute family proteins to target complementary nucleotide sequences. The resulting suppressive effect occurs via transcriptional and posttranscriptional pathways.

MicroRNAs have been detected in nearly all analyzed eukaryotes (Zhang et al. 2006), with some miRNA sequences conserved over long evolutionary distances. Additionally, miRNAs have been observed in the unicellular green alga *Chlamydomonas reinhardtii*. These findings suggest that miRNA-based gene regulation is an ancient and evolutionarily conserved mechanism (Molnar et al. 2007). Previous studies concluded that miRNAs can be distinguished from other classes of sRNAs according to their source and targets (Bartel 2004, 2009). The miRNAs originate from a hairpin-like structure formed by a relatively long, noncoding RNA generated by RNA polymerase II. The hairpin-like structure is digested by a Dicer-like enzyme, which in plants is primarily DCL1. The resulting mature miRNAs bind to an

Argonaute subfamily member and repress the expression of their target genes by inhibiting translation or degrading RNA (Bartel 2004, 2009). The miRNAs function in a homology-dependent manner to either cleave the target mRNA at a highly specific site or suppress translation. These modes of action depend largely on the complementarity between the miRNA and target RNA sequences (Bartel 2004, 2009).

Small interfering RNAs are universal effectors of gene silencing in plants, fungi, and animals. They were first observed to induce post-TGS (PTGS) in plants (i.e., RNA silencing or RNA interference) but were later confirmed to also contribute to TGS. The production of siRNAs relies on an RNA-dependent RNA polymerase to generate dsRNAs, which are recognized and cleaved by various DCLs to generate diverse siRNAs (Axtell and Merchant 2013; Fei et al. 2013). The hc-siRNAs are approximately 24 nt long and are produced by DCL3 from transposon transcripts and repeat sequences. This type of siRNA mediates TGS by promoting DNA and histone methylation to repress transposable elements via RNA-directed DNA methylation (RdDM). Meanwhile, the phasiRNAs form a plant-specific class of endogenous sRNAs (e.g., *trans*-acting siRNAs) (Vazquez et al. 2004). The phasiRNAs are usually about 21 nt long. They are generated from an mRNA-converted dsRNA by RNA-dependent RNA polymerase 6 (RDR6) and are processed by DCL4 or DCL5 (DCL3b) in rare cases (Song et al. 2012). The phasiRNAs help degrade target mRNAs in a homology-dependent manner. In contrast, the NAT-siRNAs are derived from two distinct, homologous, and interacting mRNAs that form a dsRNA that is cleaved by DCLs. Among the aforementioned sRNAs, hc-siRNAs are required for chromatin modifications, while the miRNAs, phasiRNAs, and NAT-siRNAs function mainly at the posttranscriptional level by either cleaving target transcripts or suppressing translation, although a few instances in which they mediated DNA methylations have been described (Wu et al. 2010, 2012).

15.3.2 Transcriptional Gene Silencing and RNA-Directed DNA Methylation

As described above, TGS is mainly mediated by RdDM (Baulcombe 2004), which in plants results in the de novo methylation of cytosine (C) residues. This represents a stable and heritable modification that regulates epigenetic control in eukaryotic genomes (Bond and Baulcombe 2015). In plants, the cytosine of CG, CHG, and CHH (where H can be C, A, or T) can be methylated (Daxinger et al. 2009). The initiation of de novo cytosine methylation requires 21/22-nt sRNAs and PolV but occurs independently in DNA. The 21/22-nt sRNAs can recruit a methyltransferase (DRM2) to the target site to catalyze the cytosine methylation in all sequence contexts (Zhong et al. 2014; Bond and Baulcombe 2015; Holoch and Moazed 2015). Cytosine methylation is maintained during DNA replication because of MET1 and the plant-specific chromomethylase 3. In a CG or CHG background, DNA methylation is maintained independently of sRNAs, in contrast to the maintenance of CHH within euchromatin, which requires a continuous supply of sRNA, resulting in a self-reinforcement loop involving PolIV–RdDM (Law and Jacobsen 2010).

15.3.3 Transported siRNAs Induce Gene Silencing in Remote Tissues

The fact that PTGS is non-cell autonomous and can spread from the original site of gene silencing to distant tissues was first reported at the beginning of this century. The PTGS signal can be transported over long distances through the phloem, usually from the source to the sink (Tournier et al. 2006; Liang et al. 2012; Zhang et al. 2014). Additionally, PTGS signals can be transported via plasmodesmata over relatively short distances (i.e., within 10–15 cells) (Himber et al. 2003) unless the target transcript functions as a template of RDR6, in which case the PTGS signal can spread throughout a plant (Brosnan et al. 2007). The molecular mechanism underlying the spread of PTGS reportedly involves the movement of siRNAs, which spread locally from cell to cell and over long distances through the phloem. *Arabidopsis thaliana* grafting experiments established that siRNAs of all sizes are mobile and can be transported to distant tissues in the phloem (Molnar et al. 2010; Melnyk et al. 2011a, 2011b).

Although the systemic spread of PTGS is well-characterized, a long-held belief was that TGS induced by transgene-derived siRNA is not graft-transmissible (Mlotshwa et al. 2002; Mourrain et al. 2007). However, an investigation of a transgene locus containing multiple copies of a plasmid sequence revealed that PTGS is, in fact, graft-transmissible, whereas the TGS at the same locus is not (Mourrain et al. 2007). In 2010, Molnar et al. (2010) were the first to confirm that 24-nt siRNAs are also transmissible. In an *A. thaliana* mutant in which siRNA biogenesis was blocked, the transgene-derived siRNA moved across the graft union. Although the mobile siRNAs were three orders of magnitude less abundant in the recipient tissue than in the source tissue, they were still able to efficiently initiate posttranscriptional silencing of the transgene in the recipient tissue. Furthermore, the results also provided solid evidence that a 24-nt mobile siRNA from an endogenous transcript can mediate epigenetic DNA methylations in the genome of recipient cells. Because the mobile signal is the siRNA itself, the grafting transmission of PTGS and TGS can be achieved via the transmission of siRNAs with different targets (i.e., the coding region for PTGS and the intergenic region for TGS). In a subsequent study, we observed the manifestation of the systemic TGS at an exogenous *35S:GFP* (green fluorescent protein) locus in *Nicotiana benthamiana* (Bai et al. 2011). In the same year, Melnyk et al. (2011a) described a similar phenomenon in an *A. thaliana* system (Fig. 15.3). A more recent study revealed that mobile sRNAs can regulate genome-wide DNA methylations at thousands of sites, most of which are in the non-CG context and associated with short type-1 retro elements in gene-rich genomic regions (Lewsey et al. 2016).

15.3.4 Inheritance of RNA-Directed DNA Methylation and Transcriptional Gene Silencing

The maintenance of siRNA-mediated RdDM and TGS during mitosis and meiosis is a heritable trait in plants. In *A. thaliana*, the methylation of cytosine can be maintained via meiosis for at least 30 generations (Becker et al. 2011; Schmitz et al.

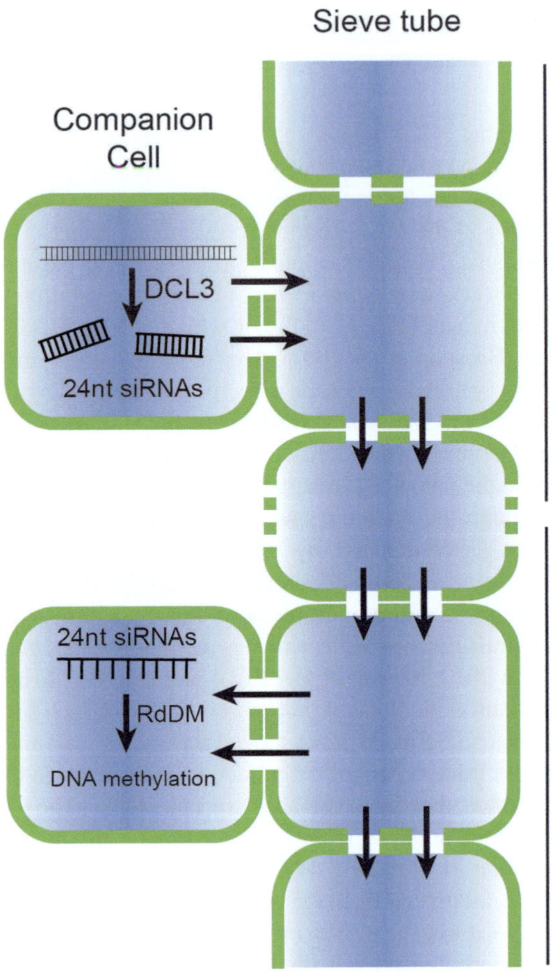

Fig. 15.3 hc-siRNAs (~24 nt siRNAs) move through phloem from source organs and mediate RdDM in the sink organs

2011). Additionally, virus-induced gene silencing (VIGS)-mediated cytosine methylation and TGS in tomato, which involve HDA6, MET1, SHH1, and PolIV, are also heritable (Blevins et al. 2014). However, cytosine methylation can also be maintained through in vitro regeneration. Vining et al. (2013) analyzed the DNA methylome of dedifferentiated poplar calli and regenerated plants and detected considerable alterations while the methylome was recovering from an in vitro regeneration. Bai et al. (2011) reported that the siRNA-mediated transcriptional silencing of an exogenous gene was maintained following an in vitro regeneration and was heritable for at least ten generations (personal communication). Interestingly, a recent study concluded that when a virus-infected plant is crossed with a non-infected plant, the methylated allele is inherited by the progeny but the gene-silenced state is not transferred to the progeny from the non-infected parent. This indicates that the DNA methylation state is a heritable trait, but the trans-acting DNA methylation-initiating factor is not (Bond and Baulcombe 2015).

15.4 Application of Transcriptional Gene Silencing for Breeding

The elucidation of the general mechanisms underlying the gene silencing process has prompted researchers to use RNAi technology to modify various plant traits. Although the co-suppression phenomenon was discovered in petunia plants (Napoli et al. 1990) in which a transgene-targeting sense RNA was introduced, it subsequently became evident that the efficient production of dsRNA for a target gene is required for the efficient induction of RNAi. A widely used method to produce dsRNA in plant cells involves transforming plants with a construct comprising an inverted repeat sequence of the target gene. Sijen et al. (2001) were the first to describe the transcriptional silencing of an endogenous gene in petunia via the introduction of exogenous inverted repeats, indicating that dsRNA can induce the transcriptional silencing of endogenous genes. In maize, TGS technology has been applied to identify and functionally characterize the regulatory components unique to transcriptional gene control (Mark Cigan et al. 2005). Additionally, TGS technology has been used to generate transgenic potato lines lacking *GBSSI* mRNA. A promoter sequence specific to three *GBSSI* alleles, but absent in one allele, was used to design an allele-specific promoter-inverted repeat construct. Two other promoter-inverted repeat constructs targeting different regions of the *GBSSI* promoter were also prepared. The gene silencing efficiencies of the different promoter sequences varied, with the highest efficiency associated with an inverted repeat construct containing the full promoter sequence (Heilersig et al. 2006). Thus, compared with the traditional PTGS approach, TGS is a powerful option for silencing specific genes belonging to families in which the members have similar coding sequences and highly diverse promoter regions.

Virus-induced gene silencing is another option for achieving TGS. It is a technology that exploits an intrinsic RNA-mediated antiviral defense mechanism in plants (Covey et al., 1997). Infected plants can inhibit the replication of the pathogenic virus by producing siRNA from the viral genome. In addition to the dsRNA produced from the intermediate generated during viral replications (Lu et al. 2003), dsRNA is also formed by the intramolecular pairing of the viral RNA (Ruiz et al. 1998; Molnar et al. 2005). In both cases, the dsRNA is recognized by DCLs, resulting in the production of viral siRNAs and the degradation of the corresponding viral mRNA or the RdDM of the target region. Earlier studies revealed that inoculations involving viral vectors carrying a DNA fragment under the control of the 35S promoter led to the transcriptional silencing of the *35S:GFP* (Jones et al. 1999) and *35S:GUS* (β-glucuronidase) (Jones et al. 2001) transgenes in systemic leaves. During VIGS, RDR6 is required for the production of secondary siRNAs that increase the efficacy of the antiviral response (Vaistij and Jones 2009). Regarding endogenous genes, Kanazawa et al. (2011) reported that the transcriptional silencing of an endogenous *CHS* in petunia can be induced by targeting dsRNA to the endogenous gene promoters using cucumber mosaic virus (CMV) as a vector. They also demonstrated that this system induces the transcriptional silencing of an endogenous tomato gene, *LeSPL-CNR*, which influences fruit ripening. Furthermore, the heritability of the TGS was confirmed based on DNA methylations and histone modifications. Because CMV is eliminated during meiosis, the transgene is not inherited by the progeny. Accordingly, CMV-based epigenetic modifications may be useful for artificially modifying the DNA methylation of crop genomes.

The VIGS approach has certain constraints when applied for breeding new cultivars. Most viruses infect a specific host, which limits their use for nonhost plants. Furthermore, although the viruses are theoretically eliminated during reproduction (except for the seed-transmissible viruses), the presence of viruses remains problematic for plants that propagate vegetatively, such as most fruit trees and some flowers and vegetables. Finally, there is a risk of generating new infectious viral strains through recombinations and mutations during genome replications (Allison et al. 1990). Thus, alternatives to VIGS will need to be developed to introduce TGS during the breeding of horticultural crops.

15.5 Grafting-Induced Epigenetic Modification

15.5.1 Theoretical Basis of Grafting-Induced Epigenetic Modification

Based on our understanding of siRNAs and TGS, we developed a novel breeding system called grafting-induced epigenetic modification (GrIEM). Bai et al. (2011) first obtained an epimutant based on GrIEM by targeting the *35S:GFP* locus in a transgenic *N. benthamiana* line. The rootstock (35SIR) producing the hairpin mRNA targeting the 35S promoter, which was controlled by a companion cell-specific promoter (i.e., Commelina yellow mottle virus promoter, CoYMV) to increase the siRNA abundance in the phloem (Matsuda et al. 2002), was generated using a standard transgenic approach (Fig. 15.4a, b). When the *35S:GFP*-expressing scion was grafted onto the 35SIR rootstock, the newly developed leaves lacked GFP around the veins. Transcriptional gene silencing was considered successful because of the presence of relatively few *GFP* transcripts along with a highly methylated target region (Fig. 15.4c).

Conversely, during a reverse grafting experiment in which the 35SIR scion was grafted onto *35S:GFP*-expressing rootstock, *GFP* was clearly silenced in the roots, especially the lateral roots, including the root apical meristem (Fig. 15.4d). These results were confirmed by Melnyk et al. (2011a). Additionally, TGS was not observed in the parental root apical meristem. Differences in the TGS between the two meristems were thought to be because lateral roots in angiosperms are initiated from the pericycle cells, which are bound to phloem cells (Lloret and Casero 2002; De Smet 2011). Therefore, lateral roots originate endogenously from tissues inside the parent root. Moreover, the founder cells that undergo TGS become the lateral root primordium, which eventually forms a lateral root because of the maintenance of TGS as described in the next section. Furthermore, the transcriptional silencing of the *35S:GFP* locus was maintained in plants that were regenerated in vitro from the GFP-eliminated region (Fig. 15.4e). The TGS was heritable for several generations (i.e., until stopping the experiment) (Fig. 15.4f).

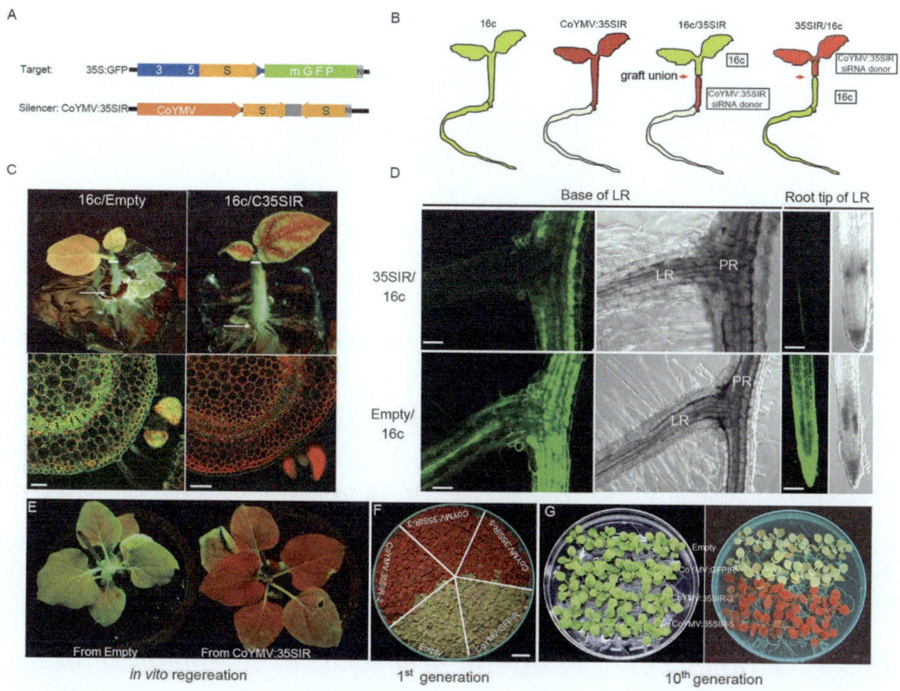

Fig. 15.4 Systemic TGS in grafting system (Modified from Bai et al. 2011). (**a**) Schematic representation of the constructs used for the target and the silencer. Note that the silencer could produce siRNA targeting the 35S promoter region (orange color). (**b**) The grafting experiments carried out in this work. 16c is the transgenic harboring 35S:GFP construct which emits green fluorescence under UV light. CoYMV:35SIR is the plant harboring silencer construct which emits red fluorescence of chlorophyll under UV light. Reciprocal grafts were performed with the CoYMV:35SIR as the rootstock and the scion separately. (**c**) Manifestation of TGS in a 16c plant grafted onto CoYMV:3SIR rootstock. GFP fluorescence in tissue surrounding the vein in newly developed leaves was lost (upper). GFP in a cross-section of the stem of a 16c scion (white bars in upper) was also silenced (lower). Photos were taken 1 week after pruning. Arrowheads indicate the graft union. Bar = 50 μm. (**d**) TGS induction in lateral roots. GFP expression in 16c root stock was suppressed completely in the lateral roots, including the root apical meristem. The roots were observed 14 days after grafting. LR = lateral root. PR = primary root. Bar = 100 μm. (**e**) The plants regenerated from the empty (left) and the silenced tissue (right). (**f**) The progeny of the regenerated plants showed GFP silencing. As a control, progeny corresponding to empty plants are shown. To illustrate the difference from PTGS, the progeny of the plants regenerated from systemically silenced leaf tissue of plants injected with CoYMV:GFPIR are also shown. Bar = 1 cm. (**g**) The progeny of the tenth generation of the regenerated plants showed GFP silencing. The transcription gene silencing of 35S:GFP is maintained after ten generations

The GrIEM system is illustrated in Fig. 15.5. First, the siRNA donor (i.e., silencer), which harbors the inverted repeat of the target sequence, is generated using a standard transgenic approach. The silencer is then grafted onto the rootstock, which is the cultivar that needs to be improved, to induce the RdDM of the target region in the lateral root system. Finally, the entire plants are obtained through an in vitro regeneration using lateral roots as explants.

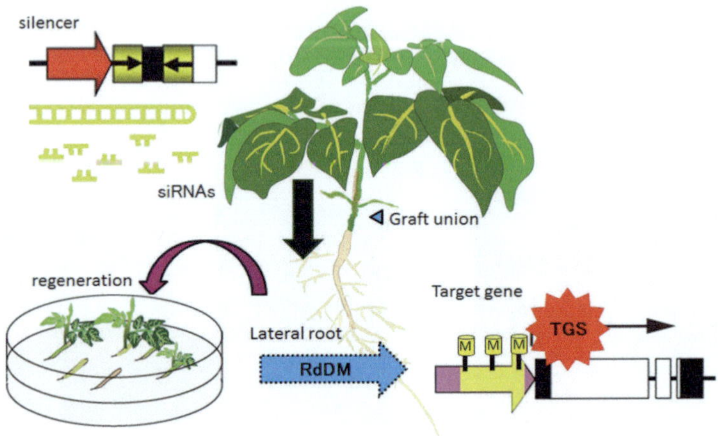

Fig. 15.5 Illustration of GrIEM system. Firstly, the siRNA donor, which harbors the inverted repeat of the target sequence, is generated using a standard transgenic approach. The silencer is then grafted onto the rootstock, which is the cultivar that needs to be improved, to induce the RdDM of the target region in the lateral root system. Finally, the entire plants are obtained through an in vitro regeneration using lateral roots as explants

15.5.2 Example of Grafting-Induced Epigenetic Modification

In our further work, we were able to test the system using potato. As potato is propagated from tuber which is formed from the underground organ, the in vitro regeneration step was not necessary. By grafting a transgenic tobacco line that produces siRNAs under the control of the 35S promoter or CoYMV promoter on transgenic or wild-type potato lines, Kasai et al. (2016) was able to induce the transcriptional silencing of the exogenous *35S:GFP* locus and the endogenous *GBSSI* locus in the potato rootstock (Fig. 15.6). Furthermore, the induced transcriptional silencing of the target genes was maintained in the progeny tubers lacking the transported siRNAs. Kasai et al. (2016) also investigated whether the off-target effect, which is observed during genome editing, occurred in the potato line with an epimutation in *GBSSI*. The methylation levels of the potential off-target sites predicted based on the potato genome (*Solanum tuberosum*, PGSCV4.03) were not significantly

Fig. 15.6 (continued) level of the target region in the potato leaves of in vitro subculturing shoots. WT, wild type; t33, 35SGBpIR line 33. Asterisks show statistically significant (**$P < 0.01$ student's t-test) differences relative to 35SGFP. Means and SD of three biological replicates are shown. (**c**) DNA methylation in the MTs formed trans-grafting plant and its stability in the progenies. Two MTs formed on a grafted plant and their progeny tubers were analyzed. Actually, the sprouts from tubers were used as analyzing materials. Asterisks show statistically significant (*$p < 0.05$; **$P < 0.01$ student's *t*-test) differences relative to WT. Means and SD of three biological replicates are shown. (**d**) Fractions of 20–24 nt siRNAs mapped to GBSSI target region in different materials. 3 wag and 7 wag are adventitious shoots in rootstock of grafted plants at 3 weeks and 7 weeks after grafting, respectively

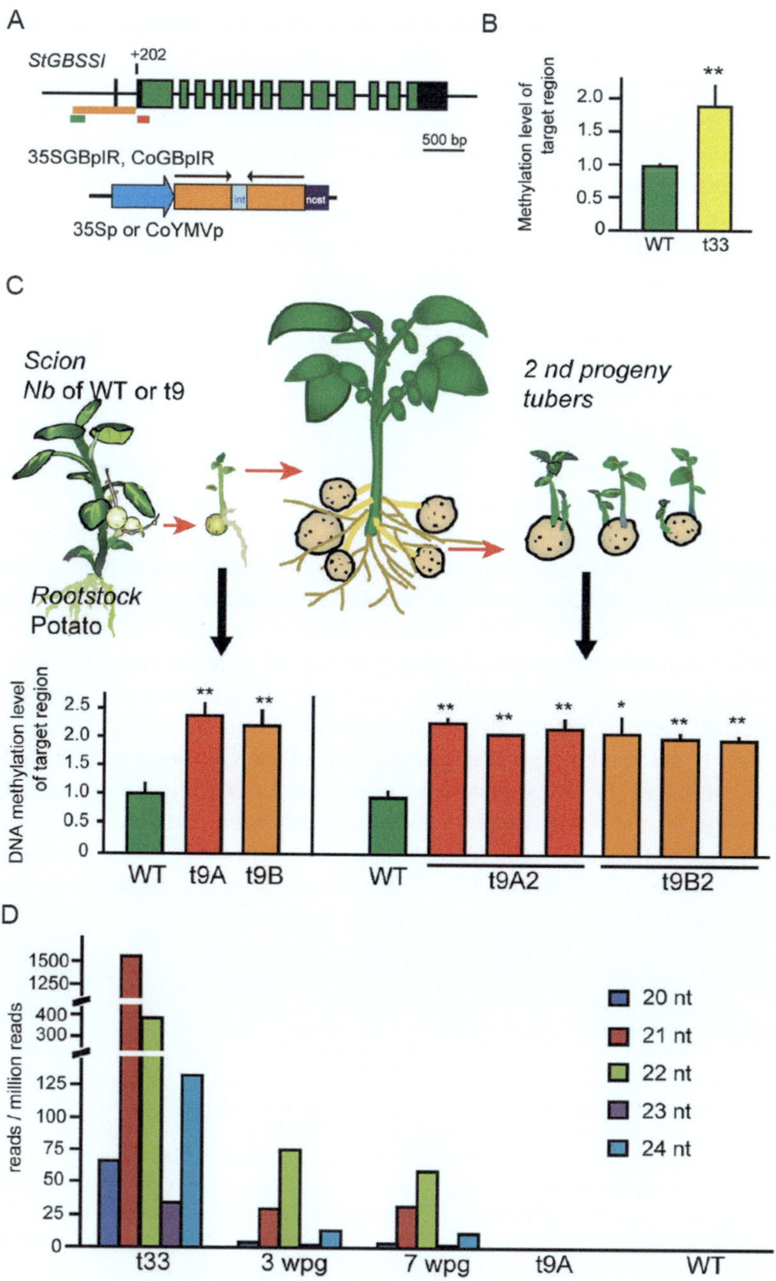

Fig. 15.6 Grafting-induced systemic TGS on endogenous *StGBSSI* gene (Modified from Kasai et al. 2016). (**a**) Schematic diagram of the GBSSI 5′-flanking region in potato "Waseshiro." Black boxes are untranslated regions, and green boxes are exons. Orange bar shows the target region. Green and red bars indicate regions used for methylation level and transcription level analyses, respectively. The construct of 35SGBpIR and CoGBpIR is drawn schematically. (**b**) Methylation

different from that of the wild-type line, indicating that there was no off-target effect, at least not in the present example (Kasai et al. 2016).

In the potato line with an epimutation in *GBSSI*, the expression of two *GBSSI* genes was suppressed, resulting in a waxy-type potato starch with a smooth pulpy texture, desirable taste, high viscosity, and a decreased tendency to retrograde compared with wild-type potato starch. This is the first report describing the artificial transcriptional silencing of an endogenous gene via siRNAs transported over long distances and represents an example of the utility of GrIEM for breeding new cultivars.

15.6 Perspective

Eight new plant breeding techniques (NPBTs) were introduced in 2012 (Lusser et al. 2012). Our GrIEM system involves a combination of two of these techniques (i.e., grafting and RdDM). Grafting as an NPBT involves culturing a chimeric plant formed between a wild-type scion and a genetically modified rootstock (Haroldsen et al. 2012). Although some specific RNAs and proteins can move through the graft union (Hannapel et al. 2013; Han et al. 2014; Thieme et al. 2016), there is no evidence that a transgene can migrate into a graft partner (Haroldsen et al. 2012). With our new technique, the scion supplies a specifically designed siRNA molecule, and the rootstock induces epigenetic changes via RdDM, which occurs naturally in plant cells. If the resulting plant is evaluated from the viewpoint of NPBTs (Heap 2013), it would likely be exempted from the restrictions imposed on genetically modified organisms. Furthermore, the epigenetically modified plant does not retain the functional siRNA and lacks recombinant DNA. Thus, GrIEM is an epigenome editing technique involving mobile RNA silencing, which may be useful for improving agriculturally important crops (Gohlke and Mosher 2015).

References

Allison R, Thompson C, Ahlquist P (1990) Regeneration of a functional RNA virus genome by recombination between deletion mutants and requirement for cowpea chlorotic mottle virus 3a and coat genes for systemic infection. Proc Natl Acad Sci 87:1820–1824

Aloni B, Karni L, Deventurero G, Levin Z, Cohen R, Katzir N, Lotan-Pompan M, Edelstein M, Aktas H, Turhan E, Joel DM, Horev C, Kapulnik Y (2008) Physiological and biochemical changes at the rootstock-scion interface in graft combinations between Cucurbita rootstocks and a melon scion. J Hortic Sci Biotechnol 83:777–783

Axtell M, Merchant S (2013) Classification and comparison of small RNAs from plants. Annu Rev Plant Biol 64:137–159

Bai S, Kasai A, Yamada K, Li T, Harada T (2011) A mobile signal transported over a long distance induces systemic transcriptional gene silencing in a grafted partner. J Exp Bot 62:4561–4570

Bartel DP (2004) MicroRNAs: genomics, biogenesis, mechanism, and function. Cell 116:281–297

Bartel DP (2009) MicroRNAs: target recognition and regulatory functions. Cell 136:215–233

Baulcombe D (2004) RNA silencing in plants. Nature 431:356–363
Becker C, Hagmann J, Muller J, Koenig D, Stegle O, Borgwardt K, Weigel D (2011) Spontaneous epigenetic variation in the *Arabidopsis thaliana* methylome. Nature 480:245–249
Blevins T, Pontvianne F, Cocklin R, Podicheti R, Chandrasekhara C, Yerneni S, Braun C, Lee B, Rusch D, Mockaitis K, Tang H, Pikaard CS (2014) A two-step process for epigenetic inheritance in *Arabidopsis*. Mol Cell 54:30–42
Bond DM, Baulcombe DC (2015) Epigenetic transitions leading to heritable, RNA-mediated de novo silencing in *Arabidopsis thaliana*. Proc Natl Acad Sci U S A 112:917–922
Brosnan CA, Mitter N, Christie M, Smith NA, Waterhouse PM, Carroll BJ (2007) Nuclear gene silencing directs reception of long-distance mRNA silencing in *Arabidopsis*. Proc Natl Acad Sci U S A 104:14741–14746
Chen ZJ (2013) Genomic and epigenetic insights into the molecular bases of heterosis. Nat Rev Genet 14:471–482
Covey SN, Al-Kaff NS, Lángara A, Turner DS (1997) Plants combat infection by gene silencing. Nature 385:781–782
Cutting JGM, Lyne MC (1993) Girdling and the reduction in Shoot Xylem Sap concentrations of Cytokinins and Gibberellins in Peach. J Hortic Sci 68:619–626
Daxinger L, Kanno T, Bucher E, van der Winden J, Naumann U, Matzke AJM, Matzke M (2009) A stepwise pathway for biogenesis of 24-nt secondary siRNAs and spreading of DNA methylation. EMBO J 28:48–57
De Smet I (2011) Lateral root initiation: one step at a time. New Phytol 193:867–873
Fei Q, Xia R, Meyers B (2013) Phased, secondary, small interfering RNAs in posttranscriptional regulatory networks. Plant Cell 25:2400–2415
Fire A, Xu SQ, Montgomery MK, Kostas SA, Driver SE, Mello CC (1998) Potent and specific genetic interference by double-stranded RNA in *Caenorhabditis elegans*. Nature 391:806–811
Gohlke J, Mosher RA (2015) Exploiting mobile RNA silencing for crop improvement. Am J Bot 102:1399–1400
Goldschmidt EE (2014) Plant grafting: new mechanisms, evolutionary implications. Front Plant Sci 5:727
Guo S, Kemphues KJ (1995) Par-1, a gene required for establishing polarity in *C. elegans* embryos, encodes a putative Ser/Thr kinase that is asymmetrically distributed. Cell 81:611–620
Han X, Kumar D, Chen H, Wu S, Kim JY (2014) Transcription factor-mediated cell-to-cell signalling in plants. J Exp Bot 65:1737–1749
Hannapel DJ, Sharma P, Lin T (2013) Phloem-mobile messenger RNAs and root development. Front Plant Sci 4:257
Harada T (2010) Grafting and RNA transport via phloem tissue in horticultural plants. Sci Hortic 125:545–550
Haroldsen VM, Chi-Ham CL, Bennett AB (2012) Transgene mobilization and regulatory uncertainty for non-GE fruit products of transgenic rootstocks. J Biotechnol 161:349–353
Heap B (2013) Europe should rethink its stance on GM crops. Nature 498:409–409
Heilersig BHJB, Loonen AEHM, Janssen EM, Wolters A-MA, Visser RGF (2006) Efficiency of transcriptional gene silencing of GBSSI in potato depends on the promoter region that is used in an inverted repeat. Mol Gen Genomics 275:437–449
Himber C, Dunoyer P, Moissiard G, Ritzenthaler C, Voinnet O (2003) Transitivity-dependent and -independent cell-to-cell movement of RNA silencing. EMBO J 22:4523–4533
Holoch D, Moazed D (2015) RNA-mediated epigenetic regulation of gene expression. Nat Rev Genet 16:71–84
Jones L, Hamilton AJ, Voinnet O, Thomas CL, Maule AJ, Baulcombe DC (1999) RNA-DNA interactions and DNA methylation in post-transcriptional gene silencing. Plant Cell 11:2291–2301
Jones L, Ratcliff F, Baulcombe DC (2001) RNA-directed transcriptional gene silencing in plants can be inherited independently of the RNA trigger and requires Met1 for maintenance. Curr Biol 11:747–757

Kalantidis K, Schumacher HT, Alexiadis T, Helm JM (2008) RNA silencing movement in plants. Biol Cell 100:13–26

Kanazawa A, Inaba J, Shimura H, Otagaki S, Tsukahara S, Matsuzawa A, Kim BM, Goto K, Masuta C (2011) Virus-mediated efficient induction of epigenetic modifications of endogenous genes with phenotypic changes in plants. Plant J 65:156–168

Kasai A, Bai S, Hojo H, Harada T (2016) Epigenome editing of potato by grafting using transgenic tobacco as siRNA donor. PLoS One 11:e0161729

Law JA, Jacobsen SE (2010) Establishing, maintaining and modifying DNA methylation patterns in plants and animals. Nat Rev Genet 11:204–220

Lewsey MG, Hardcastle TJ, Melnyk CW, Molnar A, Valli A, Urich MA, Nery JR, Baulcombe DC, Ecker JR (2016) Mobile small RNAs regulate genome-wide DNA methylation. Proc Natl Acad Sci U S A 113:E801–E810

Liang DC, White RG, Waterhouse PM (2012) Gene silencing in *Arabidopsis* spreads from the root to the shoot, through a gating barrier, by template-dependent, nonvascular, cell-to-cell movement. Plant Physiol 159:984–1000

Lloret P, Casero P (2002) Lateral root initiation. In: Plant roots. CRC Press, Boca Raton, pp 127–155. https://doi.org/10.1201/9780203909423.ch8

Lough TJ, Lucas WJ (2006) Integrative plant biology: role of phloem long-distance macromolecular trafficking. Annu Rev Plant Biol 57:203–232

Lu R, Martin-Hernandez AM, Peart JR, Malcuit I, Baulcombe DC (2003) Virus-induced gene silencing in plants. Methods 30:296–303

Lusser M, Parisi C, Plan D, Rodríguez-Cerezo E (2012) Deployment of new biotechnologies in plant breeding. Nat Biotechnol 30:231–239

Mark Cigan A, Unger-Wallace E, Haug-Collet K (2005) Transcriptional gene silencing as a tool for uncovering gene function in maize. Plant J 43:929–940

Martin A, Troadec C, Boualem A, Rajab M, Fernandez R, Morin H, Pitrat M, Dogimont C, Bendahmane A (2009) A transposon-induced epigenetic change leads to sex determination in melon. Nature 461:1135–1138

Matsuda Y, Liang G, Zhu Y, Ma F, Nelson RS, Ding B (2002) The Commelina yellow mottle virus promoter drives companion-cell-specific gene expression in multiple organs of transgenic tobacco. Protoplasma 220:51–58

Melnyk CW, Molnar A, Bassett A, Baulcombe DC (2011a) Mobile 24 nt small RNAs direct transcriptional gene silencing in the root meristems of *Arabidopsis thaliana*. Curr Biol 21:1678–1683

Melnyk CW, Molnar A, Baulcombe DC (2011b) Intercellular and systemic movement of RNA silencing signals. EMBO J 30:3553–3563

Mermigka G, Verret F, Kalantidis K (2016) RNA silencing movement in plants. J Integr Plant Biol 58:328–342

Miguel C, Marum L (2011) An epigenetic view of plant cells cultured in vitro: somaclonal variation and beyond. J Exp Bot 62:3713–3725

Mlotshwa S, Voinnet O, Mette MF, Matzke M, Vaucheret H, Ding SW, Pruss G, Vance VB (2002) RNA silencing and the mobile silencing signal. Plant Cell 14:S289–S301

Molnar A, Csorba T, Lakatos L, Varallyay E, Lacomme C, Burgyan J (2005) Plant virus-derived small interfering RNAs originate predominantly from highly structured single-stranded viral RNAs. J Virol 79:7812–7818

Molnar A, Schwach F, Studholme DJ, Thuenemann EC, Baulcombe DC (2007) miRNAs control gene expression in the single-cell alga Chlamydomonas reinhardtii. Nature 447:1126–U1115

Molnar A, Melnyk CW, Bassett A, Hardcastle TJ, Dunn R, Baulcombe DC (2010) Small silencing RNAs in plants are mobile and direct epigenetic modification in recipient cells. Science 328:872–875

Mourrain P, van Blokland R, Kooter JM, Vaucheret H (2007) A single transgene locus triggers both transcriptional and post-transcriptional silencing through double-stranded RNA production. Planta 225:365–379

Mudge K, Janick J, Scofield S, Goldschmidt EE (2009) A history of grafting. Hortic Rev 35:437–493. https://doi.org/10.1002/9780470593776.ch9

Napoli C, Lemieux C, Jorgensen R (1990) Introduction of a chimeric chalcone synthase gene into petunia results in reversible co-suppression of homologous genes in *trans*. Plant Cell 2:279–289

Pina A, Errea P (2005) A review of new advances in mechanism of graft compatibility-incompatibility. Sci Hortic 106:1–11

Ruiz MT, Voinnet O, Baulcombe DC (1998) Initiation and maintenance of virus-induced gene silencing. Plant Cell 10:937

Schmitz RJ, Schultz MD, Lewsey MG, O'Malley RC, Urich MA, Libiger O, Schork NJ, Ecker JR (2011) Transgenerational epigenetic instability is a source of novel methylation variants. Science 334:369–373

Sijen T, Vijn I, Rebocho A, van Blokland R, Roelofs D, Mol JNM, Kooter JM (2001) Transcriptional and posttranscriptional gene silencing are mechanistically related. Curr Biol 11:436–440

Song XW, Li PC, Zhai JX, Zhou M, Ma LJ, Liu B, Jeong DH, Nakano M, Cao SY, Liu CY, Chu CC, Wang XJ, Green PJ, Meyers BC, Cao XF (2012) Roles of DCL4 and DCL3b in rice phased small RNA biogenesis. Plant J 69:462–474

Stroud H, Ding B, Simon SA, Feng S, Bellizzi M, Pellegrini M, Wang G-L, Meyers BC, Jacobsen SE (2013) Plants regenerated from tissue culture contain stable epigenome changes in rice. eLife 2

Thieme CJ, Rojas-Triana M, Stecyk E, Schudoma C, Zhang WN, Yang L, Minambres M, Walther D, Schulze WX, Paz-Ares J, Scheible WR, Kragler F (2016) Endogenous *Arabidopsis* messenger RNAs transported to distant tissues. Nat Plants 2:15025

Tournier B, Tabler M, Kalantidis K (2006) Phloem flow strongly influences the systemic spread of silencing in GFP *Nicotiana benthamiana* plants. Plant J 47:383–394

Vaistij FE, Jones L (2009) Compromised virus-induced gene silencing in RDR6-deficient plants. Plant Physiol 149:1399–1407

Van der krol AR, Mur LA, Beld M, Mol JNM, Stuitje AR (1990) Flavonoid genes in petunia – addition of a limited number of gene copies may lead to a suppression of gene-expression. Plant Cell 2:291–299

Vazquez F, Vaucheret H, Rajagopalan R, Lepers C, Gasciolli V, Mallory A, Hilbert J, Bartel D, Crete P (2004) Endogenous trans-acting siRNAs regulate the accumulation of Arabidopsis mRNAs. Mol Cell 16:69–79

Vining K, Pomraning KR, Wilhelm LJ, Ma C, Pellegrini M, Di YM, Mockler TC, Freitag M, Strauss SH (2013) Methylome reorganization during *in vitro* dedifferentiation and regeneration of Populus trichocarpa. BMC Plant Biol 13:92

Wu L, Zhou H, Zhang Q, Zhang J, Ni F, Liu C, Qi Y (2010) DNA methylation mediated by a microRNA pathway. Mol Cell 38:465–475

Wu L, Mao L, Qi Y (2012) Roles of DICER-LIKE and ARGONAUTE proteins in TAS-derived small interfering RNA-triggered DNA methylation. Plant Physiol 160:990–999

Wu R, Wang X, Lin Y, Ma Y, Liu G, Yu X, Zhong S, Liu B (2013) Inter-species grafting caused extensive and heritable alterations of DNA methylation in Solanaceae plants. PLoS One 8:e61995

Zhang BH, Pan XP, Cannon CH, Cobb GP, Anderson TA (2006) Conservation and divergence of plant microRNA genes. Plant J 46:243–259

Zhang W, Kollwig G, Stecyk E, Apelt F, Dirks R, Kragler F (2014) Graft-transmissible movement of inverted-repeat-induced siRNA signals into flowers. Plant J 80:106–121

Zhong XH, Du JM, Hale CJ, Gallego-Bartolome J, Feng SH, Vashisht AA, Chory J, Wohlschlegel JA, Patel DJ, Jacobsen SE (2014) Molecular mechanism of action of plant DRM *de novo* DNA methyltransferases. Cell 157:1050–1060

Chapter 16
Theoretical and Applied Epigenetics in Plants

Yuhya Wakasa, Taiji Kawakatsu, and Fumio Takaiwa

Abstract Artificial regulation of gene expression through RNA-directed DNA methylation (RdDM)-mediated epigenome editing is one the most important and attractive next-generation technologies for plant trait improvement, often called "new plant breeding techniques" (NPBTs). RdDM can induce transcriptional gene silencing (TGS) of a target gene via modification of the cytosine methylation levels of its promoter region; thus, RdDM is useful as a method for suppression of gene expression without changing the genomic DNA sequence. Likewise, several types of strict epigenetic regulation occur at both the DNA and chromatin levels under normal growth conditions in plants. Recent studies have revealed genome-wide and organ-specific landscapes of epigenetic modifications and their close relationship to plant growth regulation. Therefore, understanding recent findings concerning epigenetic regulation in plants is very important to the future application of epigenome editing in plant breeding. In this chapter, we illustrate several aspects of theoretical and applied epigenetics in plants through discussion of recent studies.

Keywords Epigenetics · Chromatin · Histone · Methylation · New plant breeding techniques · RNA-directed DNA methylation · Transcriptional gene silencing

16.1 Mechanisms of Epigenetic Regulation in Plants

16.1.1 DNA Methylation

Cytosine DNA methylation (5mC) is a covalent modification of the fifth carbon residue of cytosine. 5mC is conserved in eukaryotes including mammals and plants, although absent in some organisms such as *Drosophila melanogaster* and *Caenorhabditis elegans* (Law and Jacobsen 2010). There are three main strategies to assay DNA methylation levels: (1) digestion of methylated/unmethylated DNA

Y. Wakasa (✉) · T. Kawakatsu · F. Takaiwa
Plant Molecular Farming Unit, Institute of Agrobiological Sciences, National Agriculture and Food Research Organization, Tsukuba, Ibaraki, Japan
e-mail: ywakasa@affrc.go.jp

S. Masuda, S. Izawa (eds.), *Applied RNA Bioscience*,
https://doi.org/10.1007/978-981-10-8372-3_16

fragments with methylation-sensitive restriction enzymes (MSREs), (2) methylated DNA immunoprecipitation (MeDIP) using an antibody against methylated cytosine, and (3) bisulfite sequencing, in which unmethylated cytosine is converted to uracil by sodium bisulfite, which is reported as thymine in sequence reads. Each strategy can be combined with high-throughput sequencing technology and extended to genome-wide analysis (Cokus et al. 2008; Down et al. 2008; Lister et al. 2008; Maunakea et al. 2010). However, bisulfite sequencing is the most comprehensive and accurate way to quantify DNA methylation levels, so it is widely used for genome-wide analysis (Urich et al. 2015).

In plants, 5mC occurs in three distinct sequence contexts: CG and CHG, which are both symmetric, and CHH (H = C, A, or T), which is asymmetric (Law and Jacobsen 2010). Based on extensive studies in *Arabidopsis thaliana*, four distinct DNA methylation pathways are known. CG methylation is catalyzed by DNA METHYLTRANSFERASE 1 (MET1) and maintained in a semiconservative manner during DNA replication (Kankel et al. 2003). CHG methylation is catalyzed by CHROMOMETHYLASE3 (CMT3), which recognizes methylation at the 9th lysine residue of the H3 tail (H3K9) (Bartee et al. 2001; Lindroth et al. 2001). CMT2, a homolog of CMT3, catalyzes CHG and CHH methylation in deep heterochromatic regions (Zemach et al. 2013; Stroud et al. 2014). DOMAINS OF REARRANGED METHYLTRANSFERASE 2 (DRM2) catalyzes cytosine methylation in all three sequence contexts, in a process referred to as RNA-directed DNA methylation (RdDM) (Cao and Jacobsen 2002; Law and Jacobsen 2010; Kawashima and Berger 2014; Cuerda-Gil and Slotkin 2016). In RdDM, 21- or 24-nucleotide (nt) small interfering RNAs (siRNAs) guide DRM2 to target regions, marked with DNA methylation and H3K9me, through association with three conserved Argonaute proteins, AGO4, AGO6, and AGO9 (Gao et al. 2010; Havecker et al. 2010; McCue et al. 2015). Non-CG DNA methylation and histone methylation (see Sect. 16.1.2) form a self-reinforcing loop, in which H3K9me controls non-CG DNA methylation and non-CG DNA methylation controls H3K9me (Stroud et al. 2014). Interplays between these pathways have been implicated by comprehensive methylome analysis of components in these pathways (Stroud et al. 2013).

DNA methylation can be actively removed by DNA demethylases. In *Arabidopsis*, DNA glycosylases DEMETER (DME), REPRESSOR OF SILENCING 1 (ROS1)/DEMETER-LIKE 1 (DML1), DML2, and DML3, are involved in DNA demethylation (Choi et al. 2002; Gong et al. 2002; Penterman et al. 2007; Ortega-Galisteo et al. 2008). In contrast to mutants of DNA methylation components, which are usually viable, *dme* is embryo lethal, highlighting the importance of active demethylation (Bartee et al. 2001; Cao and Jacobsen 2002; Choi et al. 2002; Kankel et al. 2003). DNA methylation and demethylation activities are balanced by a feedback loop between RdDM and ROS1 (Lei et al. 2015; Williams et al. 2015). *ROS1* expression is promoted by RdDM in the *ROS1* promoter region and repressed by ROS1 activity itself.

CG methylation is broadly distributed across the genome and often resides within gene bodies (transcribed regions); this type of methylation is called gene body methylation (gbM). The distributions of gbM and histone variant H2A.Z are mutually exclusive, and gbM is associated with higher gene expression (Tran et al. 2005; Zhang et al. 2006; Zilberman et al. 2007). Therefore, gbM is thought to exclude H2A.Z and allow constitutive expression (Zilberman et al. 2008; Coleman-Derr and Zilberman 2012). However, recent studies on intra- and interspecies variations of DNA methylation indicate that gbM does not have a great effect on gene expression or affect H2A.Z distribution within genes (Bewick et al. 2016; Kawakatsu et al. 2016a). Currently, the role of gbM is unclear. Co-localization of CG and non-CG methylation is a characteristic of heterochromatin and transposable elements (TEs) and contributes to gene and TE silencing (Law and Jacobsen 2010). Population-wide methylome variations are largely associated with structural variations such as TE insertion or deletion (Kawakatsu et al. 2016a) and are enriched near signaling pathway genes or immune response genes. TE transposition has shaped the epigenome of *Arabidopsis* and has introduced variation in environmental responses during diversification.

DNA methylome studies are not limited to *Arabidopsis*, currently extending to nearly 100 species (Gent et al. 2013; Project 2013; Schmitz et al. 2013a; Stroud et al. 2013; Zhong et al. 2013; Seymour et al. 2014; Ong-Abdullah et al. 2015; Ausin et al. 2016; Niederhuth et al. 2016; Takuno et al. 2016). In addition, transgenerational, populational methylome variations, tissue- and cell-type-specific methylomes, and stress-responsive methylomes have been reported (Hsieh et al. 2009; Schmitz et al. 2011; Calarco et al. 2012; Dowen et al. 2012; Ibarra et al. 2012; Schmitz et al. 2013a, b; Garg et al. 2015; Secco et al. 2015; Hsieh et al. 2016; Kawakatsu et al. 2016a, b; Narsai et al. 2016; Park et al. 2016; Wibowo et al. 2016; Hossain et al. 2017). Recent advances in single-molecule real-time sequencing enable detection of methylated cytosines from long reads without bisulfite conversion (Rand et al. 2017; Simpson et al. 2017). These technologies potentially offer a paradigm shift in DNA methylome analysis, especially in crop species with large genomes and/or multiploidy.

16.1.2 Histone Modification

Histone proteins package genomic DNA into nucleosomes, which in turn form chromatin (Roudier et al. 2009). Histones are conserved in eukaryotes. Four major histones (H2A, H2B, H3, and H4) act as core histones and H1 acts as a linker histone (Kornberg 1974; Thoma and Koller 1977; Luger et al. 1997). The histone core is an octamer complex: two H2A-H2B dimers and an H3-H4 tetramer. Approximately 147 bp of DNA wraps around each histone core and forms a nucleosome. Several histone variants share homology with major histone proteins (Deal and Henikoff 2011a). In addition, histone tails can be posttranslationally modified through methylation (me) and acetylation (ac). These modifications are implicated in flowering, leaf development, seed maturation, flower development, circadian rhythm, and chloroplast development (Deal and Henikoff 2011a; Merini and Calonje 2015; Mozgova et al. 2015). Chromatin immunoprecipitation followed by

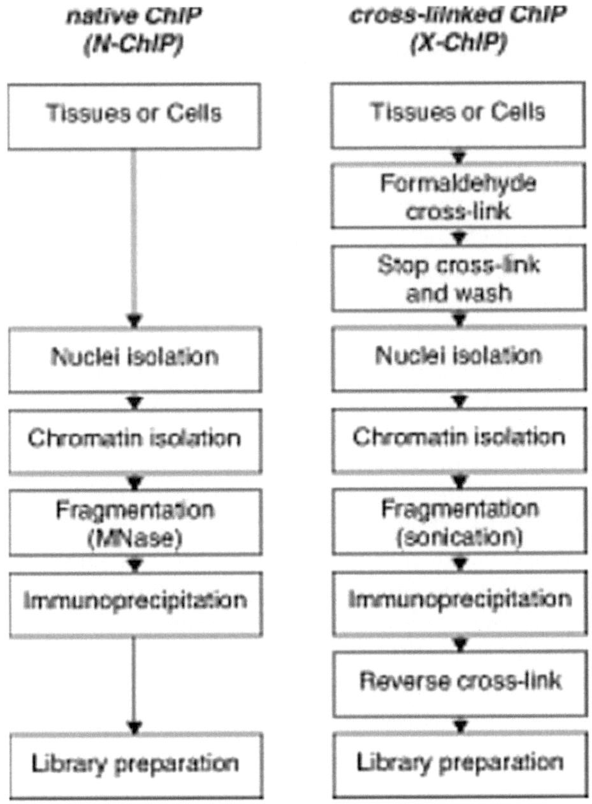

Fig. 16.1 Comparison of ChIP-seq protocols for genome-wide chromatin mark(s) distribution. Native chromatin is digested by micrococcal nuclease (MNase) in native ChIP (N-ChIP). Cross-linked chromatin is fragmented by sonication in cross-linked ChIP (X-ChIP). Nucleosomes are subjected to immunoprecipitation with specific antibodies

high-throughput sequencing (ChIP-seq) is widely used for analyzing the genome-wide distribution of histones, histone variants, and histone modifications (Luo and Lam 2014). Native chromatin digested by micrococcal nuclease (MNase) or cross-linked chromatin fragmented by sonication can be subjected to immunoprecipitation (N-ChIP [native ChIP] or X-ChIP [cross-linked ChIP]) using histone-, histone variant-, or histone modification-specific antibodies (Jackson 1978; O'Neill and Turner 1995; Barski et al. 2007; Schmid and Bucher 2007) (Fig. 16.1).

Although histone modifications are conserved in eukaryotes, their distribution patterns and functions vary. For example, mono-methylation of H3 (H3K9me1) is enriched in heterochromatin in *Arabidopsis* but is enriched at the transcription start sites (TSS) of active genes in animals (Fransz et al. 2006; Fuchs et al. 2006). H3K9 di-methylation (H3K9me2) is also enriched in heterochromatin (Turck et al. 2007). In contrast, H3K9 tri-methylation (H3K9me3) is enriched in euchromatin and is associated with active genes (Turck et al. 2007). H3K9 methylation is catalyzed by SET- and RING-associated (SRA) domain-containing SU(VAR) HOMOLOGUE 1 (SUVH1), SUVH4–6, and SET- and WIYLD-domain-containing SU(VAR)3–9 related 4 (SUVR4) (Ebbs et al. 2005; Ebbs and Bender 2006). INCREASED BONSAI METHYLATION 1 (IBM1) demethylates H3K9 (Inagaki et al. 2010).

H3K27 methylation tends to be associated with low-expression genes and tissue-specific genes (Turck et al. 2007; Roudier et al. 2011). H3K27me1 and H3K27me2 are distributed in both euchromatin and heterochromatin, and H3K27me3 is mainly observed in euchromatin (Roudier et al. 2011). H3K27me1, me2, and me3 mark distinct sets of genes. H3K27me1 and H3K27me3 are enriched in the transcribed regions of marked genes, relative to flanking regions. However, H3K27me2 levels are uniformly higher in both the transcribed and flanking regions of marked genes than in unmarked genes. H3K27me1 is catalyzed by ARABIDOPSIS TRITHORAX-RELATED PROTEIN 5 (ATXR5) and ATXR6 (Jacob et al. 2009). H3K27m3 is catalyzed by Polycomb Repressive Complex 2 (PRC2) (Margueron and Reinberg 2011). *Drosophila* PRC2 consists of four subunits: (1) enhancer of zeste (E(z)), (2) suppressor of zeste 12 (Su(z)12), (3) nucleosome-remodeling factor 55 kDa subunit (NURF55), and (4) extra sex combs (ESC). *Arabidopsis* possesses three E(z) homologs (CURLY LEAF [CLF], MEDEA [MEA], and SWINGER [SWN]), three Su(z)12 homologs (EMBRYONIC FLOWER 2 [EMF2], VERNALIZATION 2 [VRN2], and FERTILIZATION-INDEPENDENT SEED 2 [FIS2]), five NURF55 homologs (MULTI-SUBUNIT SUPPRESSOR OF IRA 1–5 [MSI1–5]), and one ESC homolog (FERTILIZATION-INDEPENDENT ENDOSPERM [FIE]) (Ach et al. 1997; Goodrich et al. 1997; Grossniklaus et al. 1998; Kenzior and Folk 1998; Luo et al. 1999; Gendall et al. 2001; Yoshida et al. 2001; Hennig et al. 2003; Chanvivattana et al. 2004; Jullien et al. 2006; Makarevich et al. 2006; Zhang et al. 2007; Jiang et al. 2008; Kim et al. 2010; Lafos et al. 2011; Pazhouhandeh et al. 2011; Derkacheva et al. 2013). PRC2 target genes especially depend on E(z) homologs.

PRC1 is also required for transcriptional repression of H3K27me3-marked genes. PRC1 can catalyze histone H2A mono-ubiquitination (H2Aub) of target genes (de Napoles et al. 2004; Wang et al. 2004). *Drosophila* PRC1 consists of four subunits (Gil and O'Loghlen 2014): (1) chromodomain protein Polycomb (Pc), (2) RING-finger protein Posterior sex comb (Psc), (3) RING-finger protein *Drosophila* RING1 (dRING1), and (4) Polyhomeotic (Ph). In *Arabidopsis*, LIKE HETEROCHROMATIN PROTEIN 1 (LHP1)/TERMINAL FLOWER 2 (TFL2) plays a similar role to Pc (Turck et al. 2007). *Drosophila* Psc possesses a RING-finger domain, a RING-finger and WD40-associated ubiquitin-like (RAWUL) domain, and a long intrinsically disordered C-terminal region (CTR). The CTR domain of Psc is involved in inhibition of nucleosome remodeling, gene repression, and chromatin compaction; however, it is missing from *Arabidopsis* Psc homologs AtBMI1A, AtBMI1B, and AtBMI1C (Sanchez-Pulido et al. 2008). EMBRYONIC FLOWER 1 (EMF1) is similar to the CTR domain and acts in a similar manner (Aubert et al. 2001; Calonje et al. 2008). AtRING1A and AtRING1B correspond to dRING1 (Schoorlemmer et al. 1997; Xu and Shen 2008). No plant homolog of Ph has yet been identified.

JUMONJI (JMJ) proteins RELATIVE OF EARLY FLOWERING 6 (REF6)/JMJ12, EARLY FLOWERING 6 (ELF6)/JMJ11, JMJ30, and JMJ32 are H3K27 demethylases (Noh et al. 2004; Lu et al. 2011; Gan et al. 2014). Notably, four C2H2 zinc finger domains of REF6 recognize a CTCTGYTY motif and guide REF6 to its binding sites to modulate H3K27me3 levels (Cui et al. 2016; Li et al. 2016).

A hierarchical model for gene repression, in which PRC2 acts upstream of PRC1, has been widely accepted (Mozgova et al. 2015). In this model, PRC2 methylates H3K27 (to H3K27me3) in target genes as the first step. Second, PRC1 is guided there through H3K27me3 recognition by Pc or LHP1, and target genes are marked with H2Aub, leading to gene repression. However, the hierarchical order of PRC2 and PRC1 is more complicated than once thought (Merini and Calonje 2015). For example, PRC1 is required for H3K27me3 at many PRC2 target genes (Kim et al. 2012; Yang et al. 2013), and PRC2 is not necessarily required for H2Aub at target genes (i.e., PRC1 recruits) (Pengelly et al. 2015). H2Aub can recruit PRC2 and promote H3K27me3 in animals (Blackledge et al. 2014; Cooper et al. 2014; Kalb et al. 2014). In addition, some PRC1 and PRC2 components interact with each other (Xu and Shen 2008; Derkacheva et al. 2013; Wang et al. 2014). Therefore, the interactions between PRC1 and PRC2 may include a positive feedback loop, direct interplay, and mutually independent mechanisms.

H3K36me3 is associated with actively expressed genes (Roudier et al. 2011). H3K36me3 is most prevalent within TSSs but is distributed throughout transcribed regions. SDG8, and possibly SDG4 and SDG25/ATXR7, catalyze H3K36me2 and H3K36me3 (Zhao et al. 2005; Cartagena et al. 2008; Xu et al. 2008; Berr et al. 2009). H3K36me3 and H3K27me3 play antagonistic roles—activation and repression—and rarely co-exist on the same histone tail (Roudier et al. 2011; Yang et al. 2014).

H3K4 methylation is mostly found in genes and other euchromatin (Roudier et al. 2011). In contrast to enhancer-associated H3K4me1 in animals, H3K4me1 in *Arabidopsis* is distributed inside transcribed regions but is less prevalent near TSSs and transcription end sites (TESs). Both H3K4me2 and H3K4me3 are enriched around TSSs but depleted around TESs. H3K4me3 is associated with highly expressed genes, whereas H3K4me1 and H3K4me2 are associated with tissue-specific genes (Roudier et al. 2011). ARABIDOPSIS TRITHORAX 1 (ATX1)/SET DOMAIN GROUP 27 (SDG27) and Complex Proteins Associated with Set1 (COMPASS)-like complex catalyze H3K4me3, and ATX2/SDG30 catalyzes H3K4me2 (Saleh et al. 2008; Jiang et al. 2011). SDG2 catalyzes H3K4me1, H3K4me2, and H3K4me3 in vitro, but in vivo, *sdg2* shows reduction only in H3K4me3 (Berr et al. 2010; Guo et al. 2010). JMJ14 and homologs of LYSINE-SPECIFIC HISTONE DEMETHYLASE 1 (LSD1), FLOWERING LOCUS D (FLD)/LSD1-like 1 (LDL1), and LDL2 are required for H3K4 demethylation (Deleris et al. 2010; Lu et al. 2010; Greenberg et al. 2013).

H3K9ac and H3K27ac are associated with active gene expression (Charron et al. 2009). Levels of both H3K9ac and H3K27ac peak near TSSs and are distributed inside gene bodies. The H3K9ac and H3K27ac target regions are largely the same but distinct from H3K27me3 target regions. H3K9ac and H3K27ac are catalyzed by histone acetyltransferase (HAT) family proteins, such as homologs of general control non-derepressible (GCN5) and TATA binding protein-associated factor 1 (TAF1) (Pandey et al. 2002; Benhamed et al. 2006). AtGCN5 also catalyzes H4K14ac. Histone deacetylase (HDAC) family proteins, such as HDA6, are responsible for histone deacetylation

(Pandey et al. 2002; Earley et al. 2006; To et al. 2011; Liu et al. 2014). HATs and HDACs act as transcriptional co-activators and co-repressors, respectively.

Combinations of histone modifications are thought to be important for the precise expression state and responsiveness of a gene. Two opposing histone marks, for example, H3K27me3 (repressing) and H3K4me3 (activating), can be co-localized in the same genomic regions (Roudier et al. 2011). As in animals, bivalent chromatin regions in plants are associated with several transcription factors (TFs) that are normally expressed at low levels but are induced at specific timing and/or in specific tissues by developmental cues (Saleh et al. 2007; Jiang et al. 2008; Berr et al. 2010; Roudier et al. 2011). It is also possible that a mixture of different cell types with different chromatin modification states could be misinterpreted as co-localization of opposing histone marks. Cell-type-specific profiling would promote further understanding not only of cell-type-specific properties but also of the combinatorial functions of histone modifications. Several recent developments show considerable promise in this area. For example, low-input ChIP-seq methods and high-throughput sequencing technologies are evolving (Adli and Bernstein 2011; Brind'Amour et al. 2015; Schmidl et al. 2015). Recently developed simple but highly efficient INTACT (Isolation of Nuclei Tagged in specific Cell Types) is feasible for cell-type-specific profiling (Deal and Henikoff 2011b).

16.1.3 Chromatin Accessibility

Transcriptional activation is primarily regulated by TF binding to regulatory DNA elements, where chromatin is open or accessible. Genome-wide chromatin accessibility can be assayed directly or indirectly through a combination of nuclease digestion and high-throughput sequencing (Meyer and Liu 2014). As in N-ChIP, MNase digests bare DNA that is not protected by nucleosomes, whereas DNase I cleaves unprotected DNA. Therefore, MNase digestion followed by sequencing (MNase-seq) identifies nucleosome positioning and indirectly detects open chromatin regions (Schones et al. 2008), whereas DNase I cleavage followed by sequencing (DNase-seq) directly detects open chromatin regions (Boyle et al. 2008). FAIRE-seq (Formaldehyde-Assisted Isolation of Regulatory Elements) also directly detects open chromatin regions by isolating un-cross-linked DNA with nucleosomes (Giresi et al. 2007). Transposase Accessible Chromatin sequencing (ATAC-seq) uses Tn5 transposase to insert sequencing-ready adaptor sequences into open chromatin regions, starting with as few as 500 cells (Buenrostro et al. 2013). Genome-wide chromatin accessibility studies have been limited in plants (Zhang et al. 2012a, 2012b; Li et al. 2014; Wu et al. 2014; Zhang et al. 2015; Lu et al. 2016). Nevertheless, these studies clearly demonstrate that the identified open chromatin regions are associated with gene expression and TF binding sites. ChIP-seq has been used for assays of genome-wide TF binding sites in vivo (Song et al. 2016). However, preparing antibodies against a wide variety of TFs or transgenic plants expressing tagged TFs with native promoters is time-, cost-, and labor-consuming. DNA affinity purification sequencing (DAP-seq) is a new technology to cost-effectively identify TF binding sites in vitro (O'Malley et al. 2016). The combination of chromatin accessibility assays and DAP-seq is expected to greatly advance our knowledge of transcriptional regulatory networks.

16.2 Application Studies on Epigenetics in Plants

16.2.1 Application of Epigenome Editing to Plant Breeding

Gene manipulation (GM) techniques have been used as molecular breeding tools to develop various GM crops with excellent traits such as resistance to insect pests, plant diseases, and specific herbicides. Furthermore, "golden rice," which accumulates provitamin A in the seed, will be practical to use in the Philippines in the near future. GM crops are at present commercially cultivated in 28 countries worldwide. The total global cultivated area of transgenic crops was estimated to have reached approximately 179.7 million hectares by 2015 (James 2016). On the other hand, conventional GM techniques are still sometimes viewed as a serious issue in numerous countries because the transgene is integrated into the genome of the target plant, and a number of people are concerned about gene flow from GM crops to the environment.

In recent years, technologies referred to as "new plant breeding techniques" (NPBTs) have been proposed as a solution for issues surrounding conventional GM crops. Genome editing (ZFN, TALEN, CRISPR/Cas9), "grafting with GM plants," "reverse breeding," *Agrobacterium* infiltration, and RdDM can all be classified as NPBTs (Lusser et al. 2012; Schaart et al. 2016). When these NPBTs are applied, it is difficult to distinguish between the newly introduced artificial mutation and natural mutations. In particular, changes to genomic DNA caused by RdDM cannot be detected by conventional molecular analysis methods such as PCR and DNA sequencing because there is no change to the DNA sequence. A key characteristic of RdDM-mediated transcriptional gene silencing (TGS) is the production of double-strand RNA (dsRNA) with homology to the promoter sequence of the target gene. The dsRNA is cleaved into 21–24 nt pieces of siRNA by DICER like (DCL) protein. In plants, these siRNAs become a cause of epigenetic modification of cytosine residues in CG, CHG, and CHH contexts into methylated cytosine (5mC). Two types of DNA-dependent RNA polymerases (Pol IV and Pol V) are necessary to advance the process of RdDM (Matzke et al. 2015). The increased methylation levels in the target gene promoter induce TGS, which is associated with changes in chromatin structure through histone modification (see Sect. 16.1.2).

Although posttranscriptional gene silencing (PTGS) has been used for functional analyses of target genes for about 20 years, TGS has an important advantage over

PTGS. Artificially induced methylation and TGS via RdDM may be preserved and inherited after removal of the trigger gene cassette; thus RdDM-mediated TGS could be considered as a type of NPBT. However, there are few research papers that discuss the relationship between the preservation or loss of TGS and the presence or absence of the trigger gene (Kasai and Kanazawa 2013).

The remainder of this chapter describes the application of epigenetic modification to improvement of plant traits through various strategies of RdDM-mediated TGS. CRISPR/Cas9-mediated epigenome editing is also briefly discussed.

16.2.2 Viral Vector-Mediated TGS in Plants

Plant viral vectors have been used to induce RdDM-mediated TGS (Fig. 16.2a). In this strategy, the plant defense response toward virus infection (called recovery) is applied to the production of dsRNA from target gene promoters. Viral vector-mediated TGS in plants has been successfully induced in both reporter genes and endogenous genes. Kanazawa and co-authors used recombinant *Cucumber mosaic virus* (CMV) to induce TGS of *Chalcone synthase-A* and *LeSPL-CNR* genes in petunia and tomato, respectively (Kanazawa et al. 2011). Both plants showed clear phenotypic changes in association with epigenetic modification of the target gene promoters (Kanazawa et al. 2011). Interestingly, these new traits have been observed in subsequent generations even though the viral vector was not detected in these progenies (Kanazawa et al. 2011). These authors further reported that 2b protein, one of the endogenous proteins derived from CMV, is useful for stable induction of RdDM because it increases the expression of TGS induction-related genes and decreases the expression of demethylation-related genes. The 2b protein functions as an RNA silencing suppressor that can inhibit PTGS and virus-induced gene silencing (VIGS) (Goto et al. 2007). These results suggest that the combination of dsRNA and the 2b protein leads to highly efficient induction of RdDM-mediated TGS in a viral vector system.

As another example, TGS induction of *Chalcone synthase-A* gene in petunia was achieved via the use of apple latent spherical virus (ALSV) as a viral vector (Kon and Yoshikawa 2014). *Tobacco rattle virus* (TRV) vectors have also been used for induction of TGS, but most of the target genes have been reporter genes such as GFP and GUS under the control of the CaMV 35S promoter (Jones et al. 2001). However, recently, Bond and Baulcombe reported that TGS induction of endogenous gene (*FLOWERING WAGENINGEN*, *FWA*) using TRV viral vector system in *Arabidopsis* and this report deeply discussed about initiation, establishment, and maintenance of TGS in endogenous gene, using *FWA* TGS *Arabidopsis* and various mutants of gene silencing-related genes (Bond and Baulcombe 2015).

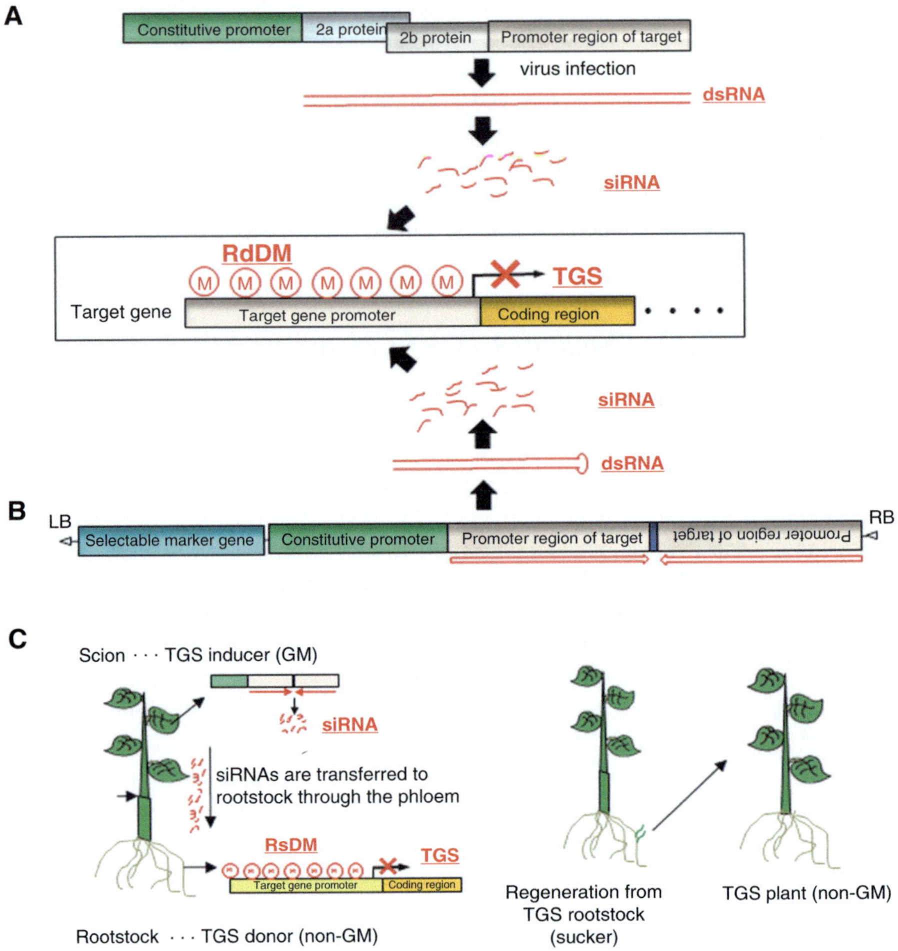

Fig. 16.2 Various strategies for induction of RdDM-mediated TGS in plants. Viral vector-mediated TGS (**a**), *Agrobacterium* (T-DNA)-mediated TGS (**b**), and grafting-mediated TGS (**c**) are shown. *M* methylation, *RB* right border, *LB* left border

16.2.3 Agrobacterium *(T-DNA)-Mediated TGS in Plants*

RdDM-mediated TGS in plants is generally induced by using T-DNA harboring gene cassettes to express dsRNA directed toward the promoter region of the target gene (Fig. 16.2b). Gene cassettes to express the dsRNA are introduced into the plant genome via *Agrobacterium*-mediated transformation. T0 plants expressing dsRNA derived from a foreign gene cassette should be treated as GM plants, whereas progenies of TGS plants after removal of T-DNA by segregation can be treated as non-GM plants in some world areas.

Although T-DNA-mediated TGS is very simple and easy to use, most research papers describe TGS of a reporter gene under the control of the CaMV 35S promoter. There are only a few reports of T-DNA-mediated TGS of endogenous genes in plants. In rice, RdDM-mediated TGS was attempted using a reporter gene (GFP under the control of the CaMV 35S promoter) and several endogenous genes. In these experiments, TGS could be easily induced for GFP; TGS of most endogenous genes could not be induced in spite of highly efficient induction of cytosine methylation of the target gene promoters (Okano et al. 2008). Although different levels of chromatin modification were observed between the CaMV 35S promoter and the endogenous gene promoter, it is not yet understood why reporter gene constructs such as CaMV 35S promoter::GFP can be silenced by TGS more easily than endogenous genes. On the other hand, successful induction of TGS of an endogenous gene by T-DNA-mediated expression of dsRNA corresponding to the target promoter region has been reported in petunia (Sijen et al. 2001), *Arabidopsis* (Deng et al. 2014), and potato (Kasai et al. 2016; Heilersig et al. 2006). However, it seems that a reproducible method for stable induction of TGS via T-DNA has not yet been found in plants.

As an alternative *Agrobacterium*-mediated strategy, T-DNA harboring a viral vector sequence is used to deliver the viral vector to plant cells. After *Agrobacterium*-mediated transformation or *Agrobacterium* infiltration, the viral vector is transferred into the nuclei of plant cells as a part of the T-DNA and can function independently as a viral vector. Viral vectors released from the T-DNA induce RdDM-mediated TGS (Ju et al. 2016).

16.2.4 Grafting-Mediated TGS in Plants

Grafting is a plant-specific strategy for inducing TGS. In vascular plants, the vascular bundle system functions to transport water, minerals, nutrients, proteins, and photosynthate from sink to source organs (or from source to sink). Some RNA molecules such as siRNA and microRNA (miRNA) are also transported by vascular bundle system. Specifically, these small RNA molecules are exclusively transported from sink to source organs through the phloem (Melnyk et al. 2011; Ham and Lucas 2017) and can move from cell to cell through the plasmodesmata (Melnyk et al. 2011; Ham and Lucas 2017). When a scion artificially expressing siRNA toward a target gene promoter (GM plant) is grafted onto a wild-type rootstock (non-GM), RdDM-mediated TGS can be induced in the rootstock via siRNA movement through the vascular bundle system and plasmodesmata. If a regenerated plantlet is obtained from the TGS rootstock, it would be transcriptionally silenced without insertion of foreign DNA into the genome. For this reason, "grafting on GM" is considered a type of NPBT (Fig. 16.2c). Bai and co-authors grafted transgenic tobacco scions expressing dsRNA directed toward parts of the CaMV 35S promoter region and under the control of the companion cell-specific *Commelina* yellow mottle virus (CoYMV) promoter onto transgenic tobacco rootstocks expressing GFP under the control of the CaMV 35S promoter. GFP fluorescence was drastically suppressed in lateral roots of the rootstock, indicating that TGS was epigenetically induced in the rootstock (Bai et al. 2011). This study showed that the movement of siRNA from scion to rootstock was more efficient than from rootstock to scion (Bai et al. 2011). These same authors have also produced an epigenetically improved potato by grafting with transgenic tobacco as the TGS inducer, resulting in modified amylose content through suppression of granule-bound starch synthase I (*GBSSI*) gene without changes in the genomic DNA sequences of the host potato (Kasai et al. 2016).

Vegetatively propagated crops such as potato and apple may have an advantage over seed-propagated crops with respect to the use of RdDM-mediated TGS because vegetatively propagated crops do not require meiosis for self-reproduction; thus, the modified methylation level may be preserved more stably than in seed-propagated crops. However, further investigation would be necessary to clarify this point.

16.2.5 CRISPR/Cas9-Mediated Epigenome Modification

The CRISPR/Cas9 system is a convenient and powerful tool for genome editing in many organisms (Cong et al. 2013; Mali et al. 2013; Fauser et al. 2014). This system is very simple, consisting of the combination of a guide RNA and Cas9 nuclease (Fig. 16.3a). A modified CRISPR/Cas9 system can be applied to epigenome editing. A nuclease-activity-disrupted Cas9 (dCas9) fused with enzymes to modify genomic DNA or histone can be used as an epigenome editing inducer (Fig. 16.3b). For example, a fusion protein consisting of nuclease-disrupted Cas9 protein and human acetyltransferase p300 successfully catalyzed the acetylation of histone H3 at target sites in human cells, resulting in the robust transcriptional activation of target genes (Hilton et al. 2015). Further modifications of CRISPR/Cas9-mediated epigenome editing will continue to be developed in animals and plants (Johnson et al. 2014; Konermann et al. 2013).

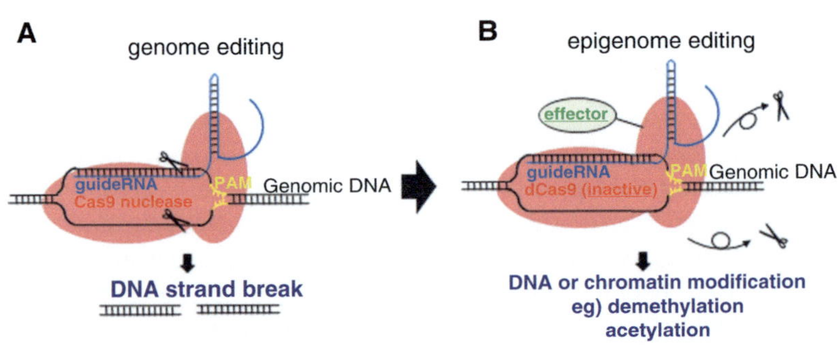

Fig. 16.3 CRISPR/Cas9-mediated epigenome editing. (**a**), Normal scheme of CRISPR/Cas9 system to induce double-strand break toward target genomic DNA. (**b**) Epigenome editing technology using fusion protein consists of effector and catalytically dead Cas9 (dCas9) lacking only endonuclease activity to induce the other modification such as demethylation or acetylation toward target genomic DNA or chromatin

16.3 Future Perspectives

This chapter describes new findings from both basic and applied studies on epigenetics in plants. Recently, the study of epigenetics has developed rapidly because of an increase in the precision of genome-wide association studies (GWAS), which have received a lot of attention in both animals and plants. At present, application of RdDM-mediated epigenome editing to plant breeding is not yet practical owing to the need to obtain stable induction of TGS toward endogenous genes and stable inheritance of the modified epigenome after removal of the trigger gene. However, we expect that many interesting findings will continue to be reported in the epigenetics field, with the result that RdDM-mediated epigenome editing will become a promising technology to produce trait-improved plants in the near future.

References

Ach RA, Taranto P, Gruissem W (1997) A conserved family of WD-40 proteins binds to the retinoblastoma protein in both plants and animals. Plant Cell 9:1595–1606

Adli M, Bernstein BE (2011) Whole-genome chromatin profiling from limited numbers of cells using nano-ChIP-seq. Nat Protoc 6:1656–1668

Aubert D, Chen L, Moon YH, Martin D, Castle LA, Yang CH, Sung ZR (2001) EMF1, a novel protein involved in the control of shoot architecture and flowering in *Arabidopsis*. Plant Cell 13:1865–1875

Ausin I, Feng S, Yu C, Liu W, Kuo HY, Jacobsen EL, Zhai J, Gallego-Bartolome J, Wang L, Egertsdotter U, Street NR, Jacobsen SE, Wang H (2016) DNA methylome of the 20-gigabase Norway spruce genome. Proc Natl Acad Sci U S A 113:E8106–E8113

Bai S, Kasai A, Yamada K, Li T, Harada T (2011) A mobile signal transported over a long distance induces systemic transcriptional gene silencing in a grafted partner. J Exp Bot 62:4561–4570

Barski A, Cuddapah S, Cui K, Roh TY, Schones DE, Wang Z, Wei G, Chepelev I, Zhao K (2007) High-resolution profiling of histone methylations in the human genome. Cell 129:823–837

Bartee L, Malagnac F, Bender J (2001) *Arabidopsis cmt3* chromomethylase mutations block non-CG methylation and silencing of an endogenous gene. Genes Dev 15:1753–1758

Benhamed M, Bertrand C, Servet C, Zhou DX (2006) *Arabidopsis* GCN5, HD1, and TAF1/HAF2 interact to regulate histone acetylation required for light-responsive gene expression. Plant Cell 18:2893–2903

Berr A, Xu L, Gao J, Cognat V, Steinmetz A, Dong A, Shen WH (2009) SET DOMAIN GROUP25 encodes a histone methyltransferase and is involved in FLOWERING LOCUS C activation and repression of flowering. Plant Physiol 151:1476–1485

Berr A, McCallum EJ, Ménard R, Meyer D, Fuchs J, Dong A, Shen WH (2010) *Arabidopsis* SET DOMAIN GROUP2 is required for H3K4 trimethylation and is crucial for both sporophyte and gametophyte development. Plant Cell 22:3232–3248

Bewick AJ, Ji L, Niederhuth CE, Willing EM, Hofmeister BT, Shi X, Wang L, Lu Z, Rohr NA, Hartwig B, Kiefer C, Deal RB, Schmutz J, Grimwood J, Stroud H, Jacobsen SE, Schneeberger K, Zhang X, Schmitz RJ (2016) On the origin and evolutionary consequences of gene body DNA methylation. Proc Natl Acad Sci U S A 113:9111–9116

Blackledge NP, Farcas AM, Kondo T, King HW, McGouran JF, Hanssen LL, Ito S, Cooper S, Kondo K, Koseki Y, Ishikura T, Long HK, Sheahan TW, Brockdorff N, Kessler BM, Koseki H, Klose RJ (2014) Variant PRC1 complex-dependent H2A ubiquitylation drives PRC2 recruitment and polycomb domain formation. Cell 157:1445–1459

Bond DM, Baulcombe DC (2015) Epigenetic transitions leading to heritable, RNA-mediated de novo silencing in *Arabidopsis thaliana*. Proc Natl Acad Sci U S A 112:917–922

Boyle AP, Davis S, Shulha HP, Meltzer P, Margulies EH, Weng Z, Furey TS, Crawford GE (2008) High-resolution mapping and characterization of open chromatin across the genome. Cell 132:311–322

Brind'Amour J, Liu S, Hudson M, Chen C, Karimi MM, Lorincz MC (2015) An ultra-low-input native ChIP-seq protocol for genome-wide profiling of rare cell populations. Nat Commun 6:6033

Buenrostro JD, Giresi PG, Zaba LC, Chang HY, Greenleaf WJ (2013) Transposition of native chromatin for fast and sensitive epigenomic profiling of open chromatin, DNA-binding proteins and nucleosome position. Nat Methods 10:1213–1218

Calarco JP, Borges F, Donoghue MT, Van Ex F, Jullien PE, Lopes T, Gardner R, Berger F, Feijo JA, Becker JD, Martienssen RA (2012) Reprogramming of DNA methylation in pollen guides epigenetic inheritance via small RNA. Cell 151:194–205

Calonje M, Sanchez R, Chen L, Sung ZR (2008) EMBRYONIC FLOWER1 participates in polycomb group-mediated AG gene silencing in *Arabidopsis*. Plant Cell 20:277–291

Cao X, Jacobsen SE (2002) Role of the arabidopsis DRM methyltransferases in de novo DNA methylation and gene silencing. Curr Biol 12:1138–1144

Cartagena JA, Matsunaga S, Seki M, Kurihara D, Yokoyama M, Shinozaki K, Fujimoto S, Azumi Y, Uchiyama S, Fukui K (2008) The *Arabidopsis* SDG4 contributes to the regulation of pollen tube growth by methylation of histone H3 lysines 4 and 36 in mature pollen. Dev Biol 315:355–368

Chanvivattana Y, Bishopp A, Schubert D, Stock C, Moon YH, Sung ZR, Goodrich J (2004) Interaction of Polycomb-group proteins controlling flowering in *Arabidopsis*. Development 131:5263–5276

Charron JB, He H, Elling AA, Deng XW (2009) Dynamic landscapes of four histone modifications during deetiolation in *Arabidopsis*. Plant Cell 21:3732–3748

Choi Y, Gehring M, Johnson L, Hannon M, Harada JJ, Goldberg RB, Jacobsen SE, Fischer RL (2002) DEMETER, a DNA glycosylase domain protein, is required for endosperm gene imprinting and seed viability in *Arabidopsis*. Cell 110:33–42

Cokus SJ, Feng S, Zhang X, Chen Z, Merriman B, Haudenschild CD, Pradhan S, Nelson SF, Pellegrini M, Jacobsen SE (2008) Shotgun bisulphite sequencing of the *Arabidopsis* genome reveals DNA methylation patterning. Nature 452:215–219

Coleman-Derr D, Zilberman D (2012) Deposition of histone variant H2A.Z within gene bodies regulates responsive genes. PLoS Genet 8:e1002988

Cong L, Ran FA, Cox D, Lin S, Barretto R, Habib N, Hsu PD, Wu X, Jiang W, Marraffini LA, Zhang F (2013) Multiplex genome engineering using CRISPR/Cas systems. Science 339:819–823

Cooper S, Dienstbier M, Hassan R, Schermelleh L, Sharif J, Blackledge NP, De Marco V, Elderkin S, Koseki H, Klose R, Heger A, Brockdorff N (2014) Targeting polycomb to pericentric heterochromatin in embryonic stem cells reveals a role for H2AK119u1 in PRC2 recruitment. Cell Rep 7:1456–1470

Cuerda-Gil D, Slotkin RK (2016) Non-canonical RNA-directed DNA methylation. Nat Plants 2:16163

Cui X, Lu F, Qiu Q, Zhou B, Gu L, Zhang S, Kang Y, Ma X, Yao Q, Ma J, Zhang X, Cao X (2016) REF6 recognizes a specific DNA sequence to demethylate H3K27me3 and regulate organ boundary formation in *Arabidopsis*. Nat Genet 48:694–699

de Napoles M, Mermoud JE, Wakao R, Tang YA, Endoh M, Appanah R, Nesterova TB, Silva J, Otte AP, Vidal M, Koseki H, Brockdorff N (2004) Polycomb group proteins Ring1A/B link ubiquitylation of histone H2A to heritable gene silencing and X inactivation. Dev Cell 7:663–676

Deal RB, Henikoff S (2011a) Histone variants and modifications in plant gene regulation. Curr Opin Plant Biol 14:116–122

Deal RB, Henikoff S (2011b) The INTACT method for cell type-specific gene expression and chromatin profiling in *Arabidopsis thaliana*. Nat Protoc 6:56–68

Deleris A, Greenberg MV, Ausin I, Law RW, Moissiard G, Schubert D, Jacobsen SE (2010) Involvement of a Jumonji-C domain-containing histone demethylase in DRM2-mediated maintenance of DNA methylation. EMBO Rep 11:950–955

Deng S, Dai H, Arenas C, Wang H, Niu HW, Chua NH (2014) Transcriptional silencing of *Arabidopsis* endogens by single-stranded RNAs targeting the promoter region. Plant Cell Physiol 55:823–833

Derkacheva M, Steinbach Y, Wildhaber T, Mozgová I, Mahrez W, Nanni P, Bischof S, Gruissem W, Hennig L (2013) *Arabidopsis* MSI1 connects LHP1 to PRC2 complexes. EMBO J 32:2073–2085

Dowen RH, Pelizzola M, Schmitz RJ, Lister R, Dowen JM, Nery JR, Dixon JE, Ecker JR (2012) Widespread dynamic DNA methylation in response to biotic stress. Proc Natl Acad Sci U S A 109:E2183–E2191

Down TA, Rakyan VK, Turner DJ, Flicek P, Li H, Kulesha E, Gräf S, Johnson N, Herrero J, Tomazou EM, Thorne NP, Bäckdahl L, Herberth M, Howe KL, Jackson DK, Miretti MM, Marioni JC, Birney E, Hubbard TJ, Durbin R, Tavaré S, Beck S (2008) A Bayesian deconvolution strategy for immunoprecipitation-based DNA methylome analysis. Nat Biotechnol 26:779–785

Earley K, Lawrence RJ, Pontes O, Reuther R, Enciso AJ, Silva M, Neves N, Gross M, Viegas W, Pikaard CS (2006) Erasure of histone acetylation by *Arabidopsis* HDA6 mediates large-scale gene silencing in nucleolar dominance. Genes Dev 20:1283–1293

Ebbs ML, Bender J (2006) Locus-specific control of DNA methylation by the *Arabidopsis* SUVH5 histone methyltransferase. Plant Cell 18:1166–1176

Ebbs ML, Bartee L, Bender J (2005) H3 lysine 9 methylation is maintained on a transcribed inverted repeat by combined action of SUVH6 and SUVH4 methyltransferases. Mol Cell Biol 25:10507–10515

Fauser F, Schiml S, Puchta H (2014) Both CRISPR/Cas-based nucleases and nickases can be efficiently for genome engineering in *Arabidopsis thaliana*. Plant J 79:348–359

Fransz P, ten Hoopen R, Tessadori F (2006) Composition and formation of heterochromatin in *Arabidopsis thaliana*. Chromosom Res 14:71–82

Fuchs J, Demidov D, Houben A, Schubert I (2006) Chromosomal histone modification patterns – from conservation to diversity. Trends Plant Sci 11:199–208

Gan ES, Xu Y, Wong JY, Goh JG, Sun B, Wee WY, Huang J, Ito T (2014) Jumonji demethylases moderate precocious flowering at elevated temperature via regulation of FLC in *Arabidopsis*. Nat Commun 5:5098

Gao Z, Liu H, Daxinger L, Pontes O, He X, Qian W, Lin H, Xie M, Lorkovic Z, Zhang S, Miki D, Zhan X, Pontier D, Lagrange T, Jin H, Matzke A, Matzke M, Pikaard C, Zhu J (2010) An RNA polymerase II- and AGO4-associated protein acts in RNA-directed DNA methylation. Nature 465:106–109

Garg R, Narayana Chevala V, Shankar R, Jain M (2015) Divergent DNA methylation patterns associated with gene expression in rice cultivars with contrasting drought and salinity stress response. Sci Rep 5:14922

Gendall AR, Levy YY, Wilson A, Dean C (2001) The VERNALIZATION 2 gene mediates the epigenetic regulation of vernalization in *Arabidopsis*. Cell 107:525–535

Gent JI, Ellis NA, Guo L, Harkess AE, Yao Y, Zhang X, Dawe RK (2013) CHH islands: de novo DNA methylation in near-gene chromatin regulation in maize. Genome Res 23:628–637

Gil J, O'Loghlen A (2014) PRC1 complex diversity: where is it taking us? Trends Cell Biol 24:632–641

Giresi PG, Kim J, McDaniell RM, Iyer VR, Lieb JD (2007) FAIRE (Formaldehyde-Assisted Isolation of Regulatory Elements) isolates active regulatory elements from human chromatin. Genome Res 17:877–885

Gong Z, Morales-Ruiz T, Ariza RR, Roldan-Arjona T, David L, Zhu JK (2002) *ROS1*, a repressor of transcriptional gene silencing in *Arabidopsis*, encodes a DNA glycosylase/lyase. Cell 111:803–814

Goodrich J, Puangsomlee P, Martin M, Long D, Meyerowitz EM, Coupland G (1997) A Polycomb-group gene regulates homeotic gene expression in *Arabidopsis*. Nature 386:44–51

Goto K, Kobori T, Kosaka Y, Natsuki T, Masuta C (2007) Characterization of silencing suppressor 2b of cucumber mosaic virus based of examination of its small RNA-bindng abilities. Plant Cell Physiol 48:1050–1060

Greenberg MV, Deleris A, Hale CJ, Liu A, Feng S, Jacobsen SE (2013) Interplay between active chromatin marks and RNA-directed DNA methylation in *Arabidopsis thaliana*. PLoS Genet 9:e1003946

Grossniklaus U, Vielle-Calzada JP, Hoeppner MA, Gagliano WB (1998) Maternal control of embryogenesis by MEDEA, a polycomb group gene in *Arabidopsis*. Science 280:446–450

Guo L, Yu Y, Law JA, Zhang X (2010) SET DOMAIN GROUP2 is the major histone H3 lysine [corrected] 4 trimethyltransferase in *Arabidopsis*. Proc Natl Acad Sci U S A 107:18557–18562

Ham BK, Lucas WJ (2017) Phloem-mobile RNAs as systemic signaling agents. Annu Rev Plant Biol 68:173–195

Havecker ER, Wallbridge LM, Hardcastle TJ, Bush MS, Kelly KA, Dunn RM, Schwach F, Doonan JH, Baulcombe DC (2010) The *Arabidopsis* RNA-directed DNA methylation argonautes functionally diverge based on their expression and interaction with target loci. Plant Cell 22:321–334

Heilersig BHJB, Loonen AEHM, Janssen EM, Wolters AA, Visser RGF (2006) Efficiency of transcriptional gene silencing of GBSSI in potato depends on the promoter region that is used in an inverted repeat. Mol Gen Genomics 275:437–449

Hennig L, Taranto P, Walser M, Schönrock N, Gruissem W (2003) *Arabidopsis* MSI1 is required for epigenetic maintenance of reproductive development. Development 130:2555–2565

Hilton IB, D'Ippolito AM, Vockley CM, Thakore PI, Crawford GE, Reddy TE, Gersbach CA (2015) Epigenome editing by a CRISPR-Cas9-based acetyltransferase activates genes from promoters and enhancers. Nat Biotechnol 33:510–517

Hossain MS, Kawakatsu T, Kim KD, Zhang N, Nguyen CT, Khan SM, Batek JM, Joshi T, Schmutz J, Grimwood J, Schmitz RJ, Xu D, Jackson SA, Ecker JR, Stacey G (2017) Divergent cytosine DNA methylation patterns in single-cell, soybean root hairs. New Phytol 214:808–819

Hsieh TF, Ibarra CA, Silva P, Zemach A, Eshed-Williams L, Fischer RL, Zilberman D (2009) Genome-wide demethylation of *Arabidopsis* endosperm. Science 324:1451–1454

Hsieh PH, He S, Buttress T, Gao H, Couchman M, Fischer RL, Zilberman D, Feng X (2016) *Arabidopsis* male sexual lineage exhibits more robust maintenance of CG methylation than somatic tissues. Proc Natl Acad Sci U S A 113:15132–15137

Ibarra CA, Feng X, Schoft VK, Hsieh TF, Uzawa R, Rodrigues JA, Zemach A, Chumak N, Machlicova A, Nishimura T, Rojas D, Fischer RL, Tamaru H, Zilberman D (2012) Active DNA demethylation in plant companion cells reinforces transposon methylation in gametes. Science 337:1360–1364

Inagaki S, Miura-Kamio A, Nakamura Y, Lu F, Cui X, Cao X, Kimura H, Saze H, Kakutani T (2010) Autocatalytic differentiation of epigenetic modifications within the *Arabidopsis* genome. EMBO J 29:3496–3506

Jackson V (1978) Studies on histone organization in the nucleosome using formaldehyde as a reversible cross-linking agent. Cell 15:945–954

Jacob Y, Feng S, LeBlanc CA, Bernatavichute YV, Stroud H, Cokus S, Johnson LM, Pellegrini M, Jacobsen SE, Michaels SD (2009) ATXR5 and ATXR6 are H3K27 monomethyltransferases required for chromatin structure and gene silencing. Nat Struct Mol Biol 16:763–768

James C (2016) 20th anniversary of the global commercialization of biotech crops (1996 to 2015) and biotech crop highlights in 2015. International Service for the Acquisition of Agri-biotech Applications (ISAAA) http://www.isaaa.org/resources/publications/briefs/51/executivesummary/pdf/B51-ExecSum-English.pdf

Jiang D, Wang Y, He Y (2008) Repression of FLOWERING LOCUS C and FLOWERING LOCUS T by the *Arabidopsis* Polycomb repressive complex 2 components. PLoS One 3:e3404

Jiang D, Kong NC, Gu X, Li Z, He Y (2011) *Arabidopsis* COMPASS-like complexes mediate histone H3 lysine-4 trimethylation to control floral transition and plant development. PLoS Genet 7:e1001330

Johnson LM, Du J, Hale CJ, Bischof S, Feng S, Chodavarapu RK, Zhong X, Marson G, Pellegrini M, Segal DJ, Patel DJ, Jacobsen SE (2014) SRA- and SET-domain-containing proteins like RNA polymerase V occupancy to DNA methylation. Nature 507:124–128

Jones L, Ratcliff F, Baulcombe DC (2001) RNA-directed transcriptional gene silencing in plants can be inherited independenly of the RNA trigger and requires Met1 for maintenance. Curr Biol 11:747–757

Ju Z, Wang L, Cao D, Zuo J, Zhu H, Fu LY, Zhu B (2016) A viral satellite DNA vector-induced transcriptional gene silencing via DNA methylation of gene promoter in *Nicotiana benthamiana*. Virus Res 223:99–107

Jullien PE, Katz A, Oliva M, Ohad N, Berger F (2006) Polycomb group complexes self-regulate imprinting of the Polycomb group gene MEDEA in *Arabidopsis*. Curr Biol 16:486–492

Kalb R, Latwiel S, Baymaz HI, Jansen PW, Müller CW, Vermeulen M, Müller J (2014) Histone H2A monoubiquitination promotes histone H3 methylation in Polycomb repression. Nat Struct Mol Biol 21:569–571

Kanazawa A, Inaba J, Shimura H, Otagaki S, Tsukahara S, Matsuzawa A, Kim BM, Goto K, Masuta C (2011) Virus-mediated efficient induction of epigenetic modifications of endogenous genes with phenotypic changes in plants. Plant J 65:156–168

Kankel MW, Ramsey DE, Stokes TL, Flowers SK, Haag JR, Jeddeloh JA, Riddle NC, Verbsky ML, Richards EJ (2003) *Arabidopsis MET1* cytosine methyltransferase mutants. Genetics 163:1109–1122

Kasai M, Kanazawa A (2013) Induction of RNA-directed DNA methylation and heritable transcriptional gene silencing as a tool to engineer novel traits in plants. Plant Biotechnol 30:233–241

Kasai A, Bai S, Hojo H, Harada T (2016) Epigenome editing of potato by grafting using transgenic tobacco as siRNA donor. PLoS One 11:e0161729

Kawakatsu T, Huang SS, Jupe F, Sasaki E, Schmitz RJ, Urich MA, Castanon R, Nery JR, Barragan C, He Y, Chen H, Dubin M, Lee CR, Wang C, Bemm F, Becker C, O'Neil R, O'Malley RC, Quarless DX, Genomes C, Schork NJ, Weigel D, Nordborg M, Ecker JR (2016a) Epigenomic diversity in a global collection of *Arabidopsis thaliana* accessions. Cell 166:492–505

Kawakatsu T, Stuart T, Valdes M, Breakfield N, Schmitz RJ, Nery JR, Urich MA, Han X, Lister R, Benfey PN, Ecker JR (2016b) Unique cell-type-specific patterns of DNA methylation in the root meristem. Nat Plants 2:16058

Kawashima T, Berger F (2014) Epigenetic reprogramming in plant sexual reproduction. Nat Rev Genet 15:613–624

Kenzior AL, Folk WR (1998) AtMSI4 and RbAp48 WD-40 repeat proteins bind metal ions. FEBS Lett 440:425–429

Kim SY, Zhu T, Sung ZR (2010) Epigenetic regulation of gene programs by EMF1 and EMF2 in *Arabidopsis*. Plant Physiol 152:516–528

Kim SY, Lee J, Eshed-Williams L, Zilberman D, Sung ZR (2012) EMF1 and PRC2 cooperate to repress key regulators of *Arabidopsis* development. PLoS Genet 8:e1002512

Kon T, Yoshikawa N (2014) Induction and maintenance of DNA methylation in promoter sequences by apple latent spherical virus-induced transcriptional gene silencing. Front Microbiol 5:595

Konermann S, Brigham MD, Trevino A, Hsu PD, Matthias H, Cong L, Platt RJ, Scott DA, Church GM, Zhang F (2013) Optical control of mammalian endogenous transcription and epigenetic states. Nature 500:472–476

Kornberg RD (1974) Chromatin structure: a repeating unit of histones and DNA. Science 184:868–871

Lafos M, Kroll P, Hohenstatt ML, Thorpe FL, Clarenz O, Schubert D (2011) Dynamic regulation of H3K27 trimethylation during *Arabidopsis* differentiation. PLoS Genet 7:e1002040

Law JA, Jacobsen SE (2010) Establishing, maintaining and modifying DNA methylation patterns in plants and animals. Nat Rev Genet 11:204–220

Lei M, Zhang H, Julian R, Tang K, Xie S, Zhu JK (2015) Regulatory link between DNA methylation and active demethylation in *Arabidopsis*. Proc Natl Acad Sci U S A 112:3553–3557

Li G, Liu S, Wang J, He J, Huang H, Zhang Y, Xu L (2014) ISWI proteins participate in the genome-wide nucleosome distribution in *Arabidopsis*. Plant J 78:706–714

Li C, Gu L, Gao L, Chen C, Wei CQ, Qiu Q, Chien CW, Wang S, Jiang L, Ai LF, Chen CY, Yang S, Nguyen V, Qi Y, Snyder MP, Burlingame AL, Kohalmi SE, Huang S, Cao X, Wang ZY, Wu K, Chen X, Cui Y (2016) Concerted genomic targeting of H3K27 demethylase REF6 and chromatin-remodeling ATPase BRM in *Arabidopsis*. Nat Genet 48:687–693

Lindroth AM, Cao X, Jackson JP, Zilberman D, CM MC, Henikoff S, Jacobsen SE (2001) Requirement of CHROMOMETHYLASE3 for maintenance of CpXpG methylation. Science 292:2077–2080

Lister R, O'Malley RC, Tonti-Filippini J, Gregory BD, Berry CC, Millar AH, Ecker JR (2008) Highly integrated single-base resolution maps of the epigenome in *Arabidopsis*. Cell 133:523–536

Liu X, Yang S, Zhao M, Luo M, Yu CW, Chen CY, Tai R, Wu K (2014) Transcriptional repression by histone deacetylases in plants. Mol Plant 7:764–772

Lu F, Cui X, Zhang S, Liu C, Cao X (2010) JMJ14 is an H3K4 demethylase regulating flowering time in *Arabidopsis*. Cell Res 20:387–390

Lu F, Cui X, Zhang S, Jenuwein T, Cao X (2011) *Arabidopsis* REF6 is a histone H3 lysine 27 demethylase. Nat Genet 43:715–719

Lu Z, Hofmeister BT, Vollmers C, DuBois RM, Schmitz RJ (2016) Combining ATAC-seq with nuclei sorting for discovery of cis-regulatory regions in plant genomes. Nucleic Acids Res 45:e41

Luger K, Rechsteiner TJ, Flaus AJ, Waye MM, Richmond TJ (1997) Characterization of nucleosome core particles containing histone proteins made in bacteria. J Mol Biol 272:301–311

Luo C, Lam E (2014) Quantitatively profiling genome-wide patterns of histone modifications in *Arabidopsis thaliana* using ChIP-seq. Methods Mol Biol 1112:177–193

Luo M, Bilodeau P, Koltunow A, Dennis ES, Peacock WJ, Chaudhury AM (1999) Genes controlling fertilization-independent seed development in *Arabidopsis thaliana*. Proc Natl Acad Sci U S A 96:296–301

Lusser M, Parisi C, Plan D, Rodriguez-Cerezo E (2012) Deployment of new biotechnologies in plant breeding. Nat Biotechnol 30:231–239

Makarevich G, Leroy O, Akinci U, Schubert D, Clarenz O, Goodrich J, Grossniklaus U, Köhler C (2006) Different Polycomb group complexes regulate common target genes in *Arabidopsis*. EMBO Rep 7:947–952

Mali P, Yang L, Esvelt KM, Aach J, Guell M, DiCarlo JE, Norville JE, Church GM (2013) RNA-guided human genome engineering via Cas9. Science 339:823–826

Margueron R, Reinberg D (2011) The Polycomb complex PRC2 and its mark in life. Nature 469:343–349

Matzke MA, Kanno T, Matzke JM (2015) RNA-directed DNA methylation: the evolution of a complex epigenetic pathway in flowering plants. Annu Rev Plant Biol 66:243–267

Maunakea AK, Nagarajan RP, Bilenky M, Ballinger TJ, D'Souza C, Fouse SD, Johnson BE, Hong C, Nielsen C, Zhao Y, Turecki G, Delaney A, Varhol R, Thiessen N, Shchors K, Heine VM, Rowitch DH, Xing X, Fiore C, Schillebeeckx M, Jones SJ, Haussler D, Marra MA, Hirst M, Wang T, Costello JF (2010) Conserved role of intragenic DNA methylation in regulating alternative promoters. Nature 466:253–257

McCue AD, Panda K, Nuthikattu S, Choudury SG, Thomas EN, Slotkin RK (2015) ARGONAUTE 6 bridges transposable element mRNA-derived siRNAs to the establishment of DNA methylation. EMBO J 34:20–35
Melnyk CW, Molnar A, Baulconbe DC (2011) Intercelluar and systemic novement of RNA silencing signals. EMBO J 30:3553–3563
Merini W, Calonje M (2015) PRC1 is taking the lead in PcG repression. Plant J 83:110–120
Meyer CA, Liu XS (2014) Identifying and mitigating bias in next-generation sequencing methods for chromatin biology. Nat Rev Genet 15:709–721
Mozgova I, Köhler C, Hennig L (2015) Keeping the gate closed: functions of the polycomb repressive complex PRC2 in development. Plant J 83:121–132
Narsai R, Secco D, Schultz MD, Ecker JR, Lister R, Whelan J (2016) Dynamic and rapid changes in the transcriptome and epigenome during germination and in developing rice (*Oryza sativa*) coleoptiles under anoxia and re-oxygenation. Plant J 89:805–824
Niederhuth CE, Bewick AJ, Ji L, Alabady MS, Kim KD, Li Q, Rohr NA, Rambani A, Burke JM, Udall JA, Egesi C, Schmutz J, Grimwood J, Jackson SA, Springer NM, Schmitz RJ (2016) Widespread natural variation of DNA methylation within angiosperms. Genome Biol 17:194
Noh B, Lee SH, Kim HJ, Yi G, Shin EA, Lee M, Jung KJ, Doyle MR, Amasino RM, Noh YS (2004) Divergent roles of a pair of homologous jumonji/zinc-finger-class transcription factor proteins in the regulation of *Arabidopsis* flowering time. Plant Cell 16:2601–2613
O'Malley RC, Huang SC, Song L, Lewsey MG, Bartlett A, Nery JR, Galli M, Gallavotti A, Ecker JR (2016) Cistrome and epicistrome features shape the regulatory DNA landscape. Cell 165:1280–1292
O'Neill LP, Turner BM (1995) Histone H4 acetylation distinguishes coding regions of the human genome from heterochromatin in a differentiation-dependent but transcription-independent manner. EMBO J 14:3946–3957
Okano Y, Miki D, Shimamoto K (2008) Small interfering RNA (siRNA) targeting of endogenous promoters induces methylation but not necessarily gene silencing, in rice. Plant J 53:65–77
Ong-Abdullah M, Ordway JM, Jiang N, Ooi SE, Kok SY, Sarpan N, Azimi N, Hashim AT, Ishak Z, Rosli SK, Malike FA, Bakar NA, Marjuni M, Abdullah N, Yaakub Z, Amiruddin MD, Nookiah R, Singh R, Low ET, Chan KL, Azizi N, Smith SW, Bacher B, Budiman MA, Van Brunt A, Wischmeyer C, Beil M, Hogan M, Lakey N, Lim CC, Arulandoo X, Wong CK, Choo CN, Wong WC, Kwan YY, Alwee SS, Sambanthamurthi R, Martienssen RA (2015) Loss of karma transposon methylation underlies the mantled somaclonal variant of oil palm. Nature 525:533–537
Ortega-Galisteo AP, Morales-Ruiz T, Ariza RR, Roldan-Arjona T (2008) *Arabidopsis* DEMETER-LIKE proteins DML2 and DML3 are required for appropriate distribution of DNA methylation marks. Plant Mol Biol 67:671–681
Pandey R, Müller A, Napoli CA, Selinger DA, Pikaard CS, Richards EJ, Bender J, Mount DW, Jorgensen RA (2002) Analysis of histone acetyltransferase and histone deacetylase families of *Arabidopsis thaliana* suggests functional diversification of chromatin modification among multicellular eukaryotes. Nucleic Acids Res 30:5036–5055
Park K, Kim MY, Vickers M, Park JS, Hyun Y, Okamoto T, Zilberman D, Fischer RL, Feng X, Choi Y, Scholten S (2016) DNA demethylation is initiated in the central cells of *Arabidopsis* and rice. Proc Natl Acad Sci U S A 113:15138–15143
Pazhouhandeh M, Molinier J, Berr A, Genschik P (2011) MSI4/FVE interacts with CUL4-DDB1 and a PRC2-like complex to control epigenetic regulation of flowering time in *Arabidopsis*. Proc Natl Acad Sci U S A 108:3430–3435
Pengelly AR, Kalb R, Finkl K, Müller J (2015) Transcriptional repression by PRC1 in the absence of H2A monoubiquitylation. Genes Dev 29:1487–1492
Penterman J, Zilberman D, Huh JH, Ballinger T, Henikoff S, Fischer RL (2007) DNA demethylation in the *Arabidopsis* genome. Proc Natl Acad Sci U S A 104:6752–6757
Project AG (2013) The *Amborella* genome and the evolution of flowering plants. Science 342:1241089

Rand AC, Jain M, Eizenga JM, Musselman-Brown A, Olsen HE, Akeson M, Paten B (2017) Mapping DNA methylation with high-throughput nanopore sequencing. Nat Methods 14:411–413
Roudier F, Teixeira FK, Colot V (2009) Chromatin indexing in *Arabidopsis*: an epigenomic tale of tails and more. Trends Genet 25:511–517
Roudier F, Ahmed I, Bérard C, Sarazin A, Mary-Huard T, Cortijo S, Bouyer D, Caillieux E, Duvernois-Berthet E, Al-Shikhley L, Giraut L, Després B, Drevensek S, Barneche F, Dèrozier S, Brunaud V, Aubourg S, Schnittger A, Bowler C, Martin-Magniette ML, Robin S, Caboche M, Colot V (2011) Integrative epigenomic mapping defines four main chromatin states in *Arabidopsis*. EMBO J 30:1928–1938
Saleh A, Al-Abdallat A, Ndamukong I, Alvarez-Venegas R, Avramova Z (2007) The *Arabidopsis* homologs of trithorax (ATX1) and enhancer of zeste (CLF) establish 'bivalent chromatin marks' at the silent AGAMOUS locus. Nucleic Acids Res 35:6290–6296
Saleh A, Alvarez-Venegas R, Yilmaz M, Le O, Hou G, Sadder M, Al-Abdallat A, Xia Y, Lu G, Ladunga I, Avramova Z (2008) The highly similar *Arabidopsis* homologs of trithorax ATX1 and ATX2 encode proteins with divergent biochemical functions. Plant Cell 20:568–579
Sanchez-Pulido L, Devos D, Sung ZR, Calonje M (2008) RAWUL: a new ubiquitin-like domain in PRC1 ring finger proteins that unveils putative plant and worm PRC1 orthologs. BMC Genomics 9:308
Schaart JG, Wiel CCM, Lotz LAP, Smulders JM (2016) Opportunities for products of new plant breeding techniques. Trends Plant Sci 21:438–449
Schmid CD, Bucher P (2007) ChIP-Seq data reveal nucleosome architecture of human promoters. Cell 131:831–832. author reply 832–833
Schmidl C, Rendeiro AF, Sheffield NC, Bock C (2015) ChIPmentation: fast, robust, low-input ChIP-seq for histones and transcription factors. Nat Methods 12:963–965
Schmitz RJ, Schultz MD, Lewsey MG, O'Malley RC, Urich MA, Libiger O, Schork NJ, Ecker JR (2011) Transgenerational epigenetic instability is a source of novel methylation variants. Science 334:369–373
Schmitz RJ, He Y, Valdes-Lopez O, Khan SM, Joshi T, Urich MA, Nery JR, Diers B, Xu D, Stacey G, Ecker JR (2013a) Epigenome-wide inheritance of cytosine methylation variants in a recombinant inbred population. Genome Res 23:1663–1674
Schmitz RJ, Schultz MD, Urich MA, Nery JR, Pelizzola M, Libiger O, Alix A, McCosh RB, Chen H, Schork NJ, Ecker JR (2013b) Patterns of population epigenomic diversity. Nature 495:193–198
Schones DE, Cui K, Cuddapah S, Roh TY, Barski A, Wang Z, Wei G, Zhao K (2008) Dynamic regulation of nucleosome positioning in the human genome. Cell 132:887–898
Schoorlemmer J, Marcos-Gutiérrez C, Were F, Martínez R, García E, Satijn DP, Otte AP, Vidal M (1997) Ring1A is a transcriptional repressor that interacts with the Polycomb-M33 protein and is expressed at rhombomere boundaries in the mouse hindbrain. EMBO J 16:5930–5942
Secco D, Wang C, Shou H, Schultz MD, Chiarenza S, Nussaume L, Ecker JR, Whelan J, Lister R (2015) Stress induced gene expression drives transient DNA methylation changes at adjacent repetitive elements. elife 4:e09343
Seymour DK, Koenig D, Hagmann J, Becker C, Weigel D (2014) Evolution of DNA methylation patterns in the Brassicaceae is driven by differences in genome organization. PLoS Genet 10:e1004785
Sijen T, Vijn I, Rebocho A, van Blokland R, Roelofs D, Mol JN, Kooter JM (2001) Transcriptional and posttranscriptional gene silencing are mechanistically related. Curr Biol 11:436–440
Simpson JT, Workman RE, Zuzarte PC, David M, Dursi LJ, Timp W (2017) Detecting DNA cytosine methylation using nanopore sequencing. Nat Methods 14:407–410
Song L, Huang SC, Wise A, Castanon R, Nery JR, Chen H, Watanabe M, Thomas J, Bar-Joseph Z, Ecker JR (2016) A transcription factor hierarchy defines an environmental stress response network. Science 354:aag1550

Stroud H, Ding B, Simon SA, Feng S, Bellizzi M, Pellegrini M, Wang GL, Meyers BC, Jacobsen SE (2013) Plants regenerated from tissue culture contain stable epigenome changes in rice. elife 2:e00354

Stroud H, Do T, Du J, Zhong X, Feng S, Johnson L, Patel DJ, Jacobsen SE (2014) Non-CG methylation patterns shape the epigenetic landscape in *Arabidopsis*. Nat Struct Mol Biol 21:64–72

Takuno S, Ran J-H, Gaut BS (2016) Evolutionary patterns of genic DNA methylation vary across land plants. Nat Plants 2:15222

Thoma F, Koller T (1977) Influence of histone H1 on chromatin structure. Cell 12:101–107

To TK, Kim JM, Matsui A, Kurihara Y, Morosawa T, Ishida J, Tanaka M, Endo T, Kakutani T, Toyoda T, Kimura H, Yokoyama S, Shinozaki K, Seki M (2011) *Arabidopsis* HDA6 regulates locus-directed heterochromatin silencing in cooperation with MET1. PLoS Genet 7:e1002055

Tran RK, Henikoff JG, Zilberman D, Ditt RF, Jacobsen SE, Henikoff S (2005) DNA methylation profiling identifies CG methylation clusters in *Arabidopsis* genes. Curr Biol 15:154–159

Turck F, Roudier F, Farrona S, Martin-Magniette ML, Guillaume E, Buisine N, Gagnot S, Martienssen RA, Coupland G, Colot V (2007) *Arabidopsis* TFL2/LHP1 specifically associates with genes marked by trimethylation of histone H3 lysine 27. PLoS Genet 3:e86

Urich MA, Nery JR, Lister R, Schmitz RJ, Ecker JR (2015) MethylC-seq library preparation for base-resolution whole-genome bisulfite sequencing. Nat Protoc 10:475–483

Wang H, Wang L, Erdjument-Bromage H, Vidal M, Tempst P, Jones RS, Zhang Y (2004) Role of histone H2A ubiquitination in Polycomb silencing. Nature 431:873–878

Wang Y, Gu X, Yuan W, Schmitz RJ, He Y (2014) Photoperiodic control of the floral transition through a distinct polycomb repressive complex. Dev Cell 28:727–736

Wibowo A, Becker C, Marconi G, Durr J, Price J, Hagmann J, Papareddy R, Putra H, Kageyama J, Becker J, Weigel D, Gutierrez-Marcos J (2016) Hyperosmotic stress memory in *Arabidopsis* is mediated by distinct epigenetically labile sites in the genome and is restricted in the male germline by DNA glycosylase activity. elife 5:e13456

Williams BP, Pignatta D, Henikoff S, Gehring M (2015) Methylation-sensitive expression of a DNA demethylase gene serves as an epigenetic rheostat. PLoS Genet 11:e1005142

Wu Y, Zhang W, Jiang J (2014) Genome-wide nucleosome positioning is orchestrated by genomic regions associated with DNase I hypersensitivity in rice. PLoS Genet 10:e1004378

Xu L, Shen WH (2008) Polycomb silencing of KNOX genes confines shoot stem cell niches in *Arabidopsis*. Curr Biol 18:1966–1971

Xu L, Zhao Z, Dong A, Soubigou-Taconnat L, Renou JP, Steinmetz A, Shen WH (2008) Di- and tri- but not monomethylation on histone H3 lysine 36 marks active transcription of genes involved in flowering time regulation and other processes in *Arabidopsis thaliana*. Mol Cell Biol 28:1348–1360

Yang C, Bratzel F, Hohmann N, Koch M, Turck F, Calonje M (2013) VAL- and AtBMI1-mediated H2Aub initiate the switch from embryonic to postgerminative growth in *Arabidopsis*. Curr Biol 23:1324–1329

Yang H, Howard M, Dean C (2014) Antagonistic roles for H3K36me3 and H3K27me3 in the cold-induced epigenetic switch at *Arabidopsis* FLC. Curr Biol 24:1793–1797

Yoshida N, Yanai Y, Chen L, Kato Y, Hiratsuka J, Miwa T, Sung ZR, Takahashi S (2001) EMBRYONIC FLOWER2, a novel polycomb group protein homolog, mediates shoot development and flowering in *Arabidopsis*. Plant Cell 13:2471–2481

Zemach A, Kim MY, Hsieh PH, Coleman-Derr D, Eshed-Williams L, Thao K, Harmer SL, Zilberman D (2013) The *Arabidopsis* nucleosome remodeler DDM1 allows DNA methyltransferases to access H1-containing heterochromatin. Cell 153:193–205

Zhang X, Yazaki J, Sundaresan A, Cokus S, Chan SW, Chen H, Henderson IR, Shinn P, Pellegrini M, Jacobsen SE, Ecker JR (2006) Genome-wide high-resolution mapping and functional analysis of DNA methylation in arabidopsis. Cell 126:1189–1201

Zhang X, Clarenz O, Cokus S, Bernatavichute YV, Pellegrini M, Goodrich J, Jacobsen SE (2007) Whole-genome analysis of histone H3 lysine 27 trimethylation in *Arabidopsis*. PLoS Biol 5:e129

Zhang W, Wu Y, Schnable JC, Zeng Z, Freeling M, Crawford GE, Jiang J (2012a) High-resolution mapping of open chromatin in the rice genome. Genome Res 22:151–162

Zhang W, Zhang T, Wu Y, Jiang J (2012b) Genome-wide identification of regulatory DNA elements and protein-binding footprints using signatures of open chromatin in *Arabidopsis*. Plant Cell 24:2719–2731

Zhang T, Zhang W, Jiang J (2015) Genome-wide nucleosome occupancy and positioning and their impact on gene expression and evolution in plants. Plant Physiol 168:1406–1416

Zhao Z, Yu Y, Meyer D, Wu C, Shen WH (2005) Prevention of early flowering by expression of FLOWERING LOCUS C requires methylation of histone H3 K36. Nat Cell Biol 7:1256–1260

Zhong S, Fei Z, Chen YR, Zheng Y, Huang M, Vrebalov J, McQuinn R, Gapper N, Liu B, Xiang J, Shao Y, Giovannoni JJ (2013) Single-base resolution methylomes of tomato fruit development reveal epigenome modifications associated with ripening. Nat Biotechnol 31:154–159

Zilberman D, Gehring M, Tran RK, Ballinger T, Henikoff S (2007) Genome-wide analysis of *Arabidopsis thaliana* DNA methylation uncovers an interdependence between methylation and transcription. Nat Genet 39:61–69

Zilberman D, Coleman-Derr D, Ballinger T, Henikoff S (2008) Histone H2A.Z and DNA methylation are mutually antagonistic chromatin marks. Nature 456:125–129

FSC
www.fsc.org
MIX
Papier aus verantwortungsvollen Quellen
Paper from responsible sources
FSC® C105338

If you have any concerns about our products, you can contact us on
ProductSafety@springernature.com

In case Publisher is established outside the EU, the EU authorized representative is:
Springer Nature Customer Service Center GmbH
Europaplatz 3, 69115 Heidelberg, Germany

Printed by Libri Plureos GmbH
in Hamburg, Germany